Surface Chemical Modification

Surface Chemical Modification

Editor

Yilei Zhang

Basel • Beijing • Wuhan • Barcelona • Belgrade • Novi Sad • Cluj • Manchester

Editor
Yilei Zhang
Department of Mechanical
Engineering, University of
Canterbury
Christchurch
New Zealand

Editorial Office
MDPI
St. Alban-Anlage 66
4052 Basel, Switzerland

This is a reprint of articles from the Special Issue published online in the open access journal *Coatings* (ISSN 2079-6412) (available at: https://www.mdpi.com/journal/coatings/special_issues/Surf_chem_mod).

For citation purposes, cite each article independently as indicated on the article page online and as indicated below:

Lastname, A.A.; Lastname, B.B. Article Title. *Journal Name* **Year**, *Volume Number*, Page Range.

ISBN 978-3-7258-0801-4 (Hbk)
ISBN 978-3-7258-0802-1 (PDF)
doi.org/10.3390/books978-3-7258-0802-1

© 2024 by the authors. Articles in this book are Open Access and distributed under the Creative Commons Attribution (CC BY) license. The book as a whole is distributed by MDPI under the terms and conditions of the Creative Commons Attribution-NonCommercial-NoDerivs (CC BY-NC-ND) license.

Contents

Adeela Rehman, Mira Park and Soo-Jin Park
Current Progress on the Surface Chemical Modification of Carbonaceous Materials
Reprinted from: *Coatings* **2019**, *9*, 103, doi:10.3390/coatings9020103 . . . . **1**

Carlos Perez-Rizquez, David Lopez-Tejedor, Laura Plaza-Vinuesa, Blanca de las Rivas, Rosario Muñoz, Jose Cumella, et al.
Chemical Modification of Novel Glycosidases from *Lactobacillus plantarum* Using Hyaluronic Acid: Effects on High Specificity against 6-Phosphate Glucopyranoside
Reprinted from: *Coatings* **2019**, *9*, 311, doi:10.3390/coatings9050311 . . . . **22**

Ruohan Zhao, Patrick Rupper and Sabyasachi Gaan
Recent Development in Phosphonic Acid-Based Organic Coatings on Aluminum
Reprinted from: *Coatings* **2017**, *7*, 133, doi:10.3390/coatings7090133 . . . . **33**

Peter Rodič, Barbara Kapun, Matjaž Panjan and Ingrid Milošev
Easy and Fast Fabrication of Self-Cleaning and Anti-Icing Perfluoroalkyl Silane Film on Aluminium
Reprinted from: *Coatings* **2020**, *10*, 234, doi:10.3390/coatings10030234 . . . . **54**

Mihaela Baibarac, Monica Daescu and Szilard N. Fejer
Optical Evidence for the Assembly of Sensors Based on Reduced Graphene Oxide and Polydiphenylamine for the Detection of Epidermal Growth Factor Receptor
Reprinted from: *Coatings* **2021**, *11*, 258, doi:10.3390/coatings11020258 . . . . **75**

Jian Sun and Quantong Yao
Fabrication of Microalloy Nitrided Layer on Low Carbon Steel by Nitriding Combined with Surface Nano-Alloying Pretreatment
Reprinted from: *Coatings* **2016**, *6*, 63, doi:10.3390/coatings6040063 . . . . **87**

Elisa Martinelli, Elisa Guazzelli, Antonella Glisenti and Giancarlo Galli
Surface Segregation of Amphiphilic PDMS-Based Films Containing Terpolymers with Siloxane, Fluorinated and Ethoxylated Side Chains
Reprinted from: *Coatings* **2019**, *9*, 153, doi:10.3390/coatings9030153 . . . . **94**

Cheuk Sing Choy, Eisner Salamanca, Pei Ying Lin, Haw-Ming Huang, Nai-Chia Teng, Yu-Hwa Pan, et al.
Argon Plasma Surface Modified Porcine Bone Substitute Improved Osteoblast-Like Cell Behavior
Reprinted from: *Coatings* **2019**, *9*, 134, doi:10.3390/coatings9020134 . . . . **108**

Ruiliang Liu and Mufu Yan
Characteristics of AISI 420 Stainless Steel Modified by Low-Temperature Plasma Carburizing with Gaseous Acetone
Reprinted from: *Coatings* **2019**, *9*, 75, doi:10.3390/coatings9020075 . . . . **121**

Chunling Zhang, Xueyan Dai, Yingnan Wang, Guoen Sun, Peihong Li, Lijie Qu, et al.
Preparation and Corrosion Resistance of ETEO Modified Graphene Oxide/Epoxy Resin Coating
Reprinted from: *Coatings* **2019**, *9*, 46, doi:10.3390/coatings9010046 . . . . **131**

Bor-Yann Chen, Yuan-Ting Tsao and Shih-Hang Chang
Cost-Effective Surface Modification of Carbon Cloth Electrodes for Microbial Fuel Cells by Candle Soot Coating
Reprinted from: *Coatings* **2018**, *8*, 468, doi:10.3390/coatings8120468 . . . . **146**

Jinglong Qu, Shufeng Yang, Hao Guo, Jingshe Li and Tiantian Wang
Synthesis of Core-Shell MgO Alloy Nanoparticles for Steelmaking
Reprinted from: *Coatings* **2018**, *8*, 161, doi:10.3390/coatings8050161 . . . . **160**

Chang Kyu Byun, Minkyung Kim and Daehee Kim
Modulating the Partitioning of Microparticles in a Polyethylene Glycol (PEG)-Dextran (DEX) Aqueous Biphasic System by Surface Modification
Reprinted from: *Coatings* **2018**, *8*, 85, doi:10.3390/coatings8030085 . . . . **169**

Rohit Khanna, Joo L. Ong, Ebru Oral and Roger J. Narayan
Progress in Wear Resistant Materials for Total Hip Arthroplasty
Reprinted from: *Coatings* **2017**, *7*, 99, doi:10.3390/coatings7070099 . . . . **178**

Review

Current Progress on the Surface Chemical Modification of Carbonaceous Materials

Adeela Rehman [1], Mira Park [2,*] and Soo-Jin Park [1,*]

1. Department of Chemistry, Inha University, 100 Inharo, Incheon 22212, Korea; adeelarehman00@gmail.com
2. Department of Organic Materials & Fiber Engineering, Chonbuk National University, Jeonju 54896, Korea
* Correspondence: wonderfulmira@jbnu.ac.kr (M.P.); sjpark@inha.ac.kr (S.-J.P.); Tel.: +82-32-876-7234 (S.-J.P.); Fax: +82-32-860-8438 (S.-J.P.)

Received: 21 December 2018; Accepted: 5 February 2019; Published: 8 February 2019

Abstract: Carbon-based materials is considered one of the oldest and extensively studied research areas related to gas adsorption, energy storage and wastewater treatment for removing organic and inorganic contaminants. Efficient adsorption on activated carbon relies heavily upon the surface chemistry and textural features of the main framework. The activation techniques and the nature of the precursor have strong impacts on surface functionalities. Consequently, the main emphasis for scientists is to innovate or improve the activation methods in an optimal way by selecting suitable precursors for desired adsorption. Various approaches, including acid treatment, base treatment and impregnation methods, have been used to design activated carbons with chemically modified surfaces. The present review article intends to deliver precise knowledge on efforts devoted by researchers to surface modification of activated carbons. Chemical modification approaches used to design modified activated carbons for gas adsorption, energy storage and water treatment are discussed here.

Keywords: activated carbon; surface modification; energy storage; clean environment

1. Introduction

Activated carbon (AC) generally refers to carbonaceous materials fabricated from various carbon-rich sources, such as wood, lignite, coal and coconut shells [1]. Activated carbons can be produced from these materials by carbonization process. During carbonization, moisture and volatile compounds are removed, leaving behind char. This char can be further activated via physical or chemical activation methods to generate highly porous activated carbons. These carbonaceous materials are widely studied as effective adsorbents owing to their porous structure, high surface area, enriched surface chemistry with the highly reactive surface functionalities. By virtue of these interesting features activated carbon is considered as a class of versatile materials with the practical applications in the fields of adsorption, catalysis, energy storage and pollutant removal from gaseous and liquid phases [2–5]. The performance of AC can be improved by tuning the textural features along with the generation of a wide range of surface functional moieties. Heteroatom doping of AC as surface functional groups is a common practice nowadays. These heteroatoms include oxygen, hydrogen, sulfur and nitrogen. By considering the surface of adsorbent, the primary functional groups which, when considered together, are essential for removal of organic and inorganic pollutants like carboxyl, carbonyl, phenols, lactones and quinones. Tailoring the surface of activated carbons produces the exceptionally high adsorption performance of these materials [6–8].

With the growing population and industrial revolution, pollutants have accumulated in air and water. Researchers are left with the challenge of developing the most efficient adsorbents with particular chemical characteristics in order to alleviate rising pollution. Nowadays, enormous research effort is devoted to modification of activated carbons, specifically their surfaces and textural features in order to meet the demands for cleaner water and air [9–12]. Many approaches have been adopted in the literature for effectively modifying surface of carbon-based materials by considering the surface chemistry and reaction mechanism, permitting higher uptake of particular contaminants by these adsorbents [13–19]. It is widely accepted that functional groups can bond directly to fused aromatic rings in hydrocarbons. Therefore, one would expect that their chemical properties resemble those of aromatic hydrocarbons. Consequently, similar chemical reactions that involve aromatic rings and functional groups can be performed, thus adsorbents with the desired specifications can be designed through thermal or chemical methods. For example, a group of researchers successfully generated weakly acidic surface functionalities on activated carbons using oxidation reactions [20–24]. Such carbons were exploited as metal adsorbents with exceptionally high affinity for metals compared to pristine carbon. Heteroatom doping and nitrogen-doping specifically, has also been well-explored and such carbons have shown enhanced catalytic activity during oxidation [25–29]. Despite the fact that activated carbon is an emerging field of interest for wastewater treatment, the exact adsorption mechanisms for organic and inorganic solutes are still vague. Recently, a π–π dispersion interaction mechanism was proposed for adsorption of organic compounds [30–32]. This mechanism suggests that aromatic species and the basal planes in carbon interact with π electrons in each system. In contrast, some experimental data claim the electron donor–acceptor mechanism governs adsorption of organic species, considering the complex formation among the adsorbed species and the surface carbonyl group in the sorbent.

Recent research has placed emphasis on modifying these physical and chemical attributes and the success achieved in the last few years regarding synthesis of surface-modified activated carbons and various modification techniques and their effects on energy storage, adsorption of pollutants from gaseous and liquid phases, is summarized in this review. Generally, carbon-based materials are activated, generating a well-defined porous structure followed by surface modification. Considering the modifying agents, the synthetic method can be classified into three main categories: (i) physical modification, (ii) chemical modification and (iii) biological modification. Herein, we have primarily reviewed surface chemical modification methods, characterization and their merits from the perspective of adsorption behavior. The characterization techniques used to analyze modified surfaces of activated carbons are acid/base titration, Fourier transform infrared spectroscopy (FT-IR), X-ray photoelectron spectroscopy (XPS) and temperature programmed desorption (TPD). For information related to detailed experimental procedures and conditions, readers are referred to full articles from the references.

2. Surface Chemical Characteristics

The surface chemistry of carbon-based materials fundamentally relies upon heterogeneity of the chemical surface, which is associated with the placement of heteroatoms on the surface [33]. Heteroatoms are generally referred to as a group of atoms other than the carbon that are present in parent matrix, for example, oxygen, nitrogen, hydrogen, sulfur and phosphorus. The nature and measure of these elements depends upon the nature of the precursors or inoculation methods used in the activation process [34–36]. By observing the nature of these heteroatoms, it was proposed that surface functionalities comprising of heteroatoms along with delocalized electrons in aromatic carbon can generate activated carbon with acidic or basic surfaces.

2.1. Acidic Surfaces

Activated carbons with acidic surfaces mainly include oxygen-containing functional groups [37–39]. Such groups are generally located on the outer surfaces or edges of the basal plane and play a major role in controlling the chemical nature of carbon. These specific outer positions are mainly considered as adsorption sites, hence the concentration of oxygen at these particular points has a great influence on the adsorption capabilities of carbon. Extensive research effort and results suggest that few groups containing oxygen atoms, including carboxylic, chromene, lactone, phenol, quinone, pyrone, carbonyl and ethers, are usually located on the carbon surfaces (Figure 1). Functional groups with carboxylic moieties are generally responsible for surface acidity [40–43]. Other surface acidic groups include carboxylic anhydrides, lactones and phenolic hydroxyls.

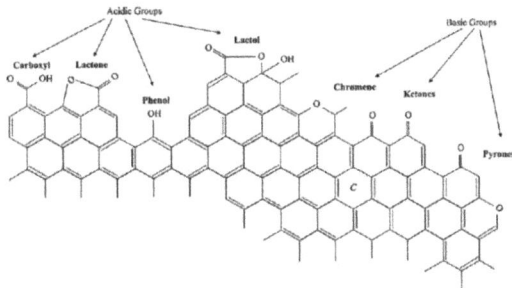

Figure 1. Acidic and basic surface functionalities on a carbon basal plane. Reprinted with permission from [44]. Copyright 2004 Elsevier Ltd.

2.2. Basic Surfaces

Two major features responsible for the basic surfaces of activated carbon are: (i) delocalized π-electrons of fused aromatic structures and (ii) basic surface functional groups (e.g., nitrogen-enriched functionalities) that can bind protons [45–47]. Previous studies explored certain functional groups containing oxygen including chromene, ketone and pyrone can also contribute to the basicity of carbon (Figure 1). On the other hand, extensive studies reveal that the basicity of activated carbons originates from resonating π-electrons in the carbon matrix. To elaborate, π-electrons in carbon layers can act as Lewis basic sites. Leon y Leon et al. [48] examined the basic surface character of two series of carbons and proved that oxygen-free carbon surfaces can efficiently adsorb protons from aqueous media. The excellent adsorption abilities are accredited to sites situated in the π-electron rich areas on the basal plane of carbon crystallites. Consequently, these regions possess Lewis basic character [49]. Similarly, chemical modification via the introduction of nitrogen-enriched functionalities on the carbon surface can facilitate adsorption of CO_2 from the atmosphere. Nitrogen moieties usually induce basic character, which can improve the interaction between the carbon surface and acidic species via dipole–dipole interactions, hydrogen bonding and covalent bonding [50,51].

3. Surface Chemical Modification of Activated Carbons

Activated carbons possess three distinct types of chemical surfaces (acidic, basic and neutral) depending on the chemical nature of surface functionalities [52,53]. These chemical characteristics can be modified by varying the functional groups on the surfaces; this will be explained in this section. Table 1 summarizes recent experiments aimed at modifying carbon surfaces to generate the desired functional groups for improved adsorption of chemical species.

Table 1. Recent research studies on modification of carbon surfaces to generate specific surface functionalities for enhanced removal of metal species.

Techniques Used	Nature of Functional Groups Induced	Applications	References
Oxidation by nitric acid	weakly acidic and non-acidic	Cr(III) and Cr(IV) removal	[54]
Oxidation by nitric and hydrofluoric acid	weakly acidic	Cu^{2+} removal	[55]
Oxidation by ammonium persulfate and sulfuric acid solution	weakly acidic	Zn^{2+} removal	[56]
Treatment using Na_2S	weakly acidic	Pb^{2+} removal	[57]
Heat treatment in H_2S	sulfur-based	Mercury (as $HgCl_2$) removal	[58]
Oxidization using nitric acid, ammonium persulphate	weakly acidic	Cu^{2+} removal	[59]
Oxidization using nitric acid	weakly acidic	Cd^{2+} removal	[60]
Treatment at ambient temperature and 900 °C in SO_2 and/or H_2S	sulfur surface complexes	Cd^{2+} removal	[61]
Treatment using hydrochloric acid	weakly acidic	Cr(VI) removal	[62]

3.1. Acid Treatment

To acidify the surface of activated carbon, the surface is generally oxidized as it enhances the acidic character and hydrophilic nature of the carbon surface by reducing the mineral content [63]. Nitric acid (HNO_3), sulfuric acid (H_2SO_4) and phosphoric acid (H_3PO_4) are widely used for this purpose [64–67]. During oxidation, oxygen-enriched functionalities are generated on the carbon surface including carboxylic, lactone and phenolic hydroxyl groups. The most frequently used activation procedures for producing oxygen-derived acidic groups are oxidation by gases and aqueous oxidants. Carbon dioxide, oxygen, steam and air are used as oxidants in gas phase oxidation. Oxidation at a low temperature leads to the generation of strong acidic functionalities (e.g., carboxylic groups), whereas an elevated temperature is responsible for the formation of weak acidic groups (e.g., phenolic). Liquid phase oxidation uses less energy as oxidation at a low temperature can introduce a higher oxygen content at surfaces compared to gas phase oxidation. Furthermore, gas phase oxidation can enhance the hydroxyl and carbonyl content on the surface of activated carbon, while liquid-phase oxidation primarily increases the concentration of carboxylic and phenolic hydroxyl groups.

From the perspective of applications, these acidic groups (i.e., oxygen-containing functional groups as proton donors) on carbon surfaces can be exploited as adsorbents of heavy metals from aqueous media via formation of metal-complexes between metal ions and negatively charged acidic groups. Numerous experimental efforts are focused on carbon surface modification by introducing acidic groups to enhance the removal of metal ions and organic species from aqueous solutions (Table 2). Some of these acidic surface functional moieties include carboxyl, quinone, carbonyl, lactone, hydroxyl and carboxylic anhydride. These specific functional groups are inclined with respect to the metal-complex formation with alkaline-earth metals and chelation with heavy metal ions. Complex formation with main group metals can be expressed as follows:

$$M^{n+} + (-COOH) \leftrightarrow (-COO)_n M + nH^+ \qquad (1)$$

Table 2. Acid treated carbons as biosorbents.

Samples	Acid Used	Species Biosorbed	References
Aquatic weeds	H_2SO_4	Cr(III), Cr(VI)	[68]
Agricultural waste	H_3PO_4, H_2O_2	Cd	[69]
Olive mill solid residue	HCl	Phenol	[70]
Activated coconut shell carbon	H_3PO_4	Zn(II)	[71]
Rice bran	HNO_3	Cd(II0, Cu(II), Pb(II), Zn(II)	[72]
Olive stone	H_2SO_4, HNO_3	Pb(II)	[73]

The reaction proceeds by cation exchange between the metal cation and the hydrogen ion attached to the carboxylic group [74]. Apart from using strong acids as oxidizing agents, such as nitric acid and sulfuric acid, other chemical oxidants such as hydrogen peroxide, acetic acid and oxygen can also generate acidic functional groups [75–77]. These conventional chemical oxidation methods can provide beneficial effects on adsorption of metal ions but they destroy the desired physical attribute of the adsorbents. The textural parameters including BET surface area pore diameter and pore volume can also be modified and the pores subsequently lose their desired original features. Many studies in the literature, including the experiment performed by Maroto-Valer et al. [78] demonstrate that nitric acid treatment reduces the specific surface area and total pore volume by 9.2% and 8.8%, respectively, where pores are blocked by newly generated oxygen groups. Similar results were also presented by Rios et al. [79] Aburub and Wurster [80] where the specific surface area of activated carbon decreased by 33.7% and 6.5%, while treating with pure nitric acid for 38 h and with an equimolar ratio mixture of nitric/sulfuric acid for 24 h. They proposed that this unprecedented reduction can be attributed to annihilation of the porous matrix within the activated carbon caused by harsh nitric acid oxidation. As these conventional methods can reduce the physical characteristics, researchers are continuously searching for alternative methods to generate acidic surface functionalities without unwanted pore blockage. One favorable technique involves the use of oxygen plasma, originally studied by Garcia et al. [81]. Their conclusions show that the technique could be used to modify the external surfaces and produced acidic groups, whereas the internal surface was nearly unaffected. The general assumption regarding plasma treatment is that the oxidizing species do not penetrate the interior of the carbon. Highly reactive mono-oxygen radicals etch the carbon surface without destroying the microporosity. Moreover, this assumption was considered true by an experimental study performed by Domingo-Garcia et al. [82] they declare that the chemical surface groups induced by oxygen plasma are located on the external surface, while those generated by treatments in aqueous media of $(NH_4)_2S_2O_8$ and H_2O_2 are located on the external and internal surfaces. Contrary to this, Lee et al. [83] used a helium–oxygen plasma produced in a reactor with dielectric-barrier discharge. During oxidation of activated carbon by this plasma, it was demonstrated that Fe^{2+} adsorption improved by a factor 3.8 after treatment. However, the specific surface area decreased by as much as 22.5% due to demolition of pore walls via gasification of carbon and pore blockage. These two isolated plasma oxidation methods reveal that not all plasma treatments can offer favorable outcomes with respect to textural specifications. Therefore, a more rigorous focus on eliminating this drawback is required in future studies. While acid treatment is a promising method for enhancing adsorption of metal ions, it weakens adsorption of organic compounds (especially phenol) from an aqueous environment. This can be attributed to the fact that acid treatment can result in the destruction of prevailing basic functionalities on the surface of activated carbon, which decreases the phenolic adsorption. This suggestion was considered in a study directed by Santiago et al. [84] where nitric acid treatment destroyed basic sites of activated carbon, resulting in a considerable decrease in phenol uptake. Furthermore, it was suggested that damaging the basic sites ultimately leads to decreased catalytic activity by phenol in the dilute phenol solutions. Nevertheless, at high phenolic concentrations, Terzyk et al. [85] found that phenol uptake improved after nitric acid treatment. This effect was explained via generation of adsorbate–adsorbate interactions at high phenolic concentration.

3.2. Base Treatment

Activated carbon with a base treatment induces a positive charge on the surface that enhances adsorption of negatively charged moieties. The most convenient way for generating porous carbons with basic surface characteristics is treatment with hydrogen or ammonia at high temperature. High temperature treatment (400–900 °C) of activated carbons with ammonia causes formation of basic nitrogen groups on the carbon surface. Nitrogen functionalities can be doped via reaction with precursors containing nitrogen (such as ammonia, nitric acid and various amines) or activation in the nitrogen-enriched environment [86–88]. Possible existing forms of the nitrogen include the following groups: amide, imide, lactame, pyrrolic and pyridinic groups. The chief role of ammonia treatment is to enhance the basicity of the activated carbon surface. Apart from the introduction of basic nitrogen atoms at high temperature, the exclusion of oxygen-based groups can significantly increase the basicity of activated carbons. These nitrogen functional groups usually induce basic character, which can boost the interaction between adsorbents and acidic species through dipole-dipole interactions, hydrogen bonding and covalent bonding.

Przepiorski [89] demonstrated the effect of heating conditions on the adsorbent during ammonia treatment. It was found that the uptake capacity towards phenol improved by as much as 29% at an optimum temperature of 700 °C. In another study, Chen et al. [90] inspected the modification of activated carbon surfaces during thermal treatment in the presence of ammonia, yielding increased perchlorate adsorption capacities from aqueous conditions without damaging the porous structure, which is beneficial for perchlorate adsorption. They demonstrated that the most promising results were obtained for samples prepared at temperatures between 650 and 700 °C with a four-fold rise in perchlorate adsorption. Economy et al. [91] used ammonia with activated carbon fibers (ACF) at higher temperatures to acquire a material with basic properties. They also claimed that such fibers were highly effective for adsorbing acidic gases like SO_2 and CO_2. Similarly, Stöhr et al. [92] evidently proved that the high performance of nitrogen-doped activated carbons prepared at high temperatures (600–900 °C). Free radicals (e.g., NH_2, NH, hydrogen and nitrogen) will start to form when carbon-based materials are kept in an atmosphere containing ammonia at elevated temperatures. These free radicals then attack the carbon surface and generate nitrogen-based functional groups [93,94]. Ammonia reactions at particular carboxylic acid groups located on the carbon surface can produce ammonium salts that generate amides and nitriles upon dehydration:

$$-COO-NH_4^+ \rightarrow -CO-NH_2 \rightarrow C\equiv N \qquad (2)$$

Substitution of OH groups with ammonia can result in the formation of amines:

$$-OH + NH_3 \rightarrow -NH_2 + H_2O \qquad (3)$$

During the reaction with ammonia at high temperatures, ether-like oxygen surface functionalities are substituted for –NH– on the carbon surface, resulting in generation of imine and pyridine moieties upon dehydrogenation [95–97]. It was found that amides, imides, imines, amines and nitriles are the predominant species at low temperature (<600 °C), while thermally stable aromatic structures such as pyrrole and pyridinic-like groups dominated at higher temperatures (>600 °C). In conclusion, basic treatment with ammonia does not require any preliminary oxidation stages, hence basicity can be induced in a facile manner. A summary of research into the modification of activated carbon with gaseous ammonia is shown in Table 3.

Table 3. Recent research studies conducted on the modification of activated carbon with gaseous ammonia.

Materials	Amination Temperature	Applications	References
Carbon adsorbents from biomass residue (almond shells)	800 °C	CO_2 adsorption	[98]
Commercial activated carbon	1000 °C	CO_2 adsorption	[99]
Commercial granular activated carbons	385 °C	Adsorption of model aromatic compounds (aniline and nitrobenzene)	[100]
Carbon materials (biomass residues, sewage sludge, pet coke)	400 °C	CO_2 adsorption	[101]
Activated carbon from peat	900 °C	Enhancement of catalytic activity of AC in oxidation reaction	[92]
Activated carbon from sulfonated styrene–divinyl-benzene copolymer	600 °C	Enhancement of molybdenum adsorption	[102]

3.3. Chemical Impregnation

It is widely accepted that the adsorption capacities of activated carbon for eliminating hazardous toxins can be significantly enhanced by impregnation with the appropriate chemicals [103]. Henning and Schafer [104] explained impregnation as a uniform distribution of chemicals in the internal surface of activated carbon. However, some researchers [105,106] associate the introduction of surface functional groups as impregnation techniques. In this context, we refer to impregnation with substances that do not affect the pH of the activated carbon surface. These impregnating materials can be metals or polymeric substances, which generally induce no significant pH changes. Henning and Schafer [84] indicated that impregnating an activated carbon surfaces should enhance the prevailing properties of the materials by improving its built-in catalytic oxidation capabilities, promote synergy among the activated carbon and impregnating agent in order to boost adsorption and augment the capacity of the material as an inert porous carrier. General industrial processing methods for producing impregnated activated carbon involve spraying virgin carbon with impregnating agents using a rotary kiln or fluidized bed. Generally, impregnating materials include hydroxides, carbonates, chromates or nitrates.

Two famous researchers (Monser and Adhoum) [107] are actively working on impregnating activated carbon with foreign compounds. In 2002, they conducted an experiment involving adsorption of copper(II), zinc(II), chromium(VI) and cyanide (CN^-) ions by activated carbons. These activated carbons were modified with tetrabutyl ammonium (TBA) and sodium diethyldithiocarbamate (SDDC). From the experimental data, it was found that TBA-modified carbons exhibit a factor 5 greater removal performance compared to pristine carbons. Similarly SDDC-modified activated carbon exhibited higher adsorption of copper (II) and zinc(II) (a factor 4 greater), as well as chromium(VI) (a factor 2 greater) than the parent activated carbon. It was concluded that the high adsorption performance is due to virgin activated carbon and the ion exchange mechanism exhibited by chemical species impregnated on the surface of activated carbon. In another study, silver and nickel-doped activated carbons were used as effective adsorbents of cyanide from an aqueous environment. From the results it was concluded that silver-doped adsorbents exhibit doubles the efficiency of nickel-doped carbons. The possible uptake mechanisms include adsorption, ion exchange with groups bearing positive charge located on the carbon surface or complex formation (e.g., $Ag(CN)^{2-}$ and $Ni(CN)_4^{2-}$). A few years later, they modified activated carbons with tetrabutyl ammonium (TBA) and copper to be exploited for efficient phthalate adsorption from industrial wastewater. Their results show that metal modification on the carbon surface exhibits a factor 2 higher adsorption compared to the parent activated carbon, while TBA-loaded AC improved phthalate uptake by a factor 1.7 compared to pristine carbon.

Recently, metal impregnation is one emerging field of research. Huang and Vane [108] reported excellent removal efficiency of arsenic from waste water. The effectiveness of iron-loaded activated carbon reveals an improved efficiency by a factor 10 compared to unloaded activated carbon. They further claimed that higher adsorption of arsenic can be attributed to arsenic complex formation involving ferrous ions. In another study, Leyva Ramos et al. [109] reported the effectiveness of aluminum-impregnated activated carbon for eliminating fluoride ions from aqueous media. Aluminum-doped carbons exhibit excellent fluoride uptake at pH 3.5. In a separate experiment, Dastgheib et al. [87] examined removal of dissolved natural organic matter (DOM) from natural water using activated carbon impregnated with iron. They concluded that the DOM uptake efficiency increases up to 120% due to iron impregnation, followed by ammonia treatment at high temperature. Further examples of chemically impregnated species on activated carbon are listed in Table 4.

Table 4. Chemical impregnated activated carbons and their potential applications.

Samples	Species Impregnated	Species Removed	References
Activated carbon	iodine and chlorine	Gas-phase elemental mercury	[110]
Granular activated carbon	sulfur	Gas-phase elemental mercury	[111]
Activated carbon	metallic silver and copper	Arsenic	[112]
Activated carbons	silver and nickel	Cyanide	[107]
Granular activated carbon	copper and silver	Cyanide	[113]

3.4. Surfactant-Modified Activated Carbons

Numerous research efforts are being devoted to surface modification of activated carbons with anionic and cationic surfactants [114–118]. A comprehensive explanation of adsorption of these surfactants on carbon surfaces is still not clear. A general assumption declares that hydrophobic interactions among surfactants and activated carbon were the primary source for surfactant adsorption on the surface of activated carbon. However, the direct implication of these surfactant-modified carbons as adsorbents for removal of pollutants from aqueous solutions is rare. An experiment conducted by Parette and Cannon [119] illustrate a new approach for modification in which granular activated carbons were doped with cationic surfactants in order to improve removal of ppb levels of perchlorate from groundwater in small-scale laboratory tests. They found that activated carbons modified with cetyl trimethyl ammonium chloride (CTAC) surfactant improved the perchlorate breakthrough time in a simulated 20–22 min empty bed by a factor 30 compared to virgin activated carbon. Few examples from the literature, for surfactant-modified carbons as adsorbents, are summarized in Table 5.

Table 5. Surfactant-modified carbons to adsorb contaminants from aqueous media.

Samples	Surfactant Used	Species Adsorbed	References
Surfactant modified activated carbon	HDTMA (hexadecyltrimethylammonium bromide) CPC (cetylpridinium chloride)	Cr(VI)	[120]
Surfactant-modified mesoporous FSM-16	cetyltrimethylammonium bromide (CTAB)	Acid dye (acid yellow and acid blue)	[121]
Activated carbon	CPC (cetylpridinium chloride)	Reactive black 5	[122]
Surfactant-modified carbon	cetyltrimethylammonium bromide (CTAB)	Cd(II)	[123]
Surfactant modified coconut coir pith	HDTMA (hexadecyltrimethylammonium bromide)	Cr(VI)	[124]

3.5. Ligand Functionalization

Recently, an innovative functionalization technique for graphitic layers of activated carbons was established by Garcia-Martin et al. [125] where N-(4-amino-1,6-dihydro-1-methyl-5-nitroso-6-oxo2-pyrimidinyl)-l-lysine (AMNLY) and N-(4-amino-1,6-dihydro-1-methyl-5-nitroso-6-oxopyrimidin-2-yl)-N-[bis(2-aminoethyl)] ethylene diamine (AMNET) were introduced on basic activated carbon with relatively low oxygen and nitrogen contents. This functionalized carbon was then used to eradicate chromate (VI) from aqueous media. They proposed that the inoculation of AMNLY and AMNET would enrich the π-electron density in the graphene layers, thus improving the adsorptive abilities of the receptors via π-dispersive and/or donor–acceptor interactions among the pyrimidine moiety in the receptors and the basic arene centers in the graphene layers. It was found that the adsorption of Cr(VI) on AMNET-supported AC was roughly amplified by a factor 1.7 as compared to virgin AC. They accredited the improved adsorptive ability to strong and selective interactions between chromate anions and NH_3^+ groups in this compound. On the other hand, impregnation of AMNLY decreased the uptake of Cr(VI) by 75%; this was attributed to the suppressed interaction among chromate anion and NH_3^+ groups as a direct consequence of the proximity of a carboxylate center with negative charge. Examples from literature are presented in Table 6.

Table 6. Ligand functionalization of activated carbons and their applications.

Samples	Ligand Functionalized	Species Adsorbed	References
Activated carbon	Benzoylthiourea	U(VI)	[126]
Carbon	5-azacytosine	U(VI)	[127]
Activated carbon	Hybrid ligands (nitric acid, thionyl chloride, ethylenediamine)	Hg(II)	[128]

4. Surface Analyses of Activated Carbons

Characterization of surface functional groups in porous carbon is complex due to the convoluted surface functionalities and inadequate understanding of their nature. The conventional and fundamental method for qualitative and quantitative determination is elemental analysis. As this method is widely used in most experiments due to its convenience, it cannot be used to predict functional groups, hence it cannot be used as an effective tool for analyzing surface chemistry. Different characterization techniques have been used to identify and confirm the presence of surface functionalities on the surfaces of activated carbon [129]. Details of some common methods are as follows.

4.1. Acid/Base Titrations

Conventional acid/base titration methods, such as those studied by Boehm [130], were used to analyze the basic and acidic functionalities of the adsorbent surface. The technique is based on neutralization of specific acids/bases by the basic or acidic functionalities on the adsorbent surface and quantification of the amount of acid or base that has reacted with these functional groups. Boehm titration is used to differentiate basic and acidic functional groups based on their neutralization abilities [131]. The number of numerous oxygen-enriched acidic sites on the activated carbon was determined under the supposition that $NaHCO_3$ can only neutralize carboxylic moieties, Na_2CO_3 would neutralize carboxylic and lactone groups and NaOH will neutralize carboxylic, phenolic and lactone groups [132]. The number of surface basic groups can be determined from the quantity of HCl used by carbon during neutralization. The quantity of consumed acid/base due to neutralization of basic/acidic functional groups on the surface of carbon can be determined using back titration with NaOH and HCl, respectively. The primary drawback of this technique is that it can only be used for samples present in large quantity. Furthermore, this method only could be used to measure the quantity of about half of the total oxygen content of activated carbon. Likewise, the total basicity takes a single value because the nature of basic surface functional groups is not well explored.

4.2. Fourier Transform Infrared Spectroscopy (FT-IR)

Infrared (IR) spectroscopy is one widely used tool to examine the surface functionalities on activated carbon. Unfortunately, IR spectra exhibit some limitations that make the spectra difficult to interpret; the measured peaks are generally a sum of interactions from distinct groups [133,134]. In addition, IR does not offer quantitative evidence regarding the existence of individual functional groups on activated carbon. These limitations of IR analyses can be reduced using Fourier transform infrared (FT-IR) spectroscopy. This technique is commonly used for qualitative determination of the chemical structure and for identifying functionalities on the surfaces of carbon materials [135–138]. The intrinsic signal-to-noise (S/N) ratio can be used to enhance the frequency resolution compared to dispersive IR spectroscopy. FT-IR spectra show the transmitted infrared intensity at various wavenumbers. A comparison of the spectra obtained before and after surface modification treatment can be used to identify which functional groups that form or decompose during the treatment [139]. Aside from the methods explained here, other generally-used techniques for illustrating the surface chemistry of carbon materials include NMR spectra [140], inverse gas chromatography (IGC) [141] and electron microscopy, including SEM and TEM [142].

4.3. Temperature Programmed Desorption (TPD)

Surface functional groups can be identified by considering the relative thermal stability of these groups. Temperature programmed desorption (TPD) is a technique for determining the concentration of functional groups that exploits the thermal stability of various groups. This method has become more famous for analyzing oxygen functional groups on the surface of activated carbon [143,144]. The TPD experiment operates on the principle that surface oxygen groups decompose at low temperatures, release CO_2 and CO at higher temperature and release H_2O and H_2 in some cases at other temperatures [145–148]. Studies in the literature show that decomposition of different groups like carboxyl and lactone generates CO_2, while decomposition of carbonyl, quinone, phenol and ether groups leads to CO production. Some other basic functional groups like pyrone and chromene decompose at high temperature to produce a CO peak. During analysis, different functional groups can be identified as the temperature is slowly increased. Functional groups corresponding to CO_2 are nearly completely utilized at high temperatures (above 1000 °C) and only a slight percentage of CO releasing groups remain on the carbon surface, which can be designated as pyrone and chromene-type functionalities. A complete picture of the surface groups can be obtained by observing the decomposition temperature, amount of devolved gases and the mechanism behind the release of a particular gaseous species (e.g., carbonyl groups devolve as CO) [149,150]. Moreover, it is difficult to interpret the TPD spectra due to the overlapping desorbed gases produced from various oxygen-containing structures. However, recognizing each surface group separately requires isolating CO and CO_2 peaks because TPD spectra exhibit high surface mobility and peaks appear as composites of CO and CO_2, particularly at high temperature [151]. Recently, researchers focused on modification methods for deconvoluting the obtained spectra with the intention of determining distinct kinds of surface oxygen structures [152].

4.4. X-Ray Photoelectron Spectroscopy (XPS)

One widely used non-destructive surface analysis technique is X-ray photoelectron spectroscopy (XPS). In this technique, particular electron binding energies of elements located at the surface are used to quantify the elemental composition and identify the chemical states of surface elements [153–155]. This spectroscopic method can identify all elements on the surface of activated carbon. The XPS technique is widely used to study carbon materials. The C1s core region can be used to identify significant changes in the nature of the carbon during surface modification, such as oxidation and acidification. From the XPS spectra, four distinct peaks can be seen at different binding energies corresponding to four different functional groups: C–C or C–H, C–O, C=O and O–C=O [156–159]. Apart from its merits, it cannot be used to distinguish functional groups with very close binding energies, such as C–C and C–H or C–O–C and C–OH. It also cannot be used to detect hydrogen atoms. However, it is one of the best tools in analytical chemistry for examining basic nitrogen functionalities. Modification of carbon surfaces with ammonia is widely explored using XPS. Nitrogen (N1s) core level studies were performed in order to obtain additional understanding of the chemical nature of surface functionalities [160,161]. Jensen and van Bekkum [162] found that ammonia modification led to the generation of amides (399.9 eV), as well as lactams and imides (399.7 eV). Mangun et al. [163] determined that pyridine (binding energy of 398.4 eV) was the only nitrogen functionality that formed during high temperature ammonia treatment of activated carbon fibers. Stöhr et al. [92] further found two N1s peaks with binding energies of 401–400 eV and 399–398 eV due to chemisorption of nitrogen during ammonia treatment. These peaks were designated as amine and nitrile and/or pyridine-like nitrogen. Some other XPS surface analysis results indicate that two basic nitrogen functional groups in coals were pyrrolic and pyridinic groups with corresponding binding energies of 400.3 and 398.7 eV, respectively [164–166].

5. Practical Applications of Surface-Modified Activated Carbons

To explore the effects induced by chemical surface modifications, researchers have exploited modified carbons in various practical applications (Table 7). Some of them are described in this section.

Table 7. Chemical modification of activated carbons and their applications.

Materials	Modification Methods	Final Outcomes	References
Corn grains	KOH activation	Increased surface area (3199 m^2/g) results in high specific capacitance (257 F/g)	[167]
Porous carbon	Nitrogen-doping with 2 wt.% of hexamine	Modification of porous carbon with nitrogen has increased the capacitance of electrodes for supercapacitor applications.	[168]
Mesoporous carbons	Nitric acid oxidation	An enhanced energy density with a highest value of 5.7 Wh/kg is obtained after oxidation.	[169]
Activated carbon	Activated carbon was prepared from eucalyptus wood with H$_3$PO$_4$ and modified by NH$_3$	Incorporation of nitrogen group in ACs increased their adsorption capacities. The CO$_2$ adsorption capacity achieved by modified carbon was 3.22 mmol/g at 1 bar.	[170]
Activated carbon	Amino/nitro groups were introduced onto the surface of the activated carbon (AC) with nitration followed by reduction.	Results showed that the contents of nitrogen on the treated samples' surface increased from 0% to 1.38 after modification. The maximum CO$_2$ adsorption capacity of the modified samples can reach 19.07 mmol/g at 298 K and 36.0 bar.	[171]
Activated carbon	Impregnation of carbon with diethanolamine, methyl diethanolamine and tetraethylene pentaamine.	Materials impregnated with diethanolamine performed best for CO$_2$ capture; the highest adsorption capacity achieved was 5.63 mmol CO$_2$/g.	[172]
Activated carbon	Highly polar carbon surfaces were generated by acid and base treatment	Two common drinking water contaminants, relatively polar methyl tertiary-butyl ether (MTBE) and relatively nonpolar trichloroethene (TCE) were successfully adsorbed by activated carbon.	[12]
Activated carbon	Acid treatment with HNO$_3$ and HCl	Acid treatment produces more active acidic surface groups such as carboxyl and lactone, resulting in a reduction in the adsorption of basic dyes.	[173]
Activated carbon	chemical treatments using HNO$_3$, H$_2$O$_2$, NH$_3$	Excellent dye adsorption performance as a result of chemical modification of activated carbon.	[174]

5.1. Surface Modified Carbons as Supercapacitors

Activated carbon with surface modified chemically are widely used as energy-storage devices [175]. Ismanto et al. [176] used cassava peel waste to produce activated carbons using simultaneous chemical (KOH) and physical (CO_2) activation methods. The obtained carbon-based materials were then subjected to surface oxidation using hydrogen peroxide (H_2O_2), nitric acid (HNO_3) and sulphuric acid (H_2SO_4) solutions. As a result of modification, no noticeable change in the textural features of the final samples was observed but the chemical characteristics of surfaces exhibit clear changes. New functionalities are introduced on the surface of activated carbon, resulting in enhanced specific capacitance in HNO_3-treated carbons. The results show that the specific capacitance increased to 264.08 $F·g^{-1}$ compared to 153.00 $F·g^{-1}$ for pristine carbon samples. In another experiment, Elmou Wahidi et al. [177] used activated carbons derived from KOH activation of argan seed shells. The modified activated carbons induced oxygen and nitrogen groups on the surface. The experimental findings show that nitrogen-enriched activated carbons exhibited the highest capacitance and retention of 355 $F·g^{-1}$ at 125 $mA·g^{-1}$ and 93% at 1 $A·g^{-1}$, respectively, compared to oxygen-doped activated carbons. These results show that surface carboxyl functionalities in oxygen-enriched activated carbons prevent electrolyte diffusion into the porous network, while the existence of nitrogen groups can produce micro-mesoporosity and excellent pseudo capacitance properties. Furthermore, Liu et al. [178] showed that HNO_3-modified porous wood carbon monolith (m-WCM) could be used in a super capacitor. The results demonstrate a significant enhancement in the electrochemical capacitive performance compared to virgin carbon materials.

5.2. Surface Modified Carbons as Efficient CO_2 Adsorbents

The literature reviewed here reveals the rapid innovation of ammonia-modified activated carbons as an alternative approach for increasing their CO_2 capture capacity. The primary attribute of surface-modified activated carbon is the introduction of nitrogen functionality, which exhibits strong basicity and can induce Lewis acid-base interactions with acidic CO_2 molecules [179–182]. This enhances the adsorption performance. However, very strong chemical interactions between CO_2 molecules and the adsorbent surface can lead to poor adsorbent regeneration and are economically unfavorable. Hence, moderate physisorption is the desired interaction for adsorbing gas molecules efficiently and releasing them when required.

5.3. Surface Modified Carbons as Organic Pollutant Adsorbents

Studies in the literature show that activated carbons are widely being used for adsorption of organic molecules compared to metals. The predominant factors contributing to the high adsorption performance of activated carbons include their excellent textural features and the generation of suitable surface functional groups. It is commonly known that oxygen-enriched acidic surfaces can reduce the adsorption of organic species from aqueous media, whereas their absence boosts the adsorption performance of AC. Weakly acidic functionalities introduced on the surface of carbons can augment the metal adsorption capacity while decreasing the adsorption of phenolic compounds in an aqueous environment. This was experimentally demonstrated by Leng and Pinto [183]. They show that phenol physisorption decreased at high concentrations of surface acidic groups, possibly due to decreased dispersive forces with the carbon basal plane. Such outcomes are consistent with the findings of Mahajan et al. [184] which showed that phenol adsorption was augmented with increased availability of π-electrons on the basal plane of carbon surfaces and decreased oxygen content at adsorption sites. Some other research groups [185,186] explained the decreased adsorption of phenol by the fact that water is preferably adsorbed on activated carbon surfaces via hydrogen bonds with oxygen moieties. This results in the formation of large clusters that hinder the movement of phenol molecules into the microporous structures. These oxygen clusters subsequently decrease the adsorptive performance by localizing free electrons in carbon. Other mechanisms [187,188] suggest the generation of two kinds of interactions among adsorbent electrons and aromatic structure of adsorbate, namely π-dispersion and electrostatic interactions. To date, nearly all experimental studies involving removal of organics from aqueous solutions used phenol and benzene-derived compounds as model studies. This can be attributed to the fact that these aromatics are frequently used in pharmaceuticals, oil refineries and pesticide manufacturing. Apart from surface chemistry, other factors such as pH of solution, temperature of solution, type of adsorbent, oxygen availability and mineral contents should be considered for efficient removal of organic contaminants from waste water.

5.4. Surface Modified Carbons as Dye Adsorbents

Over the last few decades, the textile and dye industries are making significant progress. However, removing excess dyes from waste water is still a challenging task for researchers. Various researchers, including Faria et al. [18] examined surface-modified activated carbons with 6M and 10M nitric acid and hydrogen peroxide, respectively, along with the heating at 700 °C in H_2 or N_2 atmosphere. The results show that samples prepared with H_2 treatment at high temperature exhibit excellent capacity to adsorb anionic and cationic dyes. In another interesting experiment, Orfao et al. [189] used two textile dyes (commonly known as Rifafix Red 3BN and C.I. reactive red 241) for adsorption on acid-modified activated carbon surface and basic carbon surfaces. The adsorption capabilities show that basic surfaces play a beneficial role in dye uptake due to dispersive forces between localized π-electrons in the carbon basal plane and free electrons in the dye molecules. On the other hand, acid-treated samples exhibit reduced adsorption due to the presence of repulsive forces between the oxygenated functionalities of the carbon surface and dye molecules. Conversely, these results cannot be applied to all the textile dyes. Some dyes exhibit reduced adsorption with increased basic character, such as those studied by Attia et al. [190] Here, Acid Red 73 and Acid Yellow 23 show improved uptake but Acid Blue 74 exhibits reduced uptake. However, these dyes predominantly consist of organic structures. Hence, a general hypothesis can be formulated; increasing the acidic functionalities on the surface of activated carbon should result in decreased dye uptake. This concept is strengthened by the experimental evidence obtained by Valdes et al. [40]. They found that continued exposure of activated carbon to ozone gas as an oxidizing agent transformed the basic surface sites to acidic surface sites. Consequently, the uptake of methylene blue was reduced. They further elaborated this by the fact that oxidation can release electrons from activated carbon bands, which decreases the dispersion interactions among the π-electron system in the ring structures of the dye and the graphitic planes on the carbon surface.

In another experiment, 2M nitric and hydrochloric acids were used to modify carbon surfaces. Later on, these carbons were used by Wang et al. [191] to study the uptake capacity of methylene blue. HCl and HNO_3 treatment reduced the adsorption capacity by up to 10.6% and 13.5%, respectively. They concluded a similar mechanism was responsible for these results, where acid functionalities formed on the surface of activated carbon and electrons were extracted, thus reducing the uptake of methylene blue. They further explain the activity differences resulting from the use of both acids. According to the authors, carbons treated with HCl exhibit greater uptake than carbons treated with HNO_3. This can be attributed to the fact that negative ions (Cl^-) can adsorb on positive sites on the carbon surface and can induce negative charge on the carbon surface, thus facilitating adsorption of positively charged dye molecules. However, contrary to these experiments, Jiang et al. [17] claimed that high temperature oxidation by concentrated H_2SO_4 can lead to increased methylene blue adsorption. This can be justified by the increased mesopore volumes after high temperature modification, which can result in enhanced dye adsorption.

6. Concluding Remarks and Future Outlook

In this review, an extensive study based on activated carbon modification methods, in order to improve the adsorption capacity for diverse contaminants in water and air, is presented. Different activating agents are used to induce the formation of different surface functional groups. Generally, uptake of metal ions from an aqueous environment requires an acidic treatment, while inducing basicity is highly recommended for removal of anionic pollutants, organic moieties and CO_2 from air. After examining all the possible modification methods, one finds that an acidic treatment is generally the most studied and used technique, perhaps due to its simplicity, ease of applicability, the availability of many oxidizing agents and the well-understood oxidation mechanism that has been used from many years. Currently, the main focus of researchers is to design activated carbons with modified surfaces and excellent textural parameters. In this context, many experiments were performed in order to develop economically cheap activated carbon electrodes capable of storing large amounts of energy with minimum resistance. Similarly, carbon-based materials with basic surfaces to enhance CO_2 adsorption in order to mitigate global warming, is highly needed. On the other hand, activated carbons with surfactant modification that are tailored to eliminate pollutants from aqueous media require further investigation in order to yield excellent performance as experimental findings in this area are limited. Furthermore, considering the effects of enhanced and decreased uptake of particular pollutants, selective adsorption can be induced, which provides a new pathway in the field of clean energy and environmental science. Apart from these research findings, the most important drawbacks are the cost of the modification process and treatment of the leftover hazardous chemicals used during modification. While recovering adsorbents, one should notice that chemicals used for modification should not enter the atmosphere and must be recovered. Adsorbents should be used in cyclic measurements and easy to regenerate. Keeping all these points in mind, the field requires facile, novel, simple and greener techniques for modification of activated carbons. One example of such a technique is the use of plasma treatment to generate a desired charge on the surface of carbon, making it effective for elimination of contaminants. Thus, the authors find that new, efficient modification methods should be explored as these methods improve the chemical surfaces of activated carbons while simultaneously preventing destruction of the textural characteristics.

Author Contributions: A.R. as first Author conducted the literature search. All three authors were involved in the writing and editing of the manuscript.

Funding: This research was funded by Traditional Culture Convergence Research Program through the National Research Foundation of Korea (NRF) funded by the Ministry of Science, ICT & Future Planning (2018M3C1B5052283) and the Commercialization Promotion Agency for R&D Outcomes (COMPA) funded by the Ministry of Science and ICT (MSIT) [2018_RND_002_0064, Development of 800 mAh g$^-$L pitch carbon coating materials].

Conflicts of Interest: The authors declare no conflict of interest.

References

1. Nacu, G.; Bulgariu, D.; Cristina Popescu, M.; Harja, M.; Toader Juravle, D.; Bulgariu, L. Removal of Zn (II) ions from aqueous media on thermal activated sawdust. *Desalin. Water Treat.* **2016**, *57*, 21904–21915. [CrossRef]
2. Liang, L.; Liu, C.; Jiang, F.; Chen, Q.; Zhang, L.; Xue, H.; Jiang, H.L.; Qian, J.; Yuan, D.; Hong, M. Carbon dioxide capture and conversion by an acid-base resistant metal-organic framework. *Nat. Commun.* **2017**, *8*, 1233. [CrossRef] [PubMed]
3. Hulicova-Jurcakova, D.; Seredych, M.; Lu, G.Q.; Bandosz, T.J. Combined effect of nitrogen-and oxygen-containing functional groups of microporous activated carbon on its electrochemical performance in supercapacitors. *Adv. Funct. Mater.* **2009**, *19*, 438–447. [CrossRef]
4. Georgin, J.; Dotto, G.L.; Mazutti, M.A.; Foletto, E.L. Preparation of activated carbon from peanut shell by conventional pyrolysis and microwave irradiation-pyrolysis to remove organic dyes from aqueous solutions. *J. Environ. Chem. Eng.* **2016**, *4*, 266–275. [CrossRef]
5. Hameed, B.; Tan, I.; Ahmad, A. Adsorption isotherm, kinetic modeling and mechanism of 2, 4, 6-trichlorophenol on coconut husk-based activated carbon. *Chem. Eng. J.* **2008**, *144*, 235–244. [CrossRef]
6. Ioannidou, O.; Zabaniotou, A. Agricultural residues as precursors for activated carbon production—a review. *Renew. Sustain. Energy Rev.* **2007**, *11*, 1966–2005. [CrossRef]
7. Mohamed, A.R.; Mohammadi, M.; Darzi, G.N. Preparation of carbon molecular sieve from lignocellulosic biomass: A review. *Renew. Sustain. Energy Rev.* **2010**, *14*, 1591–1599. [CrossRef]
8. Kılıç, M. Apaydın-Varol, E. Pütün, A.E. Preparation and surface characterization of activated carbons from Euphorbia rigida by chemical activation with $ZnCl_2$, K_2CO_3, NaOH and H_3PO_4. *Appl. Surf. Sci.* **2012**, *261*, 247–254. [CrossRef]

9. Ngah, W.W.; Hanafiah, M. Removal of heavy metal ions from wastewater by chemically modified plant wastes as adsorbents: a review. *Bioresour. Technol.* **2008**, *99*, 3935–3948. [CrossRef]
10. Rehman, A.; Park, S.J. Facile synthesis of nitrogen-enriched microporous carbons derived from imine and benzimidazole-linked polymeric framework for efficient CO_2 adsorption. *J. CO2 Util.* **2017**, *21*, 503–512. [CrossRef]
11. Daifullah, A.; Girgis, B. Removal of some substituted phenols by activated carbon obtained from agricultural waste. *Water Res.* **1998**, *32*, 1169–1177. [CrossRef]
12. Li, L.; Quinlivan, P.A.; Knappe, D.R. Effects of activated carbon surface chemistry and pore structure on the adsorption of organic contaminants from aqueous solution. *Carbon* **2002**, *40*, 2085–2100. [CrossRef]
13. Matsui, Y.; Fukuda, Y.; Inoue, T.; Matsushita, T. Effect of natural organic matter on powdered activated carbon adsorption of trace contaminants: characteristics and mechanism of competitive adsorption. *Water Res.* **2003**, *37*, 4413–4424. [CrossRef]
14. Chen, J.P.; Wu, S.; Chong, K.H. Surface modification of a granular activated carbon by citric acid for enhancement of copper adsorption. *Carbon* **2003**, *41*, 1979–1986. [CrossRef]
15. Karanfil, T.; Kilduff, J.E. Role of granular activated carbon surface chemistry on the adsorption of organic compounds. 1. Priority pollutants. *Environ. Sci. Technol.* **1999**, *33*, 3217–3224. [CrossRef]
16. Monser, L.; Adhoum, N. Modified activated carbon for the removal of copper, zinc, chromium and cyanide from wastewater. *Sep. Purif. Technol.* **2002**, *26*, 137–146. [CrossRef]
17. Jiang, Z.; Liu, Y.; Sun, X.; Tian, F.; Sun, F.; Liang, C.; You, W.; Han, C.; Li, C. Activated carbons chemically modified by concentrated H2SO4 for the adsorption of the pollutants from wastewater and the dibenzothiophene from fuel oils. *Langmuir* **2003**, *19*, 731–736. [CrossRef]
18. Faria, P.; Orfao, J.; Pereira, M. Adsorption of anionic and cationic dyes on activated carbons with different surface chemistries. *Water Res.* **2004**, *38*, 2043–2052. [CrossRef]
19. Dąbrowski, A.; Podkościelny, P.; Hubicki, Z.; Barczak, M. Adsorption of phenolic compounds by activated carbon—A critical review. *Chemosphere* **2005**, *58*, 1049–1070. [CrossRef]
20. Moreno-Castilla, C.; Lopez-Ramon, M.; Carrasco-Marın, F. Changes in surface chemistry of activated carbons by wet oxidation. *Carbon* **2000**, *38*, 1995–2001. [CrossRef]
21. Pradhan, B.K.; Sandle, N. Effect of different oxidizing agent treatments on the surface properties of activated carbons. *Carbon* **1999**, *37*, 1323–1332. [CrossRef]
22. Biniak, S.; Szymański, G.; Siedlewski, J.; Świątkowski, A. The characterization of activated carbons with oxygen and nitrogen surface groups. *Carbon* **1997**, *35*, 1799–1810. [CrossRef]
23. El-Hendawy, A.N.A. Influence of HNO_3 oxidation on the structure and adsorptive properties of corncob-based activated carbon. *Carbon* **2003**, *41*, 713–722. [CrossRef]
24. Daud, W.M.A.W.; Houshamnd, A.H. Textural characteristics, surface chemistry and oxidation of activated carbon. *J. Nat. Gas Chem.* **2010**, *19*, 267–279. [CrossRef]
25. Choi, C.H.; Park, S.H.; Woo, S.I. Heteroatom doped carbons prepared by the pyrolysis of bio-derived amino acids as highly active catalysts for oxygen electro-reduction reactions. *Green Chem.* **2011**, *13*, 406–412. [CrossRef]
26. Paraknowitsch, J.P.; Thomas, A. Doping carbons beyond nitrogen: an overview of advanced heteroatom doped carbons with boron, sulphur and phosphorus for energy applications. *Energy Environ. Sci.* **2013**, *6*, 2839–2855. [CrossRef]
27. Duan, X.; Sun, H.; Wang, Y.; Kang, J.; Wang, S. N-doping-induced nonradical reaction on single-walled carbon nanotubes for catalytic phenol oxidation. *ACS Catal.* **2014**, *5*, 553–559. [CrossRef]
28. Wang, Y.; Wang, X.; Antonietti, M. Polymeric graphitic carbon nitride as a heterogeneous organocatalyst: from photochemistry to multipurpose catalysis to sustainable chemistry. *Angew. Chem. Int. Ed.* **2012**, *51*, 68–89. [CrossRef]
29. Liang, J.; Jiao, Y.; Jaroniec, M.; Qiao, S.Z. Sulfur and nitrogen dual-doped mesoporous graphene electrocatalyst for oxygen reduction with synergistically enhanced performance. *Angew. Chem.* **2012**, *124*, 11664–11668. [CrossRef]
30. Moreno-Castilla, C. Adsorption of organic molecules from aqueous solutions on carbon materials. *Carbon* **2004**, *42*, 83–94. [CrossRef]
31. Villacañas, F.; Pereira, M.F.R.; Órfão, J.J.; Figueiredo, J.L. Adsorption of simple aromatic compounds on activated carbons. *J. Colloid Interface Sci.* **2006**, *293*, 128–136. [CrossRef] [PubMed]

32. Zhu, D.; Hyun, S.; Pignatello, J.J.; Lee, L.S. Evidence for π– π electron donor– acceptor interactions between π-donor aromatic compounds and π-acceptor sites in soil organic matter through pH effects on sorption. *Environ. Sci. Technol.* **2004**, *38*, 4361–4368. [CrossRef] [PubMed]
33. Li, Z.; Dai, S. Surface functionalization and pore size manipulation for carbons of ordered structure. *Chem. Mater.* **2005**, *17*, 1717–1721. [CrossRef]
34. Li, B.; Dai, F.; Xiao, Q.; Yang, L.; Shen, J.; Zhang, C.; Cai, M. Nitrogen-doped activated carbon for a high energy hybrid supercapacitor. *Energy Environ. Sci.* **2016**, *9*, 102–106. [CrossRef]
35. Kiciński, W.; Szala, M.; Bystrzejewski, M. Sulfur-doped porous carbons: synthesis and applications. *Carbon* **2014**, *68*, 1–32. [CrossRef]
36. Zhang, Y.; Park, S.-J. Incorporation of RuO_2 into charcoal-derived carbon with controllable microporosity by CO_2 activation for high-performance supercapacitor. *Carbon* **2017**, *122*, 287–297. [CrossRef]
37. Tamon, H.; Okazaki, M. Influence of acidic surface oxides of activated carbon on gas adsorption characteristics. *Carbon* **1996**, *34*, 741–746. [CrossRef]
38. Boehm, H.P. Surface oxides on carbon and their analysis: a critical assessment. *Carbon* **2002**, *40*, 145–149. [CrossRef]
39. Vinke, P.; Van der Eijk, M.; Verbree, M.; Voskamp, A.; Van Bekkum, H. Modification of the surfaces of a gasactivated carbon and a chemically activated carbon with nitric acid, hypochlorite and ammonia. *Carbon* **1994**, *32*, 675–686. [CrossRef]
40. Valdés, H.; Sánchez-Polo, M.; Rivera-Utrilla, J.; Zaror, C. Effect of ozone treatment on surface properties of activated carbon. *Langmuir* **2002**, *18*, 2111–2116. [CrossRef]
41. Toles, C.A.; Marshall, W.E.; Johns, M.M. Surface functional groups on acid-activated nutshell carbons. *Carbon* **1999**, *37*, 1207–1214. [CrossRef]
42. Prahas, D.; Kartika, Y.; Indraswati, N.; Ismadji, S. Activated carbon from jackfruit peel waste by H_3PO_4 chemical activation: pore structure and surface chemistry characterization. *Chem. Eng. J.* **2008**, *140*, 32–42. [CrossRef]
43. Wang, S.; Lu, G.M. Effects of acidic treatments on the pore and surface properties of Ni catalyst supported on activated carbon. *Carbon* **1998**, *36*, 283–292. [CrossRef]
44. Montes-Morán, M.; Suárez, D.; Menéndez, J.; Fuente, E. On the nature of basic sites on carbon surfaces: an overview. *Carbon* **2004**, *42*, 1219–1225. [CrossRef]
45. Moreno-Castilla, C.; Carrasco-Marin, F.; Utrera-Hidalgo, E.; Rivera-Utrilla, J. Activated carbons as adsorbents of sulfur dioxide in flowing air, Effect of their pore texture and surface basicity. *Langmuir* **1993**, *9*, 1378–1383. [CrossRef]
46. Lopez-Ramon, M.V.; Stoeckli, F.; Moreno-Castilla, C.; Carrasco-Marin, F. On the characterization of acidic and basic surface sites on carbons by various techniques. *Carbon* **1999**, *37*, 1215–1221. [CrossRef]
47. El-Sayed, Y.; Bandosz, T.J. Adsorption of valeric acid from aqueous solution onto activated carbons: role of surface basic sites. *J. Colloid Interface Sci.* **2004**, *273*, 64–72. [CrossRef]
48. Leon, C.L.; Solar, J.; Calemma, V.; Radovic, L.R. Evidence for the protonation of basal plane sites on carbon. *Carbon* **1992**, *30*, 797–811. [CrossRef]
49. Boehm, H. Some aspects of the surface chemistry of carbon blacks and other carbons. *Carbon* **1994**, *32*, 759–769. [CrossRef]
50. Shafeeyan, M.S.; Daud, W.M.A.W.; Houshmand, A.; Shamiri, A. A review on surface modification of activated carbon for carbon dioxide adsorption. *J Anal. Appl. Pyrolysis* **2010**, *89*, 143–151. [CrossRef]
51. Arrigo, R.; Hävecker, M.; Wrabetz, S.; Blume, R.; Lerch, M.; McGregor, J.; PJ Parrott, E.; Zeitler, J.A.; Gladden, L.F.; Knop-Gericke, A.; Schlögl, R.; Sheng Su, D. Tuning the acid/base properties of nanocarbons by functionalization via amination. *J. Am. Chem. Soc.* **2010**, *132*, 9616–9630. [CrossRef] [PubMed]
52. Kinoshita, K. *Carbon: Electrochemical and Physicochemical Properties*; John Wiley & Sons, Inc.: Hoboken, NJ, USA, 1988.
53. Wang, X.; Liu, R.; Waje, M.M.; Chen, Z.; Yan, Y.; Bozhilov, K.N.; Feng, P. Sulfonated ordered mesoporous carbon as a stable and highly active protonic acid catalyst. *Chem. Mater.* **2007**, *19*, 2395–2397. [CrossRef]
54. Aggarwal, D.; Goyal, M.; Bansal, R. Adsorption of chromium by activated carbon from aqueous solution. *Carbon* **1999**, *37*, 1989–1997. [CrossRef]
55. Yin, C.Y.; Aroua, M.K.; Daud, W.M.A.W. Review of modifications of activated carbon for enhancing contaminant uptakes from aqueous solutions. *Sep. Purif. Technol.* **2007**, *52*, 403–415. [CrossRef]

56. Álvarez-Merino, M.A.; López-Ramón, V.; Moreno-Castilla, C. A study of the static and dynamic adsorption of Zn (II) ions on carbon materials from aqueous solutions. *J. Colloid Interface Sci.* **2005**, *288*, 335–341. [CrossRef]
57. Goel, J.; Kadirvelu, K.; Rajagopal, C.; Garg, V.K. Removal of lead (II) by adsorption using treated granular activated carbon: batch and column studies. *J. Hazard. Mater.* **2005**, *125*, 211–220. [CrossRef] [PubMed]
58. Gomez-Serrano, V.; Macias-Garcia, A.; Espinosa-Mansilla, A.; Valenzuela-Calahorro, C. Adsorption of mercury, cadmium and lead from aqueous solution on heat-treated and sulphurized activated carbon. *Water Res.* **1998**, *32*, 1–4. [CrossRef]
59. Goyal, M.; Rattan, V.; Aggarwal, D.; Bansal, R. Removal of copper from aqueous solutions by adsorption on activated carbons. *Colloids Surf. A* **2001**, *190*, 229–238. [CrossRef]
60. Jia, Y.; Thomas, K. Adsorption of cadmium ions on oxygen surface sites in activated carbon. *Langmuir* **2000**, *16*, 1114–1122. [CrossRef]
61. Macias-Garcia, A.; Gomez-Serrano, V.; Alexandre-Franco, M.; Valenzuela-Calahorro, C. Adsorption of cadmium by sulphur dioxide treated activated carbon. *J. Hazard. Mater.* **2003**, *103*, 141–152. [CrossRef]
62. Park, S.J.; Jang, Y.S. Pore structure and surface properties of chemically modified activated carbons for adsorption mechanism and rate of Cr (VI). *J. Colloid Interface Sci.* **2002**, *249*, 458–463. [CrossRef] [PubMed]
63. Wu, S.; Chen, P. Modification of a commercial activated carbon for metal adsorption by several approaches. In Proceedings of the 2001 International Containment & Remediation Technology Conference and Exhibition, Orlando, FL, USA, June 2001.
64. Babel, S.; Kurniawan, T.A. Cr (VI) removal from synthetic wastewater using coconut shell charcoal and commercial activated carbon modified with oxidizing agents and/or chitosan. *Chemosphere* **2004**, *54*, 951–967. [CrossRef] [PubMed]
65. Puziy, A.; Poddubnaya, O.; Kochkin, Y.N.; Vlasenko, N.; Tsyba, M. Acid properties of phosphoric acid activated carbons and their catalytic behavior in ethyl-tert-butyl ether synthesis. *Carbon* **2010**, *48*, 706–713. [CrossRef]
66. Bedia, J.; Rosas, J.M.; Marquez, J.; Rodriguez-Mirasol, J.; Cordero, T. Preparation and characterization of carbon based acid catalysts for the dehydration of 2-propanol. *Carbon* **2009**, *47*, 286–294. [CrossRef]
67. Puziy, A.; Poddubnaya, O.; Martinez-Alonso, A.; Suárez-Garcia, F.; Tascón, J. Synthetic carbons activated with phosphoric acid: I. Surface chemistry and ion binding properties. *Carbon* **2002**, *40*, 1493–1505. [CrossRef]
68. Elangovan, R.; Philip, L.; Chandraraj, K. Biosorption of chromium species by aquatic weeds: kinetics and mechanism studies. *J. Hazard Mater.* **2008**, *152*, 100–112. [CrossRef] [PubMed]
69. Martín-Lara, M.; Pagnanelli, F.; Mainelli, S.; Calero, M.; Toro, L. Chemical treatment of olive pomace: Effect on acid-basic properties and metal biosorption capacity. *J. Hazard Mater.* **2008**, *156*, 448–457. [CrossRef] [PubMed]
70. Mameri, N.; Aioueche, F.; Belhocine, D.; Grib, H.; Lounici, H.; Piron, DL. Preparation of activated carbon from olive mill solid residue. *J. Chem. Technol. Biotechnol.* **2000**, *75*, 625–631. [CrossRef]
71. Amuda, O.; Giwa, A.; Bello, I. Removal of heavy metal from industrial wastewater using modified activated coconut shell carbon. *Biochem. Eng. J.* **2007**, *36*, 174–181. [CrossRef]
72. Montanher, S.; Oliveira, E.; Rollemberg, M. Removal of metal ions from aqueous solutions by sorption onto rice bran. *J. Hazard. Mater.* **2005**, *117*, 207–211. [CrossRef] [PubMed]
73. Blázquez, G.; Calero, M.; Ronda, A.; Tenorio, G.; Martín-Lara, M. Study of kinetics in the biosorption of lead onto native and chemically treated olive stone. *J. Ind. Eng. Chem.* **2014**, *20*, 2754–2760. [CrossRef]
74. Kikuchi, Y.; Qian, Q.; Machida, M.; Tatsumoto, H. Effect of ZnO loading to activated carbon on Pb (II) adsorption from aqueous solution. *Carbon* **2006**, *44*, 195–202. [CrossRef]
75. Barton, S.; Evans, M.; Halliop, E.; MacDonald, J. Acidic and basic sites on the surface of porous carbon. *Carbon* **1997**, *35*, 1361–1366. [CrossRef]
76. Roosen, J.; Babu, C.M.; Binnemans, K. Functionalised activated carbon for the adsorption of rare-earth elements from aqueous solutions. In Proceedings of the 4th International Symposium on Enhanced Landfill Mining (ELFM IV), Mechelen, Belgium, 5–6 February 2018; pp. 305–310.
77. Foo, G.S.; Sievers, C. Synergistic effect between defect sites and functional groups on the hydrolysis of cellulose over activated carbon. *Chem. Sus. Chem.* **2015**, *8*, 534–543. [CrossRef] [PubMed]
78. Maroto-Valer, M.M.; Dranca, I.; Lupascu, T.; Nastas, R. Effect of adsorbate polarity on thermodesorption profiles from oxidized and metal-impregnated activated carbons. *Carbon* **2004**, *42*, 2655–2659. [CrossRef]

79. Rios, R.R.A.; Alves, D.E.; Dalmázio, I.; Bento, S.F.V.; Donnici, C.L.; Lago, R.M. Tailoring activated carbon by surface chemical modification with O, S and N containing molecules. *Mater. Res.* **2003**, *6*, 129–135. [CrossRef]
80. Aburub, A.; Wurster, D.E. Phenobarbital interactions with derivatized activated carbon surfaces. *J. Colloid Interface Sci.* **2006**, *296*, 79–85. [CrossRef]
81. García, A.B.; Martínez-Alonso, A.; y Leon, C.A.L.; Tascón, J.M. Modification of the surface properties of an activated carbon by oxygen plasma treatment. *Fuel* **1998**, *77*, 613–624. [CrossRef]
82. Domingo-Garcia, M.; Lopez-Garzon, F.; Perez-Mendoza, M. Effect of some oxidation treatments on the textural characteristics and surface chemical nature of an activated carbon. *J. Colloid Interface Sci.* **2000**, *222*, 233–240. [CrossRef]
83. Lee, D.; Hong, S.H.; Paek, K.H.; Ju, W.T. Adsorbability enhancement of activated carbon by dielectric barrier discharge plasma treatment. *Surf. Coat. Tech.* **2005**, *200*, 2277–2282. [CrossRef]
84. Santiago, M.; Stüber, F.; Fortuny, A.; Fabregat, A.; Font, J. Modified activated carbons for catalytic wet air oxidation of phenol. *Carbon* **2005**, *43*, 2134–2145. [CrossRef]
85. Terzyk, A.P. Further insights into the role of carbon surface functionalities in the mechanism of phenol adsorption. *J. Colloid Interface Sci.* **2003**, *268*, 301–329. [CrossRef]
86. Jansen, R.; Van Bekkum, H. Amination and ammoxidation of activated carbons. *Carbon* **1994**, *32*, 1507–1516. [CrossRef]
87. Dastgheib, S.A.; Karanfil, T.; Cheng, W. Tailoring activated carbons for enhanced removal of natural organic matter from natural waters. *Carbon* **2004**, *42*, 547–557. [CrossRef]
88. Guerrero-Ruiz, A.; Rodriguez-Ramos, I.; Rodriguez-Reinoso, F.; Moreno-Castilla, C.; López-González, J. The role of nitrogen and oxygen surface groups in the behavior of carbon-supported iron and ruthenium catalysts. *Carbon* **1988**, *26*, 417–423. [CrossRef]
89. Przepiórski, J. Enhanced adsorption of phenol from water by ammonia-treated activated carbon. *J. Hazard. Mater.* **2006**, *135*, 453–456. [CrossRef]
90. Chen, W.; Cannon, F.S.; Rangel-Mendez, J.R. Ammonia-tailoring of GAC to enhance perchlorate removal. I: Characterization of NH_3 thermally tailored GACs. *Carbon* **2005**, *43*, 573–580. [CrossRef]
91. Economy, J.; Foster, K.; Andreopoulos, A.; Jung, H. Tailoring Carbon Fibers for Adsorbing Volatiles: Various Chemical Treatments are used to Change Surface Area, Pore Geometry and Adsorption Character Istics. *Chemtech* **1992**, *22*, 597–603.
92. Stöhr, B.; Boehm, H.; Schlögl, R. Enhancement of the catalytic activity of activated carbons in oxidation reactions by thermal treatment with ammonia or hydrogen cyanide and observation of a superoxide species as a possible intermediate. *Carbon* **1991**, *29*, 707–720. [CrossRef]
93. Boehm, H.P.; Mair, G.; Stoehr, T.; De Rincón, A.R.; Tereczki, B. Carbon as a catalyst in oxidation reactions and hydrogen halide elimination reactions. *Fuel* **1984**, *63*, 1061–1063. [CrossRef]
94. Bota, K.B.; Abotsi, G.M. Ammonia: A reactive medium for catalysed coal gasification. *Fuel* **1994**, *73*, 1354–1357. [CrossRef]
95. Meldrum, B.J.; Rochester, C.H. In situ infrared study of the surface oxidation of activated carbon in oxygen and carbon dioxide. *J. Chem. Soc. Faraday Trans.* **1990**, *86*, 861–865. [CrossRef]
96. Karanfil, T.; Schlautman, M.A.; Kilduff, J.E.; Weber, W.J. Adsorption of organic macromolecules by granular activated carbon. 2. Influence of dissolved oxygen. *Environ. Sci. Technol.* **1996**, *30*, 2195–2201. [CrossRef]
97. Park, S.-J.; Kim, B.-J. Ammonia removal of activated carbon fibers produced by oxyfluorination. *J. Colloid Interface Sci.* **2005**, *291*, 597–599. [CrossRef] [PubMed]
98. González Plaza, M.; Pevida García, C.; Arias Rozada, B.; Fermoso Domínguez, J.; Rubiera González, F.; Pis Martínez, J.J. A comparison of two methods for producing CO_2 capture adsorbents. *Energy Procedia* **2009**, *1*, 1107–1113. [CrossRef]
99. Przepiórski, J.; Skrodzewicz, M.; Morawski, A. High temperature ammonia treatment of activated carbon for enhancement of CO2 adsorption. *Appl. Surf. Sci.* **2004**, *225*, 235–242. [CrossRef]
100. Radovic, L.; Silva, I.; Ume, J.; Menendez, J.; Leon, C.L.Y.; Scaroni, A. An experimental and theoretical study of the adsorption of aromatics possessing electron-withdrawing and electron-donating functional groups by chemically modified activated carbons. *Carbon* **1997**, *35*, 1339–1348. [CrossRef]
101. Plaza, M.G.; Pevida, C.; Arias, B.; Casal, M.; Martín, C.; Fermoso, J.; Rubiera, F.; Pis, J. Different approaches for the development of low-cost CO_2 adsorbents. *J. Environ. Eng.* **2009**, *135*, 426–432. [CrossRef]

102. Abotsi, G.M.; Scaroni, A.W. Reaction of carbons with ammonia: effects on the surface charge and molybdenum adsorption. *Carbon* **1990**, *28*, 79–84. [CrossRef]
103. Adhoum, N.; Monser, L. Removal of phthalate on modified activated carbon: Application to the treatment of industrial wastewater. *Sep. Purif. Technol.* **2004**, *38*, 233–239. [CrossRef]
104. Henning, K.-D.; Schäfer, S. Impregnated activated carbon for environmental protection. *Gas Sep. Purif.* **1993**, *7*, 235–240. [CrossRef]
105. Rajaković, L.V.; Ristić, M.D. Sorption of boric acid and borax by activated carbon impregnated with various compounds. *Carbon* **1996**, *34*, 769–774. [CrossRef]
106. Şayan, E. Ultrasound-assisted preparation of activated carbon from alkaline impregnated hazelnut shell: An optimization study on removal of Cu^{2+} from aqueous solution. *Chem. Eng. J.* **2006**, *115*, 213–218. [CrossRef]
107. Adhoum, N.; Monser, L. Removal of cyanide from aqueous solution using impregnated activated carbon. *Chem. Eng. Process.* **2002**, *41*, 17–21. [CrossRef]
108. Huang, C.; Vane, L.M. Enhancing removal by a activated carbon. *J. Water Pollut. Control. Fed.* **1989**, 1596–1603.
109. Ramos, R.L.; Ovalle-Turrubiartes, J.; Sanchez-Castillo, M. Adsorption of fluoride from aqueous solution on aluminum-impregnated carbon. *Carbon* **1999**, *37*, 609–617. [CrossRef]
110. Lee, S.J.; Seo, Y.C.; Jurng, J.; Lee, T.G. Removal of gas-phase elemental mercury by iodine-and chlorine-impregnated activated carbons. *Atmos. Environ.* **2004**, *38*, 4887–4893. [CrossRef]
111. Korpiel, J.A.; Vidic, R.D. Effect of sulfur impregnation method on activated carbon uptake of gas-phase mercury. *Environ. Sci. Technol.* **1997**, *31*, 2319–2325. [CrossRef]
112. Rajaković, L.V. The sorption of arsenic onto activated carbon impregnated with metallic silver and copper. *Sep. Sci. Technol.* **1992**, *27*, 1423–1433. [CrossRef]
113. Deveci, H.; Yazıcı, E.; Alp, I.; Uslu, T. Removal of cyanide from aqueous solutions by plain and metal-impregnated granular activated carbons. *Int. J. Miner. Process.* **2006**, *79*, 198–208. [CrossRef]
114. Wu, S.H.; Pendleton, P. Adsorption of anionic surfactant by activated carbon: effect of surface chemistry, ionic strength and hydrophobicity. *J. Colloid Interface Sci.* **2001**, *243*, 306–315. [CrossRef]
115. Xiao, J.X.; Zhang, Y.; Wang, C.; Zhang, J.; Wang, C.M.; Bao, Y.X.; Zhao, Z.G. Adsorption of cationic–anionic surfactant mixtures on activated carbon. *Carbon* **2005**, *43*, 1032–1038. [CrossRef]
116. González-García, C.; González-Martín, M.; Gómez-Serrano, V.; Bruque, J.; Labajos-Broncano, L. Analysis of the adsorption isotherms of a non-ionic surfactant from aqueous solution onto activated carbons. *Carbon* **2001**, *39*, 849–855. [CrossRef]
117. González-García, C.; Gonzalez-Martin, M.; Denoyel, R.; Gallardo-Moreno, A.; Labajos-Broncano, L.; Bruque, J. Ionic surfactant adsorption onto activated carbons. *J. Colloid Interface Sci.* **2004**, *278*, 257–264. [CrossRef] [PubMed]
118. González-García, C.; González-Martín, M.; González, J.; Sabio, E.; Ramiro, A.; Gañán, J. Nonionic surfactants adsorption onto activated carbon. Influence of the polar chain length. *Powder Technol.* **2004**, *148*, 32–37. [CrossRef]
119. Parette, R.; Cannon, F.S. The removal of perchlorate from groundwater by activated carbon tailored with cationic surfactants. *Water Res.* **2005**, *39*, 4020–4028. [CrossRef]
120. Choi, H.D.; Jung, W.S.; Cho, J.M.; Ryu, B.G.; Yang, J.S.; Baek, K. Adsorption of Cr (VI) onto cationic surfactant-modified activated carbon. *J. Hazard. Mater.* **2009**, *166*, 642–646. [CrossRef]
121. Mohamed, M.M. Acid dye removal: comparison of surfactant-modified mesoporous FSM-16 with activated carbon derived from rice husk. *J. Colloid Interface Sci.* **2004**, *272*, 28–34. [CrossRef]
122. Choi, H.D.; Shin, M.C.; Kim, D.H.; Jeon, C.S.; Baek, K. Removal characteristics of reactive black 5 using surfactant-modified activated carbon. *Desalination* **2008**, *23*, 290–298. [CrossRef]
123. Nadeem, M.; Shabbir, M.; Abdullah, M.; Shah, S.; McKay, G. Sorption of cadmium from aqueous solution by surfactant-modified carbon adsorbents. *Chem. Eng. J.* **2009**, *148*, 365–370 [CrossRef]
124. Namasivayam, C.; Sureshkumar, M. Removal of chromium (VI) from water and wastewater using surfactant modified coconut coir pith as a biosorbent. *Bioresour. Technol.* **2008**, *99*, 2218–2225. [CrossRef] [PubMed]
125. García-Martín, J.; López-Garzón, R.; Godino-Salido, M.L.; Gutiérrez-Valero, M.D.; Arranz-Mascarós, P.; Cuesta, R.; Carrasco-Marín, F. Ligand adsorption on an activated carbon for the removal of chromate ions from aqueous solutions. *Langmuir* **2005**, *21*, 6908–6914. [CrossRef]

126. Zhao, Y.; Liu, C.; Feng, M.; Chen, Z.; Li, S.; Tian, G. Solid phase extraction of uranium (VI) onto benzoylthiourea-anchored activated carbon. *J. Hazard. Mater.* **2010**, *176*, 119–124. [CrossRef]
127. Song, Q.; Ma, L.; Liu, J.; Bai, C.; Geng, J.; Wang, H. Preparation and adsorption performance of 5-azacytosine-functionalized hydrothermal carbon for selective solid-phase extraction of uranium. *J. Colloid Interface Sci.* **2012**, *386*, 291–299. [CrossRef] [PubMed]
128. Zhu, J.; Deng, B.; Yang, J.; Gang, D. Modifying activated carbon with hybrid ligands for enhancing aqueous mercury removal. *Carbon* **2009**, *47*, 2014–2025. [CrossRef]
129. Zhou, J.H.; Sui, Z.J.; Zhu, J.; Li, P.; Chen, D.; Dai, Y.C.; Yuan, W.K. Characterization of surface oxygen complexes on carbon nanofibers by TPD, XPS and FT-IR. *Carbon* **2007**, *45*, 785–796. [CrossRef]
130. Boehm, H.P. Chemical Identification of Surface Groups. *Adv. Catal.* **1966**, *16*, 179–274. [CrossRef]
131. El-Sayed, Y.; Bandosz, T.J. Effect of increased basicity of activated carbon surface on valeric acid adsorption from aqueous solution activated carbon. *Phys. Chem. Chem. Phys.* **2003**, *5*, 4892–4898. [CrossRef]
132. Jia, Y.; Xiao, B.; Thomas, K. Adsorption of metal ions on nitrogen surface functional groups in activated carbons. *Langmuir* **2002**, *18*, 470–478. [CrossRef]
133. Figueiredo, J.; Pereira, M.; Freitas, M.; Orfao, J. Modification of the surface chemistry of activated carbons. *Carbon* **1999**, *37*, 1379–1389. [CrossRef]
134. Vickers, P.E.; Watts, J.F.; Perruchot, C.; Chehimi, M.M. The surface chemistry and acid–base properties of a PAN-based carbon fibre. *Carbon* **2000**, *38*, 675–689. [CrossRef]
135. Sellitti, C.; Koenig, J.; Ishida, H. Surface characterization of graphitized carbon fibers by attenuated total reflection Fourier transform infrared spectroscopy. *Carbon* **1990**, *28*, 221–228. [CrossRef]
136. Gomez-Serrano, V.; Pastor-Villegas, J.; Perez-Florindo, A.; Duran-Valle, C.; Valenzuela-Calahorro, C. FT-IR study of rockrose and of char and activated carbon. *J. Anal. Appl. Pyrolysis* **1996**, *36*, 71–80. [CrossRef]
137. Julien, F.; Baudu, M.; Mazet, M. Relationship between chemical and physical surface properties of activated carbon. *Water Res.* **1998**, *32*, 3414–3424. [CrossRef]
138. Jung, M.W.; Ahn, K.H.; Lee, Y.; Kim, K.P.; Rhee, J.S.; Park, J.T.; Paeng, K.J. Adsorption characteristics of phenol and chlorophenols on granular activated carbons (GAC). *Microchem. J.* **2001**, *70*, 123–131. [CrossRef]
139. Li, Z.; Yan, W.; Dai, S. Surface Functionalization of Ordered Mesoporous Carbons A Comparative Study. *Langmuir* **2005**, *21*, 11999–12006. [CrossRef]
140. Sakintuna, B.; Yürüm, Y. Preparation and characterization of mesoporous carbons using a Turkish natural zeolitic template/furfuryl alcohol system. *Microporous Mesoporous Mater.* **2006**, *93*, 304–312. [CrossRef]
141. Donnet, J.; Custodéro, E.; Wang, T.; Hennebert, G. Energy site distribution of carbon black surfaces by inverse gas chromatography at finite concentration conditions. *Carbon* **2002**, *40*, 163–167. [CrossRef]
142. Zielke, U.; Hüttinger, K.; Hoffman, W. Surface-oxidized carbon fibers: I. Surface structure and chemistry. *Carbon* **1996**, *34*, 983–998. [CrossRef]
143. Mayes, R.T.; Fulvio, P.F.; Ma, Z.; Dai, S. Phosphorylated mesoporous carbon as a solid acid catalyst. *Phys. Chem. Chem. Phys.* **2011**, *13*, 2492–2944. [CrossRef] [PubMed]
144. Fulvio, P.F.; Mayes, R.T.; Bauer, J.C.; Wang, X.; Mahurin, S.M.; Veith, G.M.; Dai, S. "One-pot" synthesis of phosphorylated mesoporous carbon heterogeneous catalysts with tailored surface acidity. *Catal. Today* **2012**, *186*, 12–19. [CrossRef]
145. Otake, Y.; Jenkins, R.G. Characterization of oxygen-containing surface complexes created on a microporous carbon by air and nitric acid treatment. *Carbon* **1993**, *31*, 109–121. [CrossRef]
146. Dandekar, A.; Baker, R.; Vannice, M. Characterization of activated carbon, graphitized carbon fibers and synthetic diamond powder using TPD and DRIFTS. *Carbon* **1998**, *36*, 1821–1831. [CrossRef]
147. Figueiredo, J.L.; Pereira, M.F.; Freitas, M.M.; Órfão, J.J. Characterization of active sites on carbon catalysts. *Ind. Eng. Chem. Res.* **2007**, *46*, 4110–4115. [CrossRef]
148. Rodrıguez-Reinoso, F.; Molina-Sabio, M. Textural and chemical characterization of microporous carbons. *Adv. Colloid Interface Sci.* **1998**, *76*, 271–294. [CrossRef]
149. Salame, I.I.; Bandosz, T.J. Role of surface chemistry in adsorption of phenol on activated carbons. *J. Colloid Interface Sci.* **2003**, *264*, 307–312. [CrossRef]
150. Bansal, R.; Aggarwal, D.; Goyal, M.; Kaistha, B. Influence of carbon-oxygen surface groups on the adsorption of phenol by activated carbons. *NISCAIR-CSIR* **2002**, *9*, 290–296.

151. Haydar, S.; Moreno-Castilla, C.; Ferro-García, M.; Carrasco-Marın, F.; Rivera-Utrilla, J.; Perrard, A.; Joly, J. Regularities in the temperature-programmed desorption spectra of CO_2 and CO from activated carbons. *Carbon* **2000**, *38*, 1297–1308. [CrossRef]
152. Montoya, A.; Mondragon, F.; Truong, T.N. First-principles kinetics of CO desorption from oxygen species on carbonaceous surface. *J. Phys. Chem. A* **2002**, *106*, 4236–4239. [CrossRef]
153. Turner, N.H.; Schreifels, J.A. Surface analysis: X-ray photoelectron spectroscopy and auger electron spectroscopy. *Anal. Chem.* **2000**, *72*, 99–110. [CrossRef]
154. McGuire, G.; Weiss, P.; Kushmerick, J.; Johnson, J.; Simko, S.J.; Nemanich, R.; Parikh, N.R.; Chopra, D. Surface characterization. *Anal. Chem.* **1997**, *69*, 231–250. [CrossRef]
155. Gardella, J.A., Jr. Recent developments in instrumentation for x-ray photoelectron spectroscopy. *Anal. Chem.* **1989**, *61*, 589A–600A. [CrossRef]
156. Li, K.; Ling, L.; Lu, C.; Qiao, W.; Liu, J.; Liu, L.; Mochida, I. Catalytic removal of SO_2 over ammonia-activated carbon fibers. *Carbon* **2001**, *39*, 1803–1808. [CrossRef]
157. Polovina, M.; Babić, B.; Kaluderović, B.; Dekanski, A. Surface characterization of oxidized activated carbon cloth. *Carbon* **1997**, *35*, 1047–1052. [CrossRef]
158. Bradley, R.; Ling, X.; Sutherland, I. An investigation of carbon fibre surface chemistry and reactivity based on XPS and surface free energy. *Carbon* **1993**, *31*, 1115–1120. [CrossRef]
159. Grzybek, T.; Kreiner, K. Surface changes in coals after oxidation. 1. X-ray photoelectron spectroscopy studies. *Langmuir* **1997**, *13*, 909–912. [CrossRef]
160. Boudou, J.P.; Parent, P.; Suárez-García, F.; Villar-Rodil, S.; Martínez-Alonso, A.; Tascón, J. Nitrogen in aramid-based activated carbon fibers by TPD, XPS and XANES. *Carbon* **2006**, *44*, 2452–2462. [CrossRef]
161. Severini, F.; Formaro, L.; Pegoraro, M.; Posca, L. Chemical modification of carbon fiber surfaces. *Carbon* **2002**, *40*, 735–741. [CrossRef]
162. Jansen, R.; Van Bekkum, H. XPS of nitrogen-containing functional groups on activated carbon. *Carbon* **1995**, *33*, 1021–1027. [CrossRef]
163. Mangun, C.L.; Benak, K.R.; Economy, J.; Foster, K.L. Surface chemistry, pore sizes and adsorption properties of activated carbon fibers and precursors treated with ammonia. *Carbon* **2001**, *39*, 1809–1820. [CrossRef]
164. Kelemen, S.; Gorbaty, M.; Kwiatek, P. Quantification of nitrogen forms in Argonne premium coals. *Energy Fuels* **1994**, *8*, 896–906. [CrossRef]
165. Perry, D.L.; Grint, A. Application of XPS to coal characterization. *Fuel* **1983**, *62*, 1024–1033. [CrossRef]
166. Kambara, S.; Takarada, T.; Yamamoto, Y.; Kato, K. Relation between functional forms of coal nitrogen and formation of nitrogen oxide (NOx) precursors during rapid pyrolysis. *Energy Fuels* **1993**, *7*, 1013–1020. [CrossRef]
167. Balathanigaimani, M.; Shim, W.G.; Lee, M.J.; Kim, C.; Lee, J.W.; Moon, H. Highly porous electrodes from novel corn grains-based activated carbons for electrical double layer capacitors. *Electrochem. Commun.* **2008**, *10*, 868–871. [CrossRef]
168. Candelaria, S.L.; Garcia, B.B.; Liu, D.; Cao, G. Nitrogen modification of highly porous carbon for improved supercapacitor performance. *J. Mater. Chem.* **2012**, *22*, 9884–9889. [CrossRef]
169. Wang, D.W.; Feng, L.; Min, L.; Cheng, H.-m. Improved capacitance of SBA-15 templated mesoporous carbons after modification with nitric acid oxidation. *New Carbon Mater.* **2007**, *22*, 307–314. [CrossRef]
170. Heidari, A.; Younesi, H.; Rashidi, A.; Ghoreyshi, A.A. Evaluation of CO2 adsorption with eucalyptus wood based activated carbon modified by ammonia solution through heat treatment. *Chem. Eng. J.* **2014**, *254*, 503–513. [CrossRef]
171. Zhang, C.; Song, W.; Sun, G.; Xie, L.; Wang, J.; Li, K. CO2 capture with activated carbon grafted by nitrogenous functional groups. *Energy Fuels* **2013**, *27*, 4818–4823. [CrossRef]
172. Gholidoust, A.; Atkinson, J.D.; Hashisho, Z. Enhancing CO2 adsorption via amine-impregnated activated carbon from oil sands coke. *Energy Fuels* **2017**, *31*, 1756–1763. [CrossRef]
173. Wang, S.; Zhu, Z. Effects of acidic treatment of activated carbons on dye adsorption. *Dyes Pigm.* **2007**, *75*, 306–314. [CrossRef]
174. Pereira, M.F.R.; Soares, S.F.; Órfão, J.J.; Figueiredo, J.L. Adsorption of dyes on activated carbons: influence of surface chemical groups. *Carbon* **2003**, *41*, 811–821. [CrossRef]
175. Zhai, Y.; Dou, Y.; Zhao, D.; Fulvio, P.F.; Mayes, R.T.; Dai, S. Carbon materials for chemical capacitive energy storage. *Adv. Mater.* **2011**, *23*, 4828–4850. [CrossRef] [PubMed]

176. Ismanto, A.E.; Wang, S.; Soetaredjo, F.E.; Ismadji, S. Preparation of capacitor's electrode from cassava peel waste. *Bioresour. Technol.* **2010**, *101*, 3534–3540. [CrossRef] [PubMed]
177. Elmouwahidi, A.; Zapata-Benabithe, Z.; Carrasco-Marín, F.; Moreno-Castilla, C. Activated carbons from KOH-activation of argan (Argania spinosa) seed shells as supercapacitor electrodes. *Bioresour. Technol.* **2012**, *111*, 185–190. [CrossRef] [PubMed]
178. Liu, M.C.; Kong, L.B.; Zhang, P.; Luo, Y.C.; Kang, L. Porous wood carbon monolith for high-performance supercapacitors. *Electrochim. Acta* **2012**, *60*, 443–448. [CrossRef]
179. Pevida, C.; Plaza, M.G.; Arias, B.; Fermoso, J.; Rubiera, F.; Pis, J. Surface modification of activated carbons for CO_2 capture. *Appl. Surf. Sci.* **2008**, *254*, 7165–7172. [CrossRef]
180. Maroto-Valer, M.M.; Tang, Z.; Zhang, Y. CO_2 capture by activated and impregnated anthracites. *Fuel Process. Technol.* **2005**, *86*, 1487–1502. [CrossRef]
181. Hao, G.P.; Li, W.C.; Qian, D.; Lu, A.H. Rapid synthesis of nitrogen-doped porous carbon monolith for CO_2 capture. *Adv. Mater.* **2010**, *22*, 853–857. [CrossRef]
182. Sevilla, M.; Valle-Vigón, P.; Fuertes, A.B. N-doped polypyrrole-based porous carbons for CO_2 capture. *Adv. Funct. Mater.* **2011**, *21*, 2781–2787. [CrossRef]
183. Leng, C.C.; Pinto, N. Effects of surface properties of activated carbons on adsorption behavior of selected aromatics. *Carbon* **1997**, *35*, 1375–1385. [CrossRef]
184. Mahajan, O.P.; Moreno-Castilla, C.; Walker, P., Jr. Surface-treated activated carbon for removal of phenol from water. *Sep. Sci. Technol.* **1980**, *15*, 1733–1752. [CrossRef]
185. Franz, M.; Arafat, H.A.; Pinto, N.G. Effect of chemical surface heterogeneity on the adsorption mechanism of dissolved aromatics on activated carbon. *Carbon* **2000**, *38*, 1807–1819. [CrossRef]
186. Daifullah, A.; Girgis, B. Impact of surface characteristics of activated carbon on adsorption of BTEX. *Colloids Surf. A* **2003**, *214*, 181–193. [CrossRef]
187. Rivera-Utrilla, J.; Sanchez-Polo, M.; Carrasco-Marin, F. Adsorption of 1, 3, 6-naphthalenetrisulfonic acid on activated carbon in the presence of Cd (II), Cr (III) and Hg (II) Importance of electrostatic interactions. *Langmuir* **2003**, *19*, 10857–10861. [CrossRef]
188. El-Sayed, Y.; Bandosz, T.J. Role of surface oxygen groups in incorporation of nitrogen to activated carbons via ethylmethylamine adsorption. *Langmuir* **2005**, *21*, 1282–1289. [CrossRef] [PubMed]
189. Órfão, J.; Silva, A.; Pereira, J.; Barata, S.; Fonseca, I.; Faria, P.; Pereira, M. Adsorption of a reactive dye on chemically modified activated carbons—influence of pH. *J. Colloid Interface Sci.* **2006**, *296*, 480–489. [CrossRef]
190. Attia, A.A.; Rashwan, W.E.; Khedr, S.A. Capacity of activated carbon in the removal of acid dyes subsequent to its thermal treatment. *Dyes Pigments.* **2006**, *69*, 128–136. [CrossRef]
191. Wang, S.; Zhu, Z.; Coomes, A.; Haghseresht, F.; Lu, G. The physical and surface chemical characteristics of activated carbons and the adsorption of methylene blue from wastewater. *J. Colloid Interface Sci.* **2005**, *284*, 440–446. [CrossRef]

© 2019 by the authors. Licensee MDPI, Basel, Switzerland. This article is an open access article distributed under the terms and conditions of the Creative Commons Attribution (CC BY) license (http://creativecommons.org/licenses/by/4.0/).

Article

Chemical Modification of Novel Glycosidases from *Lactobacillus plantarum* Using Hyaluronic Acid: Effects on High Specificity against 6-Phosphate Glucopyranoside

Carlos Perez-Rizquez [1], David Lopez-Tejedor [1], Laura Plaza-Vinuesa [2], Blanca de las Rivas [2], Rosario Muñoz [2], Jose Cumella [3] and Jose M. Palomo [1],*

1. Department of Biocatalysis, Institute of Catalysis (CSIC), Marie Curie 2, Campus UAM, Cantoblanco, E-28049 Madrid, Spain; c.p.rizquez@csic.es (C.P.-R.); david.lopez@csic.es (D.L.-T.)
2. Department of Microbial Biotechnology, Institute of Food Science, Technology and Nutrition (ICTAN-CSIC), José Antonio Novais 10, 28040 Madrid, Spain; laura.plaza@ictan.csic.es (L.P.-V.); blanca.r@csic.es (B.d.l.R.); r.munoz@csic.es (R.M.)
3. Institute of Medicinal Chemistry (CSIC), Juan de la Cierva 3, E-28006 Madrid, Spain; working_jose@hotmail.com
* Correspondence: josempalomo@icp.csic.es; Tel.: +34-91-585-4768

Received: 9 April 2019; Accepted: 7 May 2019; Published: 9 May 2019

Abstract: Three novel glycosidases produced from *Lactobacillus plantarum*, so called Lp_0440, Lp_2777, and Lp_3525, were isolated and overexpressed on *Escherichia coli* containing a His-tag for specific purification. Their specific activity was evaluated against the hydrolysis of *p*-nitrophenylglycosides and *p*-nitrophenyl-6-phosphate glycosides (glucose and galactose) at pH 7. All three were modified with hyaluronic acid (HA) following two strategies: A simple coating by direct incubation at alkaline pH or direct chemical modification at pH 6.8 through preactivation of HA with carbodiimide (EDC) and N-hydroxysuccinimide (NHS) at pH 4.8. The modifications exhibited important effect on enzyme activity and specificity against different glycopyranosides in the three cases. Physical modification showed a radical decrease in specific activity on all glycosidases, without any significant change in enzyme specificity toward monosaccharide (glucose or galactose) or glycoside (C-6 position free or phosphorylated). However, the surface covalent modification of the enzymes showed very interesting results. The glycosidase Lp_0440 showed low glycoside specificity at 25 °C, showing the same activity against *p*-nitrophenyl-glucopyranoside (*p*NP-Glu) or *p*-nitrophenyl-6-phosphate glucopyranoside (*p*NP-6P-Glu). However, the conjugated cHA-Lp_0440 showed a clear increase in the specificity towards the *p*NP-Glu and no activity against *p*NP-6P-Glu. The other two glycosidases (Lp_2777 and Lp_3525) showed high specificity towards *p*NP-6P-glycosides, especially to the glucose derivative. The HA covalent modification of Lp_3525 (cHA-Lp_3525) generated an enzyme completely specific against the *p*NP-6P-Glu (phosphoglycosidase) maintaining more than 80% of the activity after chemical modification. When the temperature was increased, an alteration of selectivity was observed. Lp_0440 and cHA-Lp_0440 only showed activity against *p*-nitrophenyl-galactopyranoside (*p*NP-Gal) at 40 °C, higher than at 25 °C in the case of the conjugated enzyme.

Keywords: glycosidades; phosphate-glycopyranosides; phosphoglucosidase; chemical modification; hyaluronic acid

1. Introduction

Chemical modification of enzymes has been described as a fascinating approach for changing their catalytic properties [1–5]. One of the most generic strategies, which does not require previous genetic modification of the enzyme, is focused on the incorporation of molecules on protein residues chemical groups (COOH, NH_2, OH) or by single chemical modification on the N terminus (the most reactive group at neutral pH) [3–6]. Both strategies have been successfully used to improve the activity or stability of enzymes in several cases [7–10]. However, very few examples have been reported, mainly in lipases, for the alteration of selectivity and, specifically, regioselectivity [9,10].

In this case, a particularly important application regarding the control of glycosidases regioselectivity [11,12] is their critical role in carbohydrate chemistry [13,14]. The process to control and modulate these enzymes for carbohydrate synthesis is a very important issue considering the key role that sugars play in a broad range of biological processes [15,16]. For example, the production of new sugars with different properties and the modification of natural products to enhance drugs' functional properties such as solubility, pharmacokinetics, or pharmacodynamics [17,18].

From this point of view, phosphorylated carbohydrates represent an important class of sugars very relevant in biology [19]. For example, a challenging research line is obtaining Xeno nucleic acids (XNAs), synthetic nucleic acid analogues that have a different sugar backbone in comparison to DNA or RNA and could serve as building blocks for completely new genetic systems [20]. In particular, 1,5-anhydrohexitol nucleic acid (HNA) presents a glucose-mimic core with a phosphate group in C-6 [21]. However, phosphoglycosynthetic enzymes are not that widespread and need to be synthesized for their catalytic application.

Therefore, controlling the specificity of natural specific phosphoglycosidases is a very important challenge for future applications. In this work, we present the effect on regioselectivity of the chemical modification of three different glucosidases from a bacterial source, *Lactobacillus plantarum*, with a high selectivity to *p*-nitrophenyl-6-phosphate-glycopyranosides.

One of the most interesting modification of enzymes has been achieved using polymers [22]. For example, the use of tailor-made dextran polymers showed interesting results, obtaining novel and selective biocatalysts [7–10,23].

Hyaluronic acid is a well-known, widely studied polymer because of its interesting properties. It has been used as conjugate with different enzymes in biomedical applications, such as therapeutic proteins [24,25]. Structurally, is a glycopolymer based on a linear repetitive unit of a 1,4-disaccharide constituted by a glucuronic acid and a 2-acetamido-glucosamine units. As a comparison, dextran repetitive unit is a 1,3-1,6-trisaccharide (Scheme 1). Indeed, whereas the dextran must be functionalized to be applied for coating processes, HA is already functionalized due to the abundance of carboxylic groups.

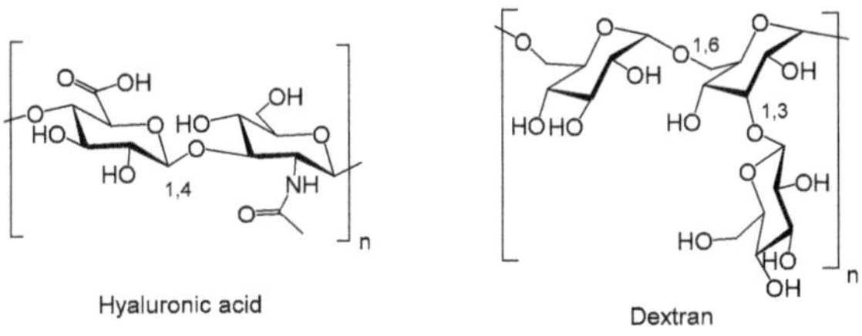

Scheme 1. Different natural polysaccharides.

Here, the covalent modification of the NH_2 groups of three glycosidases from *L. plantarum* has been performed using hyaluronic acid to alter and improve the specificity towards *p*-nitrophenyl-6-phosphate-glycopyranosides against their corresponding free glycopyranosides. For that purpose, the hydrolytic activity of the different enzymes against four different substrates, 1-*O*-(4-nitrophenyl)-β-D-glucopyranoside (*p*NP-Glu), 1-*O*-(4-nitrophenyl)-6-*O*-phosphate-β-D-glucopyranoside (*p*NP-6P-Glu), 1-*O*-(4-nitrophenyl)-β-D-galactopyranoside (*p*NP-Gal), and 1-*O*-(4-nitrophenyl)-6-*O*-phosphate-β-D-galactopyranoside (*p*NP-6P-Gal) (Scheme 2) were evaluated.

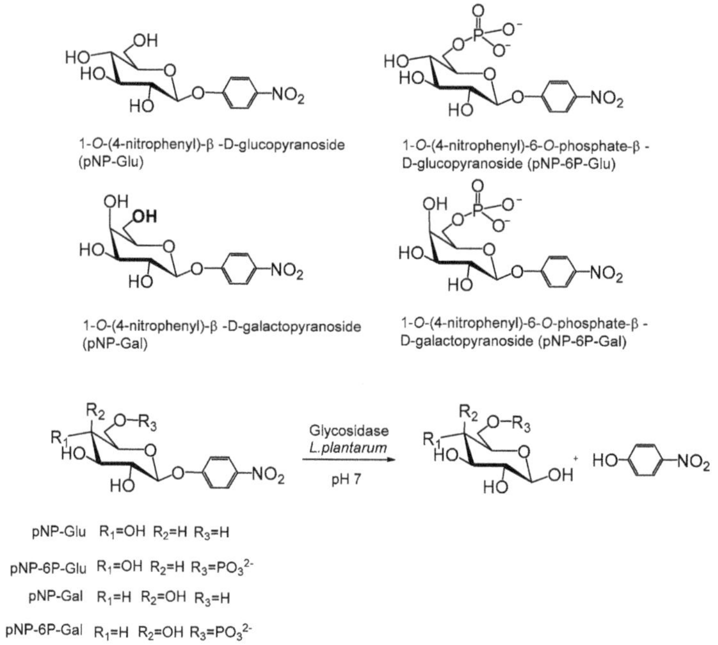

Scheme 2. Hydrolysis of *p*-nitrophenyl-glycopyranosides catalyzed by glycosidases.

2. Materials and Methods

2.1. General Description

1-*O*-(4-nitrophenyl)-β-D-glucopyranoside (*p*NP-Glu), 1-*O*-(4-nitrophenyl)-β-D-galactopyranoside (*p*NP-Gal), 1-*O*-(4-nitrophenyl)-6-*O*-phosphate-β-D-galactopyranoside (*p*NP-6P-Gal), N-hydroxysuccinimide (NHS), extra-low molecular weight hyaluronic acid 8000–15,000 Da (HA), acryl/bis-acrylamide 30% solution, ammonium persulfate (APS), *N,N,N′,N′*-tetramethylethylendiamine (TEMED), tris-base buffer, and sodium phosphate were purchased from Sigma-Aldrich (Saint Louis, MO, USA). 1-Ethyl-3-(3-dimethylaminopropyl) carbodiimide (EDC) was acquired from Tokyo Kasei (Japan). Spectrophotometric measurements were performed in a microplate reader from Biochrom (Cambridge, UK). Electrophoresis tools were acquired from Hoefer (Holliston, MA, USA).

2.2. Expression of Lp_0440, Lp_2777, and Lp_3525 Genes in Escherichia coli and Purification of the Recombinant Proteins

The Lp_0440, Lp_2777, and Lp_3525 genes from *L. plantarum* WCFS1 were amplified by polymerase chain reaction (PCR). Oligonucleotides 1666 (5′ TAACTTTAAGAAGGAGATATACAT ATGACGATTAAAGGACGAGCGTTTC) + 1667 (5′-GCTATTAATGATGATGATGATGATGCTCAATT TCGGCACCATTTGTCGC) were used to amplify Lp_0440, primers 1591 (5′-TAACTTTAAGAAGGAG ATATACATATGGCAACAACGAGTGGTTTAGA) + 1592 (5′-GCTATTAATGATGATGATGATG ATGCTTCAAATCGGCCCCATTCGTC) to amplify Lp_2777, and finally, oligonucleotides 427 (5′-CATCATGGTGACGATGACGATAAGATGTCAGAGTTCCCAGAA) + 428 (5′-AAGCTTAGTTA GCTATTATGCGTACTATTTCTTTGTCAGCCCATTATGC) to amplify the Lp_3535 gene. Genbank accession number of the enzyme genes are Lp_2777 (YP_0048900399.1); Lp_3525 (YP_004891005.1); and Lp_0440 (YP_004888459.1). Advantage HD DNA polymerase (TaKaRa, Kusatsu, Japan) was used for PCR amplifications. The purified PCR products were inserted into the pURI3 (Lp_3525 gene) or pURI3-Cter (Lp_0440 and Lp_2777 genes) vectors using a restriction enzyme- and ligation-free cloning strategy [26]. These vectors produce recombinant proteins having a six-histidine affinity tag at their N-(Lp_3525) or C-terminal (Lp_0440 and Lp_2777) ends. *Escherichia coli* DH10B cells were transformed and the recombinant plasmids obtained (pURI3-Cter-Lp_0440, pURI3-Cter-Lp_2777, and pURI3-Lp_3525) were isolated and verified by DNA sequencing, and then used to transform *E. coli* BL21(DE3) cells for expression. *E. coli* cells were grown in lysogeny broth (LB) medium containing ampicillin (100 μg/mL), until they reached an optical density at 600 nm of 0.4 and induced by adding isopropyl β-D-1-thiogalactopyranoside (IPTG) (0.4 mM final concentration). After induction, the cells were grown at 22 °C during 20 h. The induced cells were harvested by centrifugation (8000 g for 15 min at 4 °C), resuspended in phosphate buffer (50 mM, pH 7) containing 300 mM NaCl, and disrupted by French press passages (three times at 1100 psi). Insoluble fraction of the lysates was removed by centrifugation at 47,000 g for 30 min at 4 °C. The supernatant was filtered through a 0.2-μm pore-size filter and then loaded onto a Talon Superflow resin (Clontech, Mountain View, CA, USA) equilibrated in phosphate buffer (50 mM, pH 7) containing 300 mM NaCl and 10 mM imidazole to improve the interaction specificity in the affinity chromatography step. The bound enzymes were eluted using 150 mM imidazole in the same buffer. The purity of the enzymes was determined by sodium dodecyl sulfate-polyacrylamide gel electrophoresis (SDS-PAGE) in tris-glycine buffer. Fractions containing the His$_6$-tagged protein were pooled (fractions were dialyzed against 50 mM sodium phosphate buffer, 300 mM NaCl, pH 7 at 4 °C using dialysis membranes (OrDial D35-MWCO 3500, Orange Scientific, Braine-l'Alleud, Belgium) of 3.5 kDa pore diameter. Four changes of buffer were made to eliminate the imidazole present in the sample.) and analyzed. The three enzymes were defined as Lp_0440, Lp_2777, and Lp_3525.

2.3. Physical Modification of the New LpGs

First, 28 mg of HA were dissolved in 2.8 mL of water at pH 5 (final concentration 1% HA). Then, 200 μL of HA solution were added to 1 mL of a 0.05 mg/mL enzyme solution, and the mixture was left in agitation overnight. The novel biocatalysts were named HA-Lp_0440, HA-Lp_2777, or HA-Lp_3525, respectively.

2.4. Covalent Modification of Lp Glycosidases with Hyaluronic Acid (HA)

28 mg of HA were dissolved in 2.8 mL of water at pH 5 (final concentration 1% HA). To this solution, 108 mg of EDC (10 eq) and 96 mg of NHS (15 eq) were added and pH checked (must be between 5.0 and 5.5). HA activation was left in agitation for one hour. After that, 200 μL of phosphate buffer 100 mM pH 7 was incorporated to inactivate the reagents. Next, 200 μL of activated HA were added to 1 mL of a 0.05 mg/mL enzyme solution and the mixture was left in agitation overnight. The novel biocatalysts were named as cHA-Lp_0440, cHA-Lp_2777, or cHA-Lp_3525, respectively.

2.5. Native Electrophoresis Assay

For native PAGE, 12% separating gel and 4% stacking gel were used. Samples were mixed in Eppendorf tubes in 1:1 ratio with 2× sample buffer (25% glycerol, 62.5 mM tris-HCl pH 6.8 and 1% bromophenol blue) and the electrophoresis was performed in tris-HCl buffer pH 8.3 on ice at low voltage (around 120 V) and amperage, to prevent protein degradation. After that, silver staining was used [27].

2.6. Synthesis of 4-Nitropheny,6-Phosphate-β-D-Glucopyranoside (pNP-6P-Glu)

Phosphorous oxychloride (0.74 mL) was added to a solution of 4-nitrophenyl β-D-glucopyranoside (2.6 mmol) in trimethyl phosphate (6.19 mL) and stirred for 3 h at 0 °C. Then, the pH of the reaction mixture was adjusted to 7 with aqueous ammonia 30% at 0 °C. The solvents were evaporated to dryness. The solid formed was washed with MeOH (200 mL) and the liquid phase was evaporated to dryness. The residue was washed again with AcOEt (50 mL) and Et$_2$O (50 mL) and the liquid phase was evaporated under reduced pressure giving a white solid which was purified with HPLC semipreparative technology, using SunFire™ Prep C18 OBDTM (Waters corporation, Mildford, MA, USA) 5 μm, 19 mm × 150 mm column, to afford 404 mg of the glycoside 6-phosphate as white solid (41%). ^{1}H NMR (500 MHz, Methanol-d4) δ 6.68 (dd, J = 9.3, 0.8 Hz, 1H), 5.71 (dd, J = 9.2, 0.8 Hz, 1H), 3.61–3.47 (m, 1H), 2.84–2.79 (m, 1H), 2.60–2.52 (m, 1H), 2.20–2.12 (m, 1H), 1.97–1.94 (m, 2H), 1.90–1.83 (m, 1H).^{13}C NMR (126 MHz, Methanol-d4) δ 163.8, 144.0, 126.6, 117.8, 101.6, 77.7, 76.6, 74.7, 71.0, 66.7.

2.7. Enzymatic Activity Assay of Glycosidases in the Hydrolysis of p-Nitrophenyl-Glycopyranosides

Assays were monitored using a plate reader measuring at a wavelength of 405 nm. To each well, 300 μL of substrate were added and, over this, a different volume of enzyme solution was added depending of the activity of each one of them. For Lp_2777 and Lp_3525, only 10 μL (0.05 mg/mL) were needed, whereas for Lp_0440 (0.05 mg/mL), 50 μL were needed, due to its lower activity. The following substrates were tested: *p*NP-Glu, *p*NP-Gal, *p*NP-6P-Glu, and *p*NP-6P-Gal. Substrates were dissolved in phosphate buffer pH 7 to a 5 mM final concentration, except *p*NP-6P-Glu that was dissolved to a concentration of 1.25 mM. The enzyme activity of the different glycosidases was measured at room temperature and at 40 °C. For the latter purpose, the enzymes were incubated in a thermoblock that had been previously preheated to 40 °C. The sample was taken after 10 min and cooled at room temperature, then its activity was measured following the protocol described above.

Samples were measured at different times, obtaining 10 absorbance values, that were represented. Then, ΔAbs/min was obtained from the slope of the linear tendency showed by the data. Experiments were performed in triplicate.

3. Results and Discussion

3.1. Production and Purification of the Different Glycosidases

The proteins were identified based on their annotation in the database as 6-phospho-β-glucosidases and the three enzymes exhibited high sequence identity among them. The expression of Lp_3525 was previously analyzed [28] and the oligomeric state, enzymatic activity using salicin, cellobiose, and gentibiose as substrates, and crystallization of Lp_0440 was previously published [29].

The Lp_0440, Lp_2777, and Lp_3525 genes were cloned into the pURI3-Cter (Lp_0440 and Lp_2777 genes) and pURI3 (Lp_3525 gene) expression vectors by a ligation-free cloning strategy described previously [26]. The vectors incorporate the DNA sequence encoding hexahistidine at their N (Lp_3525)- or C-terminal (Lp_0440 and Lp_2777) ends to create a His-tagged fusion enzymes for further purification steps. Lp_3525 was cloned on the constructed pURI3 vector developed previously to produce proteins having a N-terminal poly-His-tag [30]. Later, a family vector from pURI3 were developed [26] where one of them, pURI3-Cter vector, allowed the production of recombinant proteins having poly-His-tag at the C-terminus, and Lp_2777 and Lp_0440 were cloned on it.

These genes were expressed under the control of an isopropyl-β-D-thiogalactopyranoside inducible promoter. Cell extracts were used to detect the presence of overproduced proteins by SDS-PAGE analysis (Figure 1). Overproduced proteins with an apparent molecular mass around 55 kDa were observed in cells harboring pURI3-Cter-0440 (Figure 1a), pURI3-Cter-2777 (Figure 1b), and pURI3-3525 (Figure 1c). The recombinant proteins (Lp_0440, Lp_2777, and Lp_3525) were purified by a metal affinity chromatography resin, and eluted with phosphate buffer (50 mM, pH 7) containing 300 mM NaCl and 150 mM imidazole. SDS-PAGE analysis (Figure 1) revealed the production of highly purified His$_6$-tagged enzymes at yield from 3.4 mg protein/L culture (Lp_3525), 4.65 mg protein/L culture (Lp_0440) to 21.37 mg protein/L culture (Lp_2777).

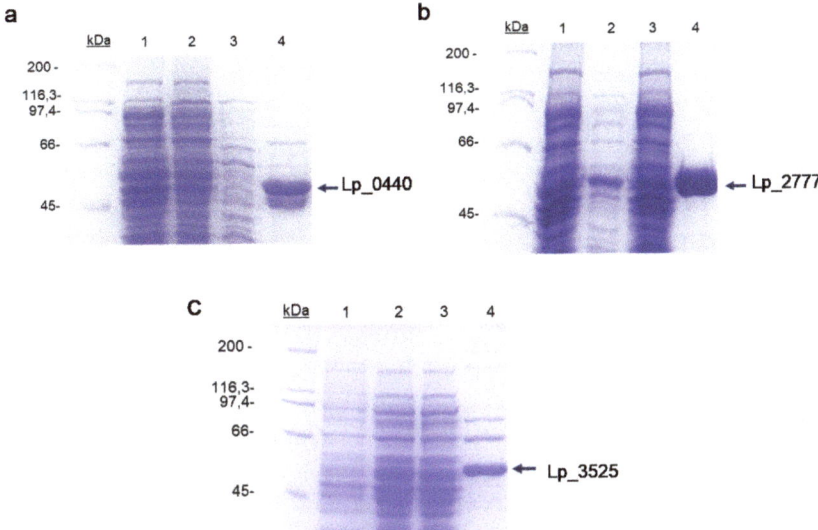

Figure 1. The sodium dodecyl sulfate-polyacrylamide gel electrophoresis (SDS-PAGE) analysis of the purification of different glycosidases from *Lactobacillus plantarum* WCFS1. (**a**) Lp_0440, (**b**) Lp_2777, and (**c**) Lp_3525. Lane 1: Soluble cell extracts of *Escherichia coli* BL21 (DE3) (pURI3-Cter for **a**,**b**) or (pURI3 for **c**). Lane 2: *E. coli* BL21 (DE3) (pURI3-Cter-0440 for (**a**), pURI3-Cter-2777 for (**b**), pURI3-Cter-3525 for (**c**). Lane 3: Flowthrough. Lane 4: Protein eluted after His affinity resin and dialysis. The gel was stained with Coomassie blue. Molecular mass markers are located on the left (SDS-PAGE Standards, Bio-Rad, Hercules, CA, USA).

3.2. Modification of Different LpGs by Hyaluronic Acid (HA)

The three different glycosidases (Lp_0440, Lp_2777, and Lp_3525) were specifically modified with hyaluronic acid (HA). The chemical modification was performed by the functionalized group COOH of the HA which was previously activated with N-hydroxysuccinimide (NHS) (Scheme 3). In all cases, the activated HA-NHS was added to the protein at pH 7, where mainly the most reactive group in the protein, the amino group in the N-terminus, reacted forming a covalent amide bond between enzyme and polymer (Scheme 3), producing the covalent HA (cHA)-glycosidase conjugates.

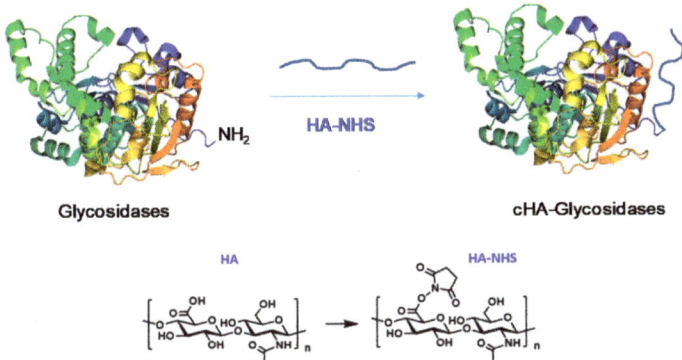

Scheme 3. The chemical modification with NHS-activated hyaluronic acid (8–14 kDa) of the different glycosidases.

Taking into consideration that the introduction of the polymer can produce alteration in the neat charge of the protein, native PAGE of these modified glycosidases at the same concentration was performed (Figure 2).

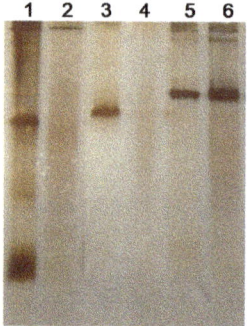

Figure 2. Native PAGE of the different modified cHA-glycosidades. Lp_0440 (Lane 1), cHA-Lp_0440 (Lane 2), Lp_2777 (Lane 3), cHA-Lp_2777 (Lane 4), Lp_3525 (Lane 5), cHA-Lp_3525 (Lane 6).

The protein bands of the conjugates cHA-Lp_0440 and cHaLp_2777 were quite blurred with a slight lower RF, whereas the intensity of the band was conserved in the conjugate cHA-Lp_3525. Both enzymes, Lp_0440 and Lp_2777 present a His$_6$-tag in the C-terminus of the protein, whereas in Lp_3525, this tag was located in the N-terminus. Considering the tridimensional structure of the enzyme Lp_0440 (Figure S1, Supplementary Materials), we have observed that both N and C-terminus are in the area in the protein. However, C-terminus is nearest to an alpha-helix in the protein, at 5.3 Å of distance to the Arg403 taken as reference whereas the N-terminus is at 16.3 Å (Figure S1). This alpha helix contains a number of polar aminoacid residues (Glu412, Asp401, Asp396). In the case of His$_6$-tag at N-terminus (Lp_3525), the amino group where the polymer is anchored is far from the protein structure (Figure S2, Supplementary Materials) whereas when the His$_6$-tag was introduced on the C-terminus (Lp_0440 and Lp_2777) the amino group on the protein from reaction is much near to the protein, being possible an interaction between the HA and tag, making maybe more difficult the electrophoretic mobility of the sample.

Also, a noncovalent strategy of the random interaction between the carboxylic groups of the polymer and amino groups on proteins (Lys) at alkaline pH was also performed. This strategy caused the interaction of the polymer on different areas of the proteins, specifically in the rich areas of Lysines (see Figure S3 in Supplementary Materials).

3.3. Glycosidase Activity, Specificity, and Regioselectivity of the Different Modified Glycosidases

The different nonmodified, physically and covalently modified glycosidases from *L. plantarum* were tested as catalysts in the hydrolysis of different *p*-nitrophenyl-glycopyranosides, non-phosphorylated (*p*NP-Glu, *p*NP-Gal) and C-6 phosphorylated (*p*NP-6P-Glu, *p*NP-6P-Gal) at 25 °C and pH 7 (Scheme 2).

The activity of the three different glycosidases were quite different depending on the used substrate (Table 1). Lp_0440 was a more promiscuous enzyme, being able to hydrolyze *p*NP-Glu, *p*NP-Gal, and *p*NP-6P-Glu. Indeed, the activity against *p*NP-Glu and *p*NP-6P-Glu was similar, and also showed two-fold less activity against *p*NP-Gal over *p*NP-Glu (Table 1, Entry 1).

Table 1. The hydrolysis of *p*-nitrophenyl-glycopyranosides catalyzed by glycosidases from *L. plantarum* at 25 °C [a].

Entry	Biocatalyst	Activity (ΔAbs/min) [a]			
		*p*NP-Glu [b]	*p*NP-6P-Glu [c]	*p*NP-Gal [b]	*p*NP-6P-Gal [b]
1	Lp_0440	39.20 ± 1.20	39.1 ± 1.10	15.6	0
2	HA-Lp_0440	5.80 ± 0.20	3.30 ± 0.10	5.1	0
3	cHA-Lp_0440	13.10 ± 0.39	1.90 ± 0.10	7	0
4	Lp_2777	0	270 ± 8	0	1.4 ± 0.07
5	HA-Lp_2777	0	0	0	0
6	cHA-Lp_2777	0	80 ± 2	0	0.5 ± 0.04
7	Lp_3525	0	335 ± 10	0	12.4 ± 0.38
8	HA-Lp_3525	0	20 ± 0.60	0	0
9	cHA-Lp_3525	0	252 ± 7.5	0	0

[a] Activity values ×10^{-3}; [b] Conditions: Substrates were dissolved in phosphate buffer pH 7 to a 5 mM final concentration, and 10 µL of enzyme solution (0.05 mg/mL) were added, except for Lp_0440, where 50 µL (0.05 mg/mL) were used; [c] Conditions: Substrates were dissolved in phosphate buffer pH 7 to a 1.25 mM final concentration, and 10 µL of enzyme solution (0.05 mg/mL) were added, except for Lp_0440, where 50 µL (0.05 mg/mL) were used.

A bioinformatic analysis of the crystal structure of Lp_0440 [28] showed three important aminoacidic residues clearly involved in the specificity and regioselectivity control (Figure 3). Ala431 is the responsible to interact with the phosphate molecule which is also stabilized, whereas the Gln22 is responsible to stabilize O4 in axial, controlling the regioselectivity between Gal and Glu.

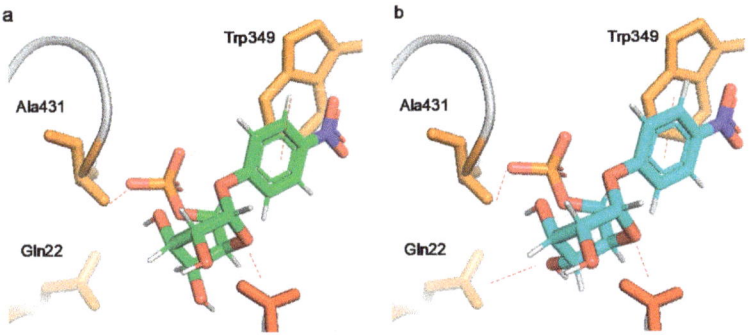

Figure 3. Lp_0440 in complex with (**a**) *p*NP-6P-Glu and (**b**) *p*NP-6P-Gal. Glu180 (active site) is marked in red, *p*NP-6P-Glu in green, and *p*NP-6P-Gal in blue. The protein structure was obtained from the Protein Data Bank (PDB code: 3qom) and the pictures were created using Pymol v. 0.99.

Lp_0440 showed 2-times better accommodation of the Glu molecules, O in equatorial position than in axial. However, the surroundings of the active site did not affect the control of the interaction in Ala431 because the location between the hydroxyl in C-6, free or phosphorylated, presented the same accommodation, resulting in the same activity. The housing of the Gln22 enables the activity against galactopyranosides.

The modification of this enzyme with HA caused changes in activity and specificity against the different glycopyranosides. The simple physical adsorption approach caused a decrease in the activity, most notably against the *p*NP-6P-Glu. However, the covalent modification with HA resulted in an alteration of the modulation, directly improving the recognition of *p*NP-Glu more than 5 times over the *p*NP-6P-Glu. It was interesting that galactose activity was almost lost in the cHA-Lp_0440 variant.

The Lp_2777 enzyme was completely specific towards the *p*-nitrophenyl-6-P-glycopyranosides, especially *p*NP-6P-Glu (almost 200 times higher activity than *p*NP-6P-Gal). No traces of activity were observed using *p*NP-Glu or *p*NP-Gal (Table 1, Entry 5). In this case, the modification with HA caused a complete loss of activity by physical adsorption (HA-Lp_2777) and did not improve the specificity of the enzyme against any of the tested substrates in the covalent approach (cHA-Lp_2777) (Table 1, Entry 6).

A very different result was achieved in the Lp_3525 (Table 1). The enzyme was very specific for pNP-6-P-glycopyranosides and, as well as Lp_2777, showed much higher activity against pNP-6P-Glu (Table 1, Entry 7). The HA coating caused a high decrease in the activity (>90%), only conserving slight activity against pNP-6P-Glu.

However, the covalent attachment of HA generated a new enzyme (cHA-Lp_3525) highly specific, recognizing exclusively the pNP-6P-Glu, showing no activity against any other tested substrate (Table 1, Entry 9). This seems to indicate that the HA modification caused the rigidification needed in the active site as hypothesized in Figure 3a.

Furthermore, temperature effect on the activity and specificity for these enzymes was evaluated (Table 2). Short-time incubation of Lp_0440 at 40 °C showed an important shift of specificity against substrates. The enzyme recognized only the pNP-Gal at this pH, with a slight increase in activity over the one shown at 25 °C. This phenomenon was also observed in the cHA-Lp_0440 (Table 2, Entry 3), hydrolyzing exclusively pNP-Gal instead of pNP-Glu. This result could be explained by a clear alteration of the Gln22 (Figure 3) by the temperature shifting the selectivity towards the O-4 in axial (galactose) than in equatorial position (glucose) on the sugar moiety. No improvements in the specificity were observed in the other two enzymes, where Lp_3525 conserved almost all the activity whereas the modified ones suffered a high decrease (Table 2).

Table 2. Hydrolysis of p-nitrophenyl-glycopyranosides catalyzed by glycosidases from L. plantarum at 40 °C [a].

Entry	Biocatalyst	Activity (ΔAbs/min) [a]			
		pNP-Glu [b]	pNP-6P-Glu [c]	pNP-Gal [b]	pNP-6P-Gal [b]
1	Lp_0440	0	0	17 ± 0.50	0
2	HA-Lp_0440	0	0	1.6 ± 0.13	0
3	cHA-Lp_0440	0	0	11.4 ± 0.34	0
4	Lp_2777	0	69 ± 2	0	3.2 ± 0.01
5	HA-Lp_2777	0	0	0	0.7 ± 0.07
6	cHA-Lp_2777	0	8.7 ± 0.26	0	0.8 ± 0.08
7	Lp_3525	0	300 ± 9	0	5.5 ± 0.17
8	HA-Lp_3525	0	0	0	0
9	cHA-Lp_3525	0	0.3 ± 0.03	0	3 ± 0.15

[a] Activity values ×10^{-3}; [b] Conditions: Substrates were dissolved in phosphate buffer pH 7 to a 5 mM final concentration, and 10 µL of enzyme solution (0.05 mg/mL) were added, except for Lp_0440, where 50 µL (0.05 mg/mL) were used; [c] Conditions: Substrates were dissolved in phosphate buffer pH 7 to a 1.25 mM final concentration, and 10 µL of enzyme solution (0.05 mg/mL) were added, except for Lp_0440, where 50 µL (0.05 mg/mL) were used.

This mild and simple strategy to modifying for the first-time glycosidases activity and specificity could be another alternative strategy than other described in the literature to modified proteins by hyaluronic acid [31–35]. Most of the methodologies described for N-terminal modification of proteins using HA (e.g., slight HA oxidation or modification of HA with an acetal spacer [31,32]) are focused on biomedical applications and only few example are applied to enzymes, mainly for stability studies [31,32].

4. Conclusions

Chemical modification with hyaluronic acid has resulted in an interesting strategy to alter and modulate the specificity of three new glycosidases from L. plantarum. A very high specificity against pNP-6P-Glu has been obtained by the chemically modified cHA-Lp_3525 phosphoglycosidase. However, a more promiscuous glycosidase as Lp_0440, it was modulated its selectivity against a high specificity towards non-phosphorylated substrate after HA covalent modification.

Therefore, these results open the door to the application of this chemical modification strategy in order to control the specificity of glycosidases. This could be relevant, for example, to obtain synthetic biocatalysts needed to produce bioactive oligosaccharides.

Supplementary Materials: The following are available online at http://www.mdpi.com/2079-6412/9/5/311/s1, Figure S1: Lp_0440 structure incorporating a His6-tag in the C-terminus; Figure S2: Differences on the hyaluronic acid (HA) chemical modification of glycosidases containing a His6-tag in the N- or C-terminus; Figure S3: Lp_0440 surface tridimensional structure.

Author Contributions: C.P.-R., D.L.-T., L.P.-V., and J.C. performed the experiments; B.d.l.R. and R.M. prepared the enzymes, supervised and partially wrote the manuscript; J.M.P. supervised, conceived and designed the experiments; and J.M.P., C.P.-R., and D.L.-T. wrote the manuscript.

Funding: This work was sponsored by the Spanish Government (Project No. AGL2017-84614-C2-2-R) and CSIC Project No. CSIC-PIE 201880E011). We also thank the Ministry of Education, Youth and Sports of the Community of Madrid and the European Social Fund for a contract to C.P.-R (PEJD-2017PRE/SAL-3762) in the program of Youth Employment and the Youth Employment Initiative (YEI) 2017.

Acknowledgments: The authors thank the support by the Spanish National Research Council (CSIC).

Conflicts of Interest: The authors declare no conflict of interest.

References

1. Spicer, C.D.; Davis, B.G. Selective chemical protein modification. *Nat. Commun.* **2014**, *5*, 4740. [CrossRef]
2. MacDonald, J.I.; Munch, H.K.; Moore, T.; Francis, M.B. One-step site-specific modification of native proteins with 2-pyridinecarboxyaldehydes. *Nat. Methods* **2015**, *11*, 326–331. [CrossRef] [PubMed]
3. Chan, A.O.Y.; Ho, C.M.; Chong, H.C.; Leung, Y.C.; Huang, J.S.; Wong, M.K.; Che, C.M. Modification of N-terminal α-amino groups of peptides and proteins using ketenes. *J. Am. Chem. Soc.* **2012**, *134*, 2589–2598. [CrossRef] [PubMed]
4. Lai, Z.W.; Petrera, A.; Schilling, O. Protein amino-terminal modifications and proteomic approaches for N-terminal profiling. *Chem. Boil.* **2015**, *24*, 71–79. [CrossRef] [PubMed]
5. Filice, M.; Romero, O.; Guisán, J.M.; Palomo, J.M. trans,trans-2,4-Hexadiene incorporation on enzymes for site-specific immobilization and fluorescent labeling. *Org. Biomol. Chem.* **2011**, *9*, 5535–5540. [CrossRef] [PubMed]
6. Wu, B.; Wijma, H.J.; Song, L.; Rozeboom, H.J.; Poloni, C.; Tian, Y.; Arif, M.I.; Nuijens, T.; Quaedflieg, P.J.L.M.; Szymanski, W.; et al. Versatile peptide C-terminal functionalization via a computationally engineered peptide amidase. *ACS Catal.* **2016**, *6*, 5405–5414. [CrossRef]
7. Díaz-Rodríguez, A.; Davis, B.G. Chemical modification in the creation of novel biocatalysts. *Chem. Boil.* **2011**, *15*, 211–219. [CrossRef]
8. Romero, O.; Filice, M.; de las Rivas, B.; Carrasco-Lopez, C.; Klett, J.; Morreale, A.; Hermoso, J.A.; Guisan, J.M.; Abian, O.; Palomo, J.M. Semisynthetic peptide-lipase conjugates for improved biotransformations. *Chem. Commun.* **2012**, *48*, 9053–9055. [CrossRef]
9. Gutarra, M.L.E.; Romero, O.; Abian, O.; Torres, F.A.G.; Freire, D.M.G.; Castro, A.M.; Guisan, J.M.; Palomo, J.M.; Freire, D.M.G. Enzyme surface glycosylation in the solid phase: Improved activity and selectivity of Candida antarctica lipase B. *Chem. Cat. Chem.* **2011**, *3*, 1902–1910. [CrossRef]
10. Callaghan, C.; Redmond, M.; Filice, M.; Alnoch, R.C.; Mateo, C.; Palomo, J.M. Biocatalytic process optimization for the production of high-added-value 6-O-hydroxy and 3-O-hydroxy glycosyl building blocks. *Chem. Cat. Chem.* **2017**, *9*, 2536–2543. [CrossRef]
11. Perugino, G.; Trincone, A.; Rossi, M.; Moracci, M. Oligosaccharide synthesis by glycosynthases. *Trends Biotechnol.* **2004**, *22*, 31–37. [CrossRef]
12. Ajisaka, K.; Yamamoto, Y. Control of the regioselectivity in the enzymatic syntheses of oligosaccharides using glycosidases. *Trends Glycosci. Glycotechnol.* **2002**, *14*, 1–11. (In Japanese) [CrossRef]
13. García, C.; Hoyos, P.; Hernáiz, M.J. Enzymatic synthesis of carbohydrates and glycoconjugates using lipases and glycosidases in green solvents. *Biocatal. Biotrans.* **2018**, *36*, 131–140. [CrossRef]
14. Harit, V.K.; Ramesh, N.G. Amino-functionalized iminocyclitols: Synthetic glycomimetics of medicinal interest. *RSC Adv.* **2016**, *6*, 109528–109607. [CrossRef]

15. Seeberger, P.H.; Werz, D.B. Synthesis and medical applications of oligosaccharides. *Nat. Cell Boil.* **2007**, *446*, 1046–1051. [CrossRef]
16. Murrey, H.E.; Hsieh-Wilson, L.C. The chemical neurobiology of carbohydrates. *Chem. Rev.* **2008**, *108*, 1708–1731. [CrossRef]
17. Campbell, C.T.; Yarema, K.J. Large-scale approaches for glycobiology. *Genome Boil.* **2005**, *6*, 236. [CrossRef]
18. Filice, M.; Ubiali, D.; Fernandez-Lafuente, R.; Fernandez-Lorente, G.; Guisán, J.M.; Palomo, J.M.; Terreni, M. A chemo-biocatalytic approach in the synthesis of β-*O*-naphtylmethyl-*N*-peracetylated lactosamine. *J. Mol. Catal. B Enzym.* **2008**, *52*, 106–112. [CrossRef]
19. Chen, L.; Huang, G. The antioxidant activity of derivatized cushaw polysaccharides. *Int. J. Boil. Macromol.* **2019**, *128*, 1–4. [CrossRef]
20. Pinheiro, V.B.; Taylor, A.I.; Cozens, C.; Abramov, M.; Renders, M.; Zhang, S.; Chaput, J.C.; Wengel, J.; Peak-Chew, S.-Y.; McLaughlin, S.H.; et al. Synthetic genetic polymers capable of heredity and evolution. *Science* **2012**, *336*, 341–344. [CrossRef]
21. Herdewijn, P. Nucleic acids with a six-membered 'carbohydrate' mimic in the backbone. *Chem. Biodivers.* **2010**, *7*, 1–59. [CrossRef]
22. Wright, T.A.; Page, R.C.; Konkolewicz, D. Polymer conjugation of proteins as a synthetic post-translational modification to impact their stability and activity. *Polym. Chem.* **2019**, *10*, 434–454. [CrossRef]
23. Romero, O.; Rivero, C.W.; Guisan, J.M.; Palomo, J.M.; Leite, L. Novel enzyme-polymer conjugates for biotechnological applications. *PeerJ* **2013**, *1*, e27. [CrossRef]
24. Montagner, I.M.; Merlo, A.; Carpanese, D.; Pietà, A.D.; Mero, A.; Grigoletto, A.; Loregian, A.; Renier, D.; Campisi, M.; Zanovello, P.; et al. A site-selective hyaluronan-interferonα2a conjugate for the treatment of ovarian cancer. *J. Control. Release* **2016**, *236*, 79–89. [CrossRef]
25. Huang, G.; Huang, H. Application of hyaluronic acid as carriers in drug delivery. *Drug Deliv.* **2018**, *25*, 766–772. [CrossRef]
26. Curiel, J.A.; Rivas, B.D.L.; Mancheño, J.M.; Muñoz, R. The pURI family of expression vectors: A versatile set of ligation independent cloning plasmids for producing recombinant His-fusion proteins. *Protein Expr. Purif.* **2011**, *76*, 44–53. [CrossRef]
27. Chevallet, M.; Luche, S.; Rabilloud, T. Silver staining of proteins in polyacrylamide gels. *Nat. Protoc.* **2006**, *1*, 1852–1858. [CrossRef]
28. Marasco, R.; Muscariello, L.; Varcamonti, M.; De Felice, M.; Sacco, M. Expression of the bglH gene of *Lactobacillus plantarum* is controlled by carbon catabolite repression. *J. Bacteriol.* **1998**, *180*, 3400–3404.
29. Michalska, K.; Tan, K.; Li, H.; Hatzos-Skintges, C.; Bearden, J.; Babnigg, G.; Joachimiak, A. GH1-family 6-P-β-glucosidases from human microbiome lactic acid bacteria. *Acta Crystallogr. Sect. D Boil. Crystallogr.* **2013**, *69*, 451–463. [CrossRef]
30. De las Rivas, B.; Curiel, J.A.; Mancheño, J.M.; Muñoz, R. Expression vectors for enzyme restriction- and ligation-independent cloning for producing recombinant His-fusion proteins. *Biotechnol. Prog.* **2007**, *23*, 680–686. [CrossRef]
31. Mero, A.; Pasqualin, M.; Campisi, M.; Renier, D.; Pasut, G. Conjugation of hyaluronan to proteins. *Carbohydr. Polym.* **2013**, *92*, 2163–2170. [CrossRef]
32. Yang, J.-A.; Kim, E.-S.; Kwon, J.H.; Kim, H.; Shin, J.H.; Yun, S.H.; Choi, K.Y.; Hahn, S.K. Transdermal delivery of hyaluronic acid–Human growth hormone conjugate. *Biomaterials* **2012**, *33*, 5947–5954. [CrossRef]
33. Mero, A.; Campisi, M. Hyaluronic acid bioconjugates for the delivery of bioactive molecules. *Polymers* **2014**, *6*, 346–369. [CrossRef]
34. Martínez-Sanz, E.; Ossipov, D.A.; Hilborn, J.; Larsson, S.; Jonsson, K.B.; Varghese, O.P. Bone reservoir: Injectable hyaluronic acid hydrogel for minimal invasive bone augmentation. *J. Control. Release* **2011**, *152*, 232–240. [CrossRef]
35. Yang, J.-A.; Park, K.; Jung, H.; Kim, H.; Hong, S.W.; Yoon, S.K.; Hahn, S.K. Target specific hyaluronic acid–Interferon alpha conjugate for the treatment of hepatitis C virus infection. *Biomaterials* **2011**, *32*, 8722–8729. [CrossRef]

© 2019 by the authors. Licensee MDPI, Basel, Switzerland. This article is an open access article distributed under the terms and conditions of the Creative Commons Attribution (CC BY) license (http://creativecommons.org/licenses/by/4.0/).

Review

Recent Development in Phosphonic Acid-Based Organic Coatings on Aluminum

Ruohan Zhao, Patrick Rupper * and Sabyasachi Gaan *

Additives and Chemistry Group, Advanced Fibers, Empa, Swiss Federal Laboratories for Materials Science and Technology, Lerchenfeldstrasse 5, 9014 St. Gallen, Switzerland; ruohan.zhao@empa.ch
* Correspondence: patrick.rupper@empa.ch (P.R.); sabyasachi.gaan@empa.ch (S.G.);
   Tel.: +41-587-657-559 (P.R.); +41-587-657-611 (S.G.)

Received: 11 July 2017; Accepted: 21 August 2017; Published: 23 August 2017

Abstract: Research on corrosion protection of aluminum has intensified over the past decades due to environmental concerns regarding chromate-based conversion coatings and also the higher material performance requirements in automotive and aviation industries. Phosphonic acid-based organic and organic-inorganic coatings are increasingly investigated as potential replacements of toxic and inefficient surface treatments for aluminum. In this review, we have briefly summarized recent work (since 2000) on pretreatments or coatings based on various phosphonic acids for aluminum and its alloys. Surface characterization methods, the mechanism of bonding of phosphonic acids to aluminum surface, methods for accessing the corrosion behavior of the treated aluminum, and applications have been discussed. There is a clear trend to develop multifunctional phosphonic acids and to produce hybrid organic-inorganic coatings. In most cases, the phosphonic acids are either assembled as a monolayer on the aluminum or incorporated in a coating matrix on top of aluminum, which is either organic or organic-inorganic in nature. Increased corrosion protection has often been observed. However, much work is still needed in terms of their ecological impact and adaptation to the industrially-feasible process for possible commercial exploitation.

Keywords: phosphonic acids; corrosion protection; aluminum; sol-gel; coatings

1. Introduction

Aluminum (Al) and its alloys have been widely used in engineering applications because of their higher strength to weight ratio, ductility, formability, and lower costs. In many applications, such as in aircraft, automobiles, and structural parts in buildings, its corrosion resistance becomes very important. Aluminum, by itself, is relatively stable to corrosion because it readily oxidizes to form a passive protective oxide layer on its surface which is robust and does not simply flake off. However, the galvanic corrosion of aluminum is very critical and is complicated in various alloys and under acidic pH. An excellent review on corrosion protection of aluminum prior to the year 2000 can be found in the literature [1]. The need to develop chromate-free treatments for aluminum is becoming increasingly important. Many organic and inorganic protective coatings are being investigated by researchers with the potential advantages and shortcomings [1]. Organic-inorganic conversion coatings are quite attractive due to the simplicity of their application and possible favorable interaction with organic layers in hybrid materials, which offer enhanced corrosion resistance and mechanical performance [2].

A variety of phosphonic acids is commonly used to modify the surfaces of metals and their oxides for their corrosion protection, stabilization as nano-particles, adhesion improvement to organic layers, hydrophobization, hydrophilization, etc. They are excellent chelating agents and bind very strongly to metals resulting in different 1D to 3D metal organic frameworks (MOFs), also called metal phosphonates [3,4]. They form hydrolytically-stable bonds with the metals and provide excellent coverage compared to many thiols, silanes, and carboxylic acids. A phosphonate corrosion inhibitor adsorbs well on the metal surface, reducing its solubility in aqueous media and, thus, decreases the area of active metal surface and increases the activation energy to hydrolysis. Such phosphonic acids can be a simple molecule where the phosphonic acid is attached to an alkyl or aromatic group. Self-assembly of such phosphonic acids on aluminum surface is well studied where the binding of these molecules to aluminum can be monodentate, bidentate, or tridentate [5–7]. More recently there is an increased impetus to develop functional phosphonic acids, which can not only bind to the metal surface but also provide linkage to an organic matrix in multicomponent systems [8,9]. In some cases, phosphonic acid acts as a linker between aluminum and a hydrophobic polymerizable group (pyrrole) [10] or chemically-stable protective material (graphene oxide) [11]. Such molecules are applied on the surface of aluminum via dip coatings, spray from aqueous solutions, or are incorporated in a polymeric matrix (adhesives, glues, or paints) which is then coated on the metal surface. A novel ultrasonic assisted deposition (USAD) method has also been developed to coat phosphate films on aluminum [12]. This application procedure is believed to improve the interaction of the phosphonic acid to the aluminum surface.

Functional phosphonic acids with hydrophobic aliphatic groups or fluorinated groups can increase the hydrophobicity of the metal surface, thereby acting as barrier to aqueous solutions and improving its corrosion protection. In some cases the functional groups of phosphonic acids react with organic layers (adhesives) and provide a stable barrier against aqueous environments.

The following sections first briefly summarize the methods for characterization of phosphonic acids on aluminum surfaces and several techniques used to evaluate the corrosion behavior, in order to understand the following chapter, which deals with the application of various types of phosphonic acids on aluminum, together with their physico-chemical behavior relevant to corrosion protection and adhesion to organic layers. Thereby, a distinction has been made between the aluminum surface treatment with phosphonic acids or phosphates as a pre-coating and surface coatings dissolved in paints that contain phosphonic acids.

2. Characterization of the Phosphonic Acid-Modified Aluminum Surfaces

Self-assembled monolayers (SAM) are ordered molecular assemblies and are commonly used to modify aluminum surfaces. They are spontaneously formed by the adsorption of molecules with head groups that show affinity to a specific substrate [13]. Phosphonic acids form SAMs on aluminum surfaces by condensation reactions of the acid functional group with basic surface-bound alumino-hydroxyl species. Thereby, the phosphonic acid head group shows a strong affinity to the aluminum substrate, creating an alumino-phosphonate linkage via a strong Al–O–P chemical bond:

$$R\text{-}PO(OH)_2 + Al\text{-}OH \rightarrow R\text{-}(OH)OP\text{-}O\text{-}Al + H_2O \quad [5\text{-}7].$$

To investigate a successful formation of the organophosphorus layers, including information about the properties of the films (surface morphology, structural ordering, density and uniformity, binding mode to the aluminum substrate, as well as stability), different analytical methods have been applied in the literature. A detailed characterization of the layers is necessary to understand their intended function (for instance corrosion protection, adhesion promotion, (non)-wettability, surface passivation). Table 1 summarizes the various analytical techniques that have been used in the last 20 years, or so, to study the organization and chemistry of organophosphorus compounds on an aluminum substrate. Various characterization techniques have been categorized by us into three groups: Surface morphology; presence, composition, and stability of layers; orientation of molecules and binding mode to aluminum, and are all briefly explained in the remainder of this section.

Table 1. Commonly-employed surface characterization techniques.

Characterization	Technique	Literature
Surface morphology	Scanning electron microscopy (SEM)	[9,14,15]
	Atomic force microscopy (AFM)	[8,15]
Presence, composition and stability of layer	X-ray photoelectron spectroscopy (XPS)	[6–8,11,16–21]
	Auger electron spectroscopy (AES)	[22]
	Fourier-transform infrared spectroscopy (FTIR)	[8,11,14,16,17,23,24]
	Time-of-flight secondary ion mass spectrometry (ToF-SIMS)	[6]
	Water contact angle measurements (WCA)	[5,7,10,16,25,26]
Orientation of molecules and binding mode to aluminum	Angle-resolved X-ray photoelectron spectroscopy (ARXPS)	[7,10,27,28]
	Infrared reflection absorption spectroscopy at grazing angle (IRRAS)	[5,12,16,29–33]
	X-ray photoelectron spectroscopy (XPS)	[20,28,34]
	Solid-state ^{31}P nuclear magnetic resonance (NMR)	[12,31–33]

2.1. Surface Morphology

Microscopy measurements, i.e., optical microscopy, scanning electron microscopy (SEM) and atomic force microscopy (AFM), were used to investigate the surface morphology of the organophosphorus self-assembled monolayers onto the aluminum substrate. Physical characteristics, like uniformity and homogeneity of the surface phosphatization, were obtained by comparing the coated with the bare aluminum substrate using SEM [14]. AFM has been used to obtain information about the surface smoothness [8,15].

2.2. Presence, Composition, and Stability of Layer

A variety of surface sensitive analytical techniques has been used to investigate the presence of the layer and its composition. X-ray photoelectron spectroscopy (XPS) and Auger electron spectroscopy (AES) are both methods which determine the elemental composition in the first couple of nanometers of a surface and, hence, were often applied to verify the presence of phosphorus, indicating a successful organophosphorus coating chemically-bound to aluminum [6,8,11,16,17,19–22]. As a general preparative step, the samples were usually rinsed/washed after the coating process to remove the residual chemicals and physically adsorbed organophosphorus compounds. In addition to the determination of the presence of the organophosphorus coating, XPS was also used to study the concentration of the organic molecules on the aluminum surface [18]. The experimental phosphorus to aluminum concentration ratio allowed the quantification of the organophosphorus molecule surface concentration and modelling of the XPS data helped to determine the thickness of the monolayer [18]. Thereby, the ratio of signals from the coating and from the underlying aluminum substrate was recorded, from which the coating thickness can be derived. Depth profiling via argon ion sputtering in combination with XPS was also used to determine the coating thickness (especially for thicker layers). The comparison between experimental and theoretical values for the carbon-to-phosphorus ratio was used as proof that chemically-uniform organophosphorus films formed on aluminum [7]. The recording of characteristic binding energies allowed to follow the individual steps in the decomposition of the phosphonates on the aluminum surface [20]. Table 2 summarizes XPS binding energies found in the literature for organophosphorus layers on metal surfaces.

Another analytical technique often used to study the presence and composition of organophosphorus layers on aluminum is reflectance Fourier-transform infrared spectroscopy (FTIR), often measured at a grazing angle to increase the surface sensitivity. FTIR was applied to investigate the adsorption of the coating by detecting specific vibrational modes involving phosphorus bonds [8,11,14,16,17,23,24]. Table 3 contains a summary of frequencies of vibrational modes involving the phosphorus atom found in the literature for self-assembled organophosphorus monolayers on aluminum.

Table 2. XPS binding energies relevant to phosphonic acids modified aluminum surface.

Assignment	Binding Energy (eV)	Literature
Al 2p		[7,16,20,35,36]
metal	72.3–73.3	
Al-oxide, Al-hydroxide	74.2–76.2	
Al-phosphonate	~75.5	
P 2p		[16,34,37]
phosphonates	133.3–134.2	
P 2s		[20]
P–C, P=O, P–O	192.5–192.7	
O 1s		[16,28,34,35,38,39]
Al–oxide	530.6–531.1	
Al–hydroxide	532.3–532.4	
water	534–535	
P=O, P–O$^-$, P–O–Metal	531.4–532.1	
P–OH, P–OR	532.6–534.3	

Table 3. Infrared vibrational modes involving the phosphorus atom for the self-assembled organophosphorus layer on aluminum.

Assignment	Peak Position (cm^{-1})	Literature
P–O sym. stretch	910–960	[5,24,27,30,37]
P–O asym. stretch	1000–1040	
P=O stretch	1100–1250	[8,11,12,16,24,27,30–32,37]
PO$_2^-$ sym. stretch	1000–1070	[37]
PO$_2^-$ asym. stretch	~1160	
PO$_3^{2-}$ sym. stretch	960–1060	[5,8,24,29,37]
PO$_3^{2-}$ asym. stretch	1115–1140	
P–OH stretch	2500–2750	[16,27,31]

Time-of-flight secondary ion mass spectrometry (ToF-SIMS), as a complementary technique to XPS, has also been used to demonstrate the presence of the organophosphorus layer, thereby investigating selective functionalization on aluminum and glass regions of a surface by mapping analysis [6]. In addition, surface plasmon resonance spectroscopy (SPR) was applied to examine the kinetics of the adsorption process [27]. Thereby, it was found that the adsorption process of phosphonic acids on aluminum starts very quickly and reaches a plateau after some minutes.

Water contact angle (WCA) measurements were used to investigate the hydrophilicity and hydrophobicity of the surface. A change compared to a reference untreated aluminum surface indicated the successful formation of an organophosphorus layer. In addition, WCA measurements were also applied to study the orientation of the molecules (see also below) because in well-ordered monolayers, the access of the water drop to the aluminum surface is reduced [7]. From the investigation of the temporal behavior of WCA, information about surface coverage and chain ordering was obtained [5]. Finally, the stability of the adsorbed layers was determined by recording advancing and receding contact angles by repeated cycling [10,26].

2.3. Orientation of Molecules and Binding Mode to Aluminum

In order to characterize the binding (orientation and mode) of the organophosphorus molecules to an aluminum surface, mostly three methods have been used in the literature: Infrared reflection absorption spectroscopy (IRRAS) at grazing angle (often applying polarized infrared radiation), X-ray photoelectron spectroscopy (XPS) at a fixed electron take-off angle as well as angle-resolved (ARXPS) and solid-state ^{31}P nuclear magnetic resonance (NMR) spectroscopy.

The most expected orientation with the phosphonic acid group reacting with the surface hydroxyl groups of the aluminum substrate and the hydrocarbon tail and/or terminal functional groups on top was proven by angle-resolved XPS. The carbon (from the tail) to phosphorus (from the head group) intensity ratio was determined at low and high electron take-off angles, representing two different information depths [7,10,27,28]. In addition, using infrared reflection absorption spectroscopy (IRRAS) and comparing the intensities of vibrations polarized perpendicular or parallel to the hydrocarbon backbone, information about the orientation of the organophosphorus molecule was also obtained [5].

The phosphonic group can bind to the aluminum atom via a direct P–O–Al bond in different modes, i.e., mono-, bi-, or tridentate, depending on whether one, two or all three oxygen atoms from the phosphonic group are involved, respectively. It has been found that both the deposition mode (for instance stirring, sonication), as well as the structure of the phosphonates, play a role [12]. Additionally, the evolution from tridentate to lower binding modes as film formation proceeds (the increase of coverage can change the chain ordering) was observed in the literature [5]. These behaviors might explain the variety of binding modes reported in the literature for organophosphorus layers on aluminum.

Table 3 summarizes the frequencies of infrared vibrational modes found in the literature for self-assembled organophosphorus layers on aluminum. By observing which vibrational bands are present and which are not, certain binding modes have been assigned (mostly bidentate [16,29] and tridentate [12,32,33], and some monodentate [30]). For instance, a missing P–OH stretching mode together with the presence of PO_3^{2-} stretching modes (deprotonation of the phosphonic acid group) in the IR spectrum indicates the formation of a bidentate binding. Mixtures of binding modes have also been found [5,30,31].

As a second technique to investigate the binding mode of organophosphorus molecules to aluminum, X-ray photoelectron spectroscopy (XPS) was used. Table 2 summarizes the XPS binding energies found in the literature for self-assembled organophosphorus layers on aluminum. As can be seen from Table 2, due to overlapping bands, neither the aluminum nor the phosphorus signal allows to clearly differentiate between various binding modes of phosphorus to aluminum. It has to be noted that a trend in the experimental binding energy to increase for the structure $[PO_n(OR)_m]^{y-}$ is observed as the ratio of $n:m$ of "free" O ligands (n) to covalently bound OR ligands (m) is stepwise changed from 4:0 to 3:1 to 2:2 and to 1:3. Therefore, also the XPS phosphorus binding energy has been used to assign the binding mode (i.e., mono-, bi-, or tridentate) [34]. However, a final conclusion cannot be drawn just from the P 2p signal and further experimental evidence is needed. The O 1s signal is of particular interest regarding the assessment of the type of bond between organophosphorus and the aluminum substrate. As the binding energy for P=O and P–O–Metal is different from the one for P–OH and P–OR (see Table 2), the ratio of the intensities at these two binding energies was used to determine the binding mode. In agreement with other studies using IR (see above), a mixed monodentate and bidentate mode was found [34]. As a third method, the linewidth and shifts of peaks in ^{31}P NMR spectra have been used to study the bonds between the aluminum and the oxygen atoms of the organophosphorus molecule and to assign a specific binding mode by comparison with literature data [12,31].

3. Corrosion Evaluation Methods

The corrosion resistance of the coated aluminum surface is evaluated mainly by four different ways: Electrochemical measurements, conventional measurements, spectral analysis and surface analysis, and are briefly summarized in Table 4. In most cases treated aluminum substrates are coated with an epoxy layer to study its adhesion and corrosion performance. The corrosion resistance of aluminum substrates is highly related to the microstructure, chemical composition, electrochemical behavior of the coating, and to the interface between coating/metal. Therefore, the measurements mentioned in Table 4 are mostly comprehensive and are used to gather more information in order to analyze the corrosion behavior from different perspectives.

Table 4. Common methods for corrosion measurements.

Measurements	Examples	Literature
Electrochemical measurements	Electrochemical noise method (ENM) Potentiodynamic polarization measurements Electrochemical impedance spectroscopy (EIS)	[22] [14,15,17,22,30,40,41]
Localized electrochemical techniques	Scanning Kelvin Probe (SKP) and Scanning Vibrating Electrode Technique (SVET)	[42]
Conventional measurements	Immersion test Salt spray test Weight loss treatment	[14,23,41] [8,27] [43]
Spectral analysis	Infrared spectroscopy (IR)	[12]
Wettability	Contact angle measurements	[10,19]

3.1. Electrochemical Measurements

Electrochemical changes occurring during the coating failure can be detected via an electrical signal which could provide information of metal corrosion and coating property alterations. Quantitative and semi-quantitative evaluation of coatings could be achieved by electrochemical measurements, such as EIS, ENM, and hydrogen permeation current method.

3.1.1. Electrochemical Impedance Spectroscopy (EIS)

EIS is a non-destructive measurement, which can provide time dependent surface information in a corrosive medium. A small amplitude sinusoidal alternating signal was added in the coating/primer/metal system. Through analyzing the impedance spectrum and the admittance spectrum, an equivalent circuit model, shown in Figure 1, was built to evaluate electrochemical information of the coating system. During the electrolyte penetration through the coating, C_{coat} increases and R_{coat} decreases. It is confirmed that a surface with an impedance at the low frequency below 10^7 $\Omega \cdot cm^2$ is considered a poor protective barrier [44].

EIS is a nondestructive method and provides fast detection and comprehensive information on coating corrosive behavior. The G106-89 standard has been developed for EIS measurement. However, EIS only provides general information on corrosion behavior of the surface, which means it cannot help understand the mechanism of the corrosion start point [45].

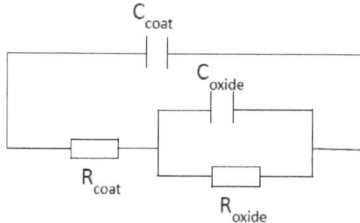

Figure 1. An equivalent circuit model for epoxy coated aluminum. R_{oxide} is the resistance of oxide aluminum layer, C_{oxide} is the capacitance of oxide aluminum layer, R_{coat} is the resistance of the coating, and C_{coat} is the capacitance of the coating. (Reproduced from [41] with permission; Copyright 2004 Elsevier).

3.1.2. Electrochemical Noise Method (ENM)

Electrochemical noise can be described as naturally-occurring fluctuations in potential and current around a mean value in the electrochemical cell [46]. The parameters voltage noise (σ_v) and current noise (σ_i) can be derived from the fluctuations. The noise resistance R_n, which can be calculated by the Ohms Law, is used to evaluate the corrosion resistance (see Equation (1)):

$$R_n = \frac{\sigma_v}{\sigma_i} \tag{1}$$

The higher the value of R_n is, the better the corrosion resistance of the coating [22]. When the value of noise resistance R_n is less than 10^6 $\Omega \cdot cm^2$, the coating shows poor corrosion resistance. When R_n is more than 10^8 $\Omega \cdot cm^2$, the coating exhibits good corrosion protection. The value between 10^6 and 10^8 $\Omega \cdot cm^2$ indicates an intermediate level of corrosion resistance [47].

ENM is an electrically non-intrusive sensitive method, which requires only a few minutes for a single measurement [48]. Recently, it has become the main measurement method for determining the metal corrosion rate, the localized corrosion process, and is widely used in industry for corrosion detection.

3.1.3. Potentiodynamic Polarization Measurements

Potentiodynamic anodic polarization can characterize the metal corrosion behavior by its current potential relationship. It can be used to determine the function and the type of inhibitor. The sample potential is scanned slowly in the positive direction, which means it forms an oxide coating during the test. The passivation tendencies and inhibitors influence can be easily studied by this method. Meanwhile, the corrosion behavior of different coatings or metals can be compared on a rational basis, and it can be used as a pretest to give a suitable corrosive condition range for further long-term measurements. The general corrosion behavior of coated substrates can be evaluated by the corrosion potential E_{corr}. Specifically, the more positive E_{corr} and the lower the current values, the better the corrosion resistance [22].

3.1.4. Localized Electrochemical Techniques

For many corrosion phenomena, high resolution instead of average data about the behavior of an electrochemically-active surface is required (for instance to differentiate small local anodes and cathodes on a metal surface) [42]. A scanning Kelvin probe (SKP), as a non-contact and non-destructive method, allows measuring and mapping the local potential and current difference on the microscale. The scanning vibrating electrode technique (SVET) enables to spatially characterize corrosion activity by measuring the potential gradients in the electrolyte due the presence of anodic and cathodic areas on a metal surface. These localized techniques find applications in different corrosion investigations like effect of microstructure and finish on corrosion initiation, localized corrosion phenomena, and detection of electrochemically-active pin-hole defects in coatings.

3.2. Conventional Measurements

Conventional corrosion measurement methods are inexpensive and easy to perform. Different national and international standards have been developed which include immersion, salt spray, damp heat, gas corrosion, and filiform corrosion tests. They are commonly used in industry. However, they are not suitable for the corrosion kinetics study or the corrosion mechanism study as they provide qualitative results, usually involving long test cycles and poor repeatability.

3.2.1. Immersion Test

The coated substrates are directly immersed in the corrosive medium. After a certain exposure to corrosion medium, the damage or corrosion of the coating is observed to evaluate corrosion resistance. The immersion tests are widely used for different coatings in different corrosion media. The common immersion tests are water resistance test, salt water resistance test, acid resistance test, and various organic solvent resistance tests. One can find numerous national immersion tests specific to a country. Some international ASTM standards for immersion tests are B895-05, G44, and G110.

3.2.2. Salt Spray Test

The salt spray test is the most classic and widely-used method to evaluate the corrosion resistance of coatings. The first international salt spray standard, ASTM B117, was recognized in 1939. The coated substrates are scratched first and then the test is carried out in a closed test chamber, where salt water (5% NaCl solution, pH 6.5–7.2) is sprayed using pressurized air at 35 ± 2 °C. Later standards include ISO9227, JIS Z 2371, and ASTM G85. To modify the corrosion condition, acetic acid-salt spray (ASS), copper-accelerated acetic acid-salt spray (CASS), prohesion cycle test as well as various modified versions were established.

3.3. Spectral Analysis

The decomposition of a polymer in the coating is one of the main reason that causes coating failures and metal corrosion, which would lead to further deposition. The chemical changes of the polymer and the inorganic corrosion products can be quantitatively detected by, for example, FTIR, infrared microscopy, or laser Raman spectroscopy (LRS). They can be used to study the corrosion performance and corrosion mechanism because of the highly-precise positioning and quantification [45]. The detail about spectral analysis has been described in Sections 2.2 and 2.3.

3.4. Surface Analysis

The various properties of the coating are highly related to its microstructure, chemical composition, and its bonding condition at the interface of coating/metal. The surface changes during the corrosion process were characterized using the same analytical methods as already described in Section 2.1.

4. Aluminum Surface Treatments with Phosphonic Acids

The following sections describe the various types of phosphonic acids and some phosphates which have been used to modify the surface of aluminum with the main objective to improve its corrosion protection and/or adhesion promotion to organic layers. A list of phosphonic acids commonly used to improve corrosion protection of aluminum is shown in Table 5. Some of these phosphonic acids have simple structures (PPA or MSAP), others are more functional phosphonic acids with terminal groups like vinyl (VPA), amino (APP), and pyrrol (Cn-Ph-P).

Table 5. Chemical name, structure, and abbreviation of organophosphonic acids.

Chemical Name	Structure	Abbreviation	Literature
Phenylphosphonic acid		PPA	[14,22]
Vinylphosphonic acid		VPA	[49]
1,12-dodecyldiphosphonic acid		DDP	[8]
Amino trimethylene phosphonic acid		ATMP	[15]

Table 5. *Cont.*

Chemical Name	Structure	Abbreviation	Literature
Ethylenediamine tetra methylene phosphonic acid		EDTPO	[17,30,40]
1,2-diaminoethanetetrakis-methylenephosphonic acid		DETAPO	[23]
Monostearyl acid phosphate		MSAP	[12]
Phosphoric acid mono-(12-hydroxy-dodecyl) ester		–	[41]
(12-ethylamino-dodecyl)-phosphonic acid		–	[27]
Aminopropyl phosphonate		APP	[9,11]
ω-(3-phenylpyrrol-1-ylalkyl) phosphonic acid		Cn-Ph-P	[10]
ω-(2,5-Dithienylpyrrol-1-yl-alky) phosphonic acid		SNS-n-P	[10]
2-(Phosphonooxy) benzoic acid		Fosfosal	[14]
Phosphonosuccinic acid		PPSA	[14]
(1H,1H, 2H,2H-heptadecafluorodec-1-yl) phosphonic acid		HDF-PA	[19]

Table 5. Cont.

Chemical Name	Structure	Abbreviation	Literature
Fluoro alkyl phosphates	R_x PO(OH)$_y$ R = CF$_3$(CF$_2$CF$_2$)$_z$(CH$_2$CH$_2$O) x = 1, 2 or 3, x + y = 3, z = 1 to 7	Zonyl UR	[12]
Poly (vinyl phosphonic acid)		PvPA	[24]

4.1. Phosphonic Acids Used as Pre-Coating

4.1.1. Phenylphosphonic Acid (PPA)

In an earlier work, it was found that a hydrophobic sol-gel film can provide good corrosion resistance to aluminum [50]. Subsequently, the same researchers incorporated organic phenyl phosphonic acid as an anion in a hydrophobic sol-gel film to further enhance the protection of aluminum against pitting corrosion. It was expected that such combinations of phosphonic acid and sol-gel coatings would be beneficial in corrosion protection. Various sol-gel coatings with two different anions were coated on the aluminum substrate. The formulations of such coatings used in this research are shown in Table 6.

Table 6. Various formulations of sol-gel coatings.

Sol-Gel Chemistry	PTMOS	PTMOS	PTMOS + TMA	PTMOS	TEOS	TEOS	TEOS + TMA
Anion	–	PPA	PPA	TBPA	–	PPA	PPA

Phenyltrimethoxysilane (PTMOS), tetraethyl orthosilicate (TEOS), N-trimethoxysilylpropyl-N,N,N-trimethyl ammonium chloride (TMA), tert-butylphosphonic acid (TBPA).

Aluminum rods embedded in a Teflon sheath (99.999%) and aluminum plates (5050-H24) were used as substrates and coated using various sol-gel formulations (Table 6) via dip-coating. The substrates were then submerged in 100 ppm NaCl solution at 25 °C for 20 min and subjected to ENM to study the initiation and propagation of corrosion at the interface of aluminum plates and coating. It is considered that the higher the value of R_n, the better the corrosion resistance of the coating. The results showed that PTMOS + PPA sol-gel coatings exhibited the best protection (R_n = 6.0 kΩ) compared to uncoated aluminum (R_n = 1.5 kΩ) and PTMOS coatings (R_n = 4.0 kΩ).

In potentiodynamic polarization measurements, aluminum in the form of rods was used. These rods were exposed to 100 ppm NaCl solution for 30 min before the measurements. For these measurements, as mentioned before, the more positive E_{corr} and the lower the current values, the better the corrosion resistance. PTMOS + PPA showed the highest E_{corr} value and a low current. In contrast, the TEOS-based film showed a negative E_{corr} value, which indicated the poor corrosion resistance. Even addition of TMA in TEOS film accelerated the corrosion process. Potentiodynamic polarization measurements of samples for electrodeposition process showed slight enhancement of corrosion inhibition. Elemental analysis was used to detected phosphorus (due to PPA) in the films. For PTMOS based film, phosphorus was detected and it was hypothesized that the anion of PPA and phenyl group of PTMOS tends to form stable π-interactions in the sol-gel system [22].

4.1.2. Vinylphosphonic Acid (VPA)

Cooling of an aluminum alloy plate heat exchanger by seawater needs superior corrosion resistance. Trifluoroethylene polymers were preferred to form a fluorocarbon resin coating on aluminum alloy because of high adhesion to the organic phosphonic acid primer coating and high corrosion resistance. Vinylphosphonic acid (VPA) was used as the organic phosphonic acid primer because of its handle and its superior adhesion effect. The aluminum alloy (3003) was first anodically oxidized and then immersed in VPA aqueous solution (10 g/L) at 65 °C for 10 s or 120 s, followed by fluorocarbon resin coating and dried at 50 °C for 24 h. For the corrosion test, one hundred 1 mm^2 cross-cuts were formed on the coating, and a tape was attached on it and then peeled by the method specified in JIS K 5600-5-6. For samples with coatings without an organic phosphonic acid primer, all hundred cross-cuts of the coating were peeled off (0/100) upon tape peeling. However, for the coating with VPA (10 s or 120 s), all 100 cross-cuts remained unpeeled (100/100), which indicated an excellent anti-corrosion property [49].

4.1.3. 1,12-Dodecyldiphosphonic Acid (DDP)

Self-assembled monolayers (SAMs) consisting of zirconium phosphate and their derivatives were used as the substitute to chromating on aluminum for ecological reasons [8]. Aluminum alloy (1100) was pretreated with 0.4 vol % of 3-aminopropyltriethoxysilane (APTES) in toluene by chemical vapor deposition. APTES is believed to prevent the formation of defects and provide protection against the harsh subsequent phosphonation treatment (pH ≤ 2). Subsequently, the substrate was immersed in 0.2 mol POCl$_3$ and γ-collidine of acetonitrile solution. After ultrasonically rinsing with acetonitrile, the substrate was dipped in 5 mmol/dm^3 zirconyl chloride ZrCl$_2$·8H$_2$O of ethyl alcohol-aqueous solution to form a zirconium phosphate (ZrP) layer. Ultimately the substrate was suspended in 2 mmol/dm^3 DDP of acetone-aqueous solution to form the three layered Zr-DDP coating as shown in Scheme 1.

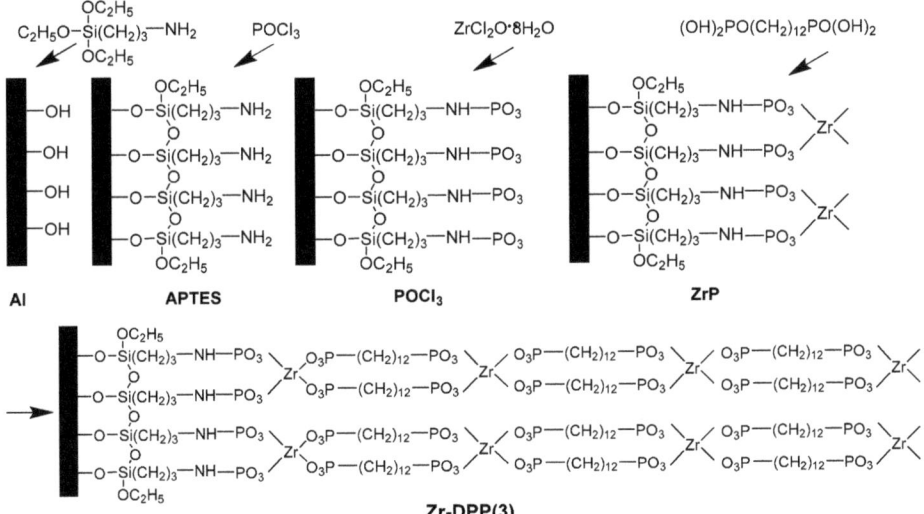

Scheme 1. Multilayer zirconium phosphate treatment consisting of DDP. (Reproduced from [8] with permission; Copyright 2003 Elsevier).

The Zr-DDP multi-layers film was tested for corrosion resistance by exposure to 5 wt % NaCl solution at 308 K for 48 h in the salt-spray test. The corrosion rate of multi-layer Zr-DDP coating is 0% (indicating no corrosion at all) and showed superior anti-corrosion properties compared to APTES coating (100%), POCl$_3$ coating (80%), and conventional ZrP coating (100%). After a 72 h salt spray test, Zr-DDP films were corroded with a corrosion rate of 60% whereas the other treatments showed 100% corrosion rate. The authors concluded that Zr-DDP films constructed by SAMs are promising surface finishing treatments to replace the conventional phosphating or chromating treatments [8].

4.1.4. Amino Trimethylene Phosphonic Acid (ATMP)

ATMP is known to have anti-scale performance due to its excellent chelating ability, and a low threshold inhibition and lattice distortion process. ATMP was used as an inhibitor to improve the anti-corrosion properties of aluminum alloy (AA2024-T3) [15]. The deposition bath was made by mixing ATMP in distilled water, tetraethylorthosilicate (TEOS), and ethanol in a 6:4:90 ($v/v/v$) ratio. In this work, the effect of three different pretreatments (acetic acid, acetic acid-NaOH, NaOH) and ATMP concentration on the corrosion protection were studied. SEM and energy-dispersive X-ray spectroscopy (EDS) analysis of the treated aluminum surface confirmed that acetic acid pretreatment of the aluminum surface was the most efficient pretreatment. The authors assumed that acetic acid pretreatment increased surface area of the alloy matrix, which led to more formation of favorable metalosiloxane and metal-phosphonic bonds to the surface.

Three different ATMP concentrations of deposition baths were investigated. The corrosion behavior of the treated aluminum was tested against 0.05 M NaCl solution for 48 h. EIS was used to assess the corrosion resistance, as mentioned in Section 3.1.1. High R_{HF} (the resistance of the high frequency related to the coating layer) value and low CPE_{HF} (the capacitance of the high frequency related to the coating layer) value indicated higher corrosion resistance. Results of the TEOS coating containing ATMP (5.00×10^{-4} M) showed R_{HF} (kOhm·cm^2): 81.4, CPE_{HF} (µF·cm^{-2}·s^{n-1}): 16.2, whereas increasing or decreasing the ATMP concentration had an adverse effect on corrosion protection. In contrast, TEOS coating without ATMP resulted in R_{HF} (kOhm·cm^2): 5.06, CPE_{HF} (µF·cm^{-2}·s^{n-1}): 160.5, indicating no corrosion resistance at all. The authors believe that the corrosion protection was caused by the strong chemical bonding of phosphonic groups to the aluminum substrate [15].

4.1.5. Ethylenediamine Tetra Methylene Phosphonic Acid (EDTPO)

Modification of aluminum alloy (AA2024) with a well-known anticorrosive epoxy-polyamide paint in combination with vinyltrimethoxysilane (VTMS)/tetraethylorthosilicate (TEOS) nanocoating was used to improve its corrosion resistance and adhesion to organic layers. EDTPO, as a catalyst, was incorporated into VTMS/TEOS to form a sol-gel coating on the substrate surface and is believed to improve the formation of Al–O–P bonds which is beneficial for anticorrosive properties [17]. The concentration of EDTPO in the sol-gel deposition solution was 3.75×10^{-5} mol·L^{-1}. EIS measurement was carried out in 3.5% NaCl aqueous solution. Substrates coated with epoxy and sol-gel film with or without EDTPO showed good corrosion resistance: The EIS results remained constant at $10^{9.5}$ Ω·cm^{-2} up to 300 days, which indicated an excellent corrosion resistance. Whereas the values for the epoxy coating without sol-gel film started to decrease after 160 days and 300 days later, coating resistance was only $10^{7.9}$ Ω·cm^2. A cyclic accelerated corrosion test in 3.5% NaCl solution for 60 days was evaluated by ASTM D-1654. The results showed that the failure area of the coating with EDTPO was 0.8%, which indicates an excellent corrosion resistance compared to the coating without EDTPO (1.3%) or the coating without any sol-gel primer (32%). After accelerated corrosion, the pull-off test (UNE-EN-ISO 4624) showed that the adhesion reduction for the coating with EDTPO to be 0%. However, the adhesion of coating without EDTPO showed a reduction of 15%, and the adhesion of the coating without any sol-gel primer reduced to 40%. It was concluded that the EDTPO-modified silane coating (VTMS/TEOS) has excellent adhesion and anti-corrosion properties because of its barrier properties [17].

In a similar work, two different concentrations of EDTPO (3.75×10^{-5} and 3.75×10^{-4} mol·L^{-1}) were incorporated in TEOS to form sol-gel deposition solutions, which are named TEOS/EDTPO 10^{-5} and TEOS/EDTPO 10^{-4}, respectively. Aluminum alloy (AA2024-T3) panels were immersed in these formulations to form an anti-corrosion film. As mentioned in Section 3.1.1, a high R_{HF} value and low CPE_{HF} value indicates a corrosion resistance. After seven days of soaking in the 0.05 mol·L^{-1} NaCl solution at 25 °C, EIS test results of TEOS/EDTPO 10^{-4} was R_{HF} (Ω·cm^2): 248.3, CPE_{HF} (μF·cm^{-2}·s^{n-1}): 1.74 and for TEOS/EDTPO 10^{-5} it was R_{HF} (Ω·cm^2): 177.4, CPE_{HF} (μF·cm^{-2}·s^{n-1}): 15.8. Thus, it was clear that TEOS/EDTPO 10^{-4} could provide a better corrosion resistance. The substrate coated with TEOS/EDTPO 10^{-4} was further investigated by soaking it in 0.05 mol·L^{-1} NaCl solution at 70 °C with various immersion times. After a 24 h exposure, the impedance modulus is lower than the measurements for 25 °C. Nevertheless, the corrosion resistance of the EDTPO containing coatings was still approximately six times higher than for TEOS-only coatings. In this work, it was also confirmed that EDTPO enhanced the corrosion resistance of the sol-gel coating even at a higher temperature [30].

In a separate work done by the same authors, the corrosion behavior of EDTPO and ATMP was compared in TEOS sol-gel coating on the aluminum alloy (AA1100) surface. Substrates were dip-coated in two different solutions: 3.75 × 10^{-4} mol·L^{-1} EDTPO in TEOS ethanol solution and 5.00 × 10^{-4} mol·L^{-1} ATMP in TEOS ethanol solution. EIS was used to evaluate the corrosion of coated substrates in a 0.05 mol·L^{-1} NaCl solution for seven days. The corrosion resistance of TEOS/EDTPO was R_{HF} (Ω·cm^2): 141.0, CPE_{HF} (μF·cm^{-2}·s^{n-1}): 2.45. For TEOS/ATMP it was R_{HF} (Ω·cm^2): 113.7, CPE_{HF} (μF·cm^{-2}·s^{n-1}): 1.65. It was concluded that both EDTPO and ATMP could provide considerable corrosion protection for AA1100. However, the corrosion resistance of EDTPO-containing coatings was higher than for ATMP-containing coatings [40].

4.1.6. 1,2-Diaminoethanetetrakis-Methylenephosphonic Acid (DETAPO)

The adhesion and corrosion performance of silane sol-gel film coating on aluminum has been reported. Especially ATMP and EDTPO, as the phosphonic catalysts in a sol-gel system, have been well studied for the protection against corrosion [15,17,30,40]. Phosphonic derivative DETAPO, which has a similar structure has also been tested because of higher concentration of O=P–OH group in its structure compared to ATMP and EDTPO [23]. In this work, aluminum alloy (AA2024-T3) was coated with VTMS/TEOS sol-gel film containing DETAPO and the epoxy-polyamide resin. A cyclic accelerated corrosion test in 3.5% NaCl for 45 days was performed according to the ASTM D1654 method. The results showed that the failure area of DETAPO modified coating was around 10–11%, compared to the 35% failure area for the epoxy coating without any silane film. Pull-off tests after accelerated corrosion showed that adhesion reduction of DETAPO-modified silane coating was around 62.5–52.9%, on the other hand, the coating without any silane film reduced to 34.7% adhesion. It was clear that the DETAPO-modified coating can provide an outstanding corrosion resistance for aluminum surface [23].

4.1.7. (12-Ethylamino-Dodecyl)-Phosphonic Acid

Self-assembled monolayers (SAMs) were used to form thin organic adhesion promoter layers to replace the chromating process on aluminum. (12-ethylamino-dodecyl)-phosphonic acid was chosen as a model functional organic phosphonic acid in this work because of its phosphonic acid anchor group which acts as a reactive group for the aluminum surface, its aliphatic part acting as a hydrophobic spacer, and its amino head group serving as a reactive group to organic material [27]. The aluminum alloy (AlMg) was dip-coated in the 10^{-3} M (12-ethylamino-dodecyl)-phosphonic acid solution of ethanol/water (3:1 vol %). The surface analysis by FTIR and surface plasmon resonance spectroscopy (SPR) confirmed that organophosphonic acid could spontaneously adsorb on the aluminum surface and subsequently form oriented layers. The adhesion promotion and corrosion resistance of the aluminum surface were confirmed by the acetic acid salt spray-test (ASS-test, DIN 50021) and filiform test (DIN 50024). Coated panels were scratched and exposed in a climate chamber of NaCl solution and acetic acid. After 1200 h exposure, the infiltrations for the amino phosphonic acid coating was less than 1 mm, which indicates excellent corrosion resistance compared to 8 mm infiltration for the uncoated substrate [27].

4.1.8. Aminopropyl Phosphonate (APP)

To replace harmful chromate conversion layers, environmentally-friendly aminopropyl phosphonate (APP), aminopropyl silane (APS), and hexamethyldisiloxane (HMDSO) were used to coat aluminum alloy (6016) [9]. As known, filiform corrosion (FFC) usually occurs at the interface of a polymer and an aluminum alloy. Hence, the corrosion resistance and adhesion promotion of these coatings were studied to find a suitable pretreatment for automotive applications. The influence of the rolling direction on filiform corrosion was also investigated. Substrates were dipped in 1 mmol APP solution with pH 7 for 1 h followed by an epoxy coating. APS was coated on substrates with the same dip-coating method and, for HMDSO, a plasma coating technique was used. Before the corrosion test, substrates were scribed perpendicularly or parallel to the rolling direction. The corrosion test was carried out by soaking scratched substrates in a vessel with concentrated HCl vapor for 20 min. FFC was evaluated based on filament initiating time and the number of filaments observed in the digital microscope. From visual observation, it could be confirmed that filaments grew predominantly along the rolling direction, which indicated that the rolling direction had an influence on the interfacial bonding. The FFC severities were as follows: APS > etching > APP > HMDSO plasma. The adhesion of the epoxy adhesive to the substrate was analyzed by a peeling test, and the results follow a trend for different coatings: HMDSO plasma > APS > APP > etching. It was concluded that the adhesion of the coating had a great influence on filament morphology and further on filament propagation, but an increase in adhesion of the coating was not a guarantee of resistance to FFC [9].

The corrosion performance of aluminum powder was enhanced by a novel functionalization with graphene oxide (GO) with phosphonic acid as a linker [11]. APP was used as the "link" agent to connect graphene oxide (GO) with aluminum powder. APP was added in a suspension of GO and deionized water. After stirring, refluxing, dialyzing, and drying, GO-APP was added in the aluminum powder suspension to form the oxide modified aluminum powder (GO-Al). Corrosion performance was tested in dilute hydrochloric acid. Since corrosion of aluminum generated hydrogen, corrosion behavior could be evaluated by detecting the amount and the yield time of hydrogen gas. The results showed that for the Al/HCl system, the hydrogen was detected after 100 min. However, in the GO-Al/HCl system, hydrogen was detected only after 140 min, which confirms enhanced the anti-corrosive performance of GO-Al. Through XPS, FTIR, field emission scanning electron microscopy (FE-SEM), and EDS, it could be confirmed that the flaky aluminum particle was successfully covalently bonded and covered by GO. In conclusion, GO as a barrier coating was well connected with epoxy through APP [11].

4.1.9. ω-(3-Phenylpyrrol-1-ylalkyl) Phosphonic Acid (Cn-Ph-P)

Cn-Ph-P was used to build SAMs on the aluminum surface (Al/Al$_2$O$_3$ chemical vapor deposition on polished, p-doped Si wafers) because its phosphonic acid group could anchor to the aluminum surface and its pyrrole group could be used to achieve an in situ surface polymerization with further monomers [10]. The substrates were immersed in a Cn-Ph-P solution for different time intervals (5 min, 1 h, and 24 h) to form SAMs on the surface. After that, the pyrrole monomer was polymerized with the free terminal pyrrole group of the SAM by oxidants, like sodium or ammonium peroxidisulphate, followed by the growth of polypyrrole (PPY) on the surface. The contact angle was used to evaluate the adsorption behavior. It was inferred that a more hydrophobic surface (higher contact angles) was caused by phosphonic acid group reacting with surface and the terminal polymerizable group present on top of it. The surface contact angle rose up after adsorption of CnPhP, which indicated that CnPhP was oriented on the surface. The contact angles of C12PhP were slightly higher than C10PhP because of the stronger van der Waals interaction between the alkyl chains. C12PhP showed excellent stability after eight repeated cycling using the Wilhelmy method [10].

4.1.10. ω-(2,5-Dithienylpyrrol-1-yl-alkyl) Phosphonic Acid (SNS-n-P)

SNS-n-P as a model functional phosphonic acid was used to study the influence of varying alkyl chain length (SNS10P, SNS4P) to the adsorption on the aluminum surface (Al/Al_2O_3 chemical vapor deposition on polished, p-doped Si wafers). The contact angle of SNS10P coated surface did not change after ten repeated cycles by the Wilhelmy method, which indicated that SNS10P had a very stable bond to the surface. However, the contact angle of SNS4P coating decreased significantly with the increasing number of repeated cycles. Thus, this comparison of analogue structures SNS10P and SNS4P showed that the alkyl chain length of the structures influences the bonding at the surface [10].

4.1.11. (1H,1H, 2H,2H-Heptadecafluorodec-1-yl) Phosphonic Acid (HDF-PA)

Trichlorosilane or metal-ligand coordination has mostly been used to engineer surface wetting properties on metals. However, they were highly moisture sensitive due to hydrolysis and self-condensation of silanes [51–53]. Therefore, HDF-PA was employed to coat different self-assembled monolayers (SAMs) on metal oxide surfaces because phosphonic acid-functionalized molecules were stable in water, and built a stable metal–ligand coordination and hetero-condensation between phosphonate and substrate surfaces. Substrates with a 30 nm thick Al_2O_3 film were dip-coated in 2 mM HDF-PA isopropyl alcohol for 1 h. After the water flow test, changes in surface contact angle were analyzed to evaluate the reliability of the chemical bond between SAMs and substrate surface. (Heptadecafluoro-1,1,2,2-tetrahydrodecyl) trichlorosilane (HDF-S) was also studied as a comparison. The bare Al_2O_3 films showed contact angle values of $70.0° \pm 1.5°$, the contact angles of HDF-PA and HDF-S-coated Al_2O_3 films were enhanced to $99.9° \pm 1.0°$ and $102.7° \pm 2.4°$. After exposing the surface to 5 L of water droplets, contact angles of HDF-S coating remained around $102.7°$, whereas the contact angles of HDF-PA coating reduced from $99.9°$ to $69.3°$. The results showed that HDF-S formed a more stable bond than HDF-PA, which may be due to the cross-linked siloxane network on the surface. A low-temperature (<150 °C) thermal annealing process was used to improve the HDF-PA bond formation. The results showed that after the water flow test, contact angles for 100 °C and 150 °C annealed substrates were stable at around 101 °C. Therefore, the durability of HDF-PA coatings could be improved by an additional thermal annealing process at 100–150 °C [19].

4.1.12. Poly (Vinyl Phosphonic Acid) (PvPA)

PvPA was coated as a thin polymeric interfacial layer on aluminum alloy (AA 1050) surface by dip-coating and compared to the poly (acrylic acid) (PAA), poly (ethylene-alt-maleic anhydride) (PEMah) coated aluminum surface. After that, an epoxy (Resolution Epikote 1001) adhesive was coated on the modified substrate. The PvPA-based system showed poor adhesion performance in the pull-off test compared to PEMah, because PvPA forms a weakly-cured epoxy/polymer interphase [24].

4.2. Phosphates Used as Pre-Coating

4.2.1. Monostearyl Acid Phosphate (MSAP)

A novel ultrasonic assisted deposition (USAD) method was used to coat phosphate films on aluminum and compared with films formed by mild mechanical stirring [12]. MSAP was used in this study and its impact on the adsorption behavior of these films was evaluated. Simultaneously, the corrosion resistance and adhesion properties of the coatings were studied as well. Infrared spectroscopy (IR) and NMR were used to analyze the stability of the MSAP film, after soaking the substrates for 5 h at 100 °C in a reflux distillation column containing water and steam. It was observed in NMR spectra that the USAD substrates had resolved peaks at -7 and -14 ppm, whereas the chemical shift of the stirring coated substrates occurred at 0 ppm. Based on the previous NMR studies for alkylphosphate films on metal oxide surfaces [54–58], it was confirmed that the peaks at -7 and -14 ppm meant bidentate and tridentate interactions between the aluminum oxide surface and the phosphate. It was concluded that the USAD treatment favored tridentate interaction between MSAP and aluminum oxide [12].

4.2.2. Phosphoric Acid Mono-(12-Hydroxy-Dodecyl) Ester

From earlier studies, it was clear that cooperated pretreatments of phosphoric acid anodic oxidation (PAA) and an adhesion promoter (AP) (phosphoric acid mono alkyl ester) could enhance the adhesion and durability of the aluminum/epoxy bonding and thus could be considered as an alternative to the chromium-based pretreatment for aluminum [59]. In a more recent work, aluminum substrates were first anodized in a phosphoric acid solution to create PAA. Then the PPA substrates were dip-coated in 10^{-3} M aqueous solution of phosphoric acid mono-(12-hydroxy-dodecyl) ester for 5 min at room temperature [41]. EIS and the floating roller peel test were used to study the corrosion performance of PAA and PAA/AP systems. After 48 h soaking in 3% NaCl solution at 80 °C, the EIS results showed that the PAA and PAA/AP pretreated surfaces had a strong influence on the crosslinking of the epoxy at the interphase. For the floating roller peel test, one rigid and one flexible coated substrate was first bonded with Delo-Duopox 1891 adhesive. The joints were then corroded according to accelerated method VDA 621-415. The peel test was performed based on DIN EN 1464. The results indicated that after aging, PAA/AP treated substrates showed higher peel strength (around 3.3 N/mm) than PAA treated substrates (around 2.8 N/mm). It was confirmed that AP could enhance the cross-linking of the epoxy coat. The authors suggested that the AP layer was well-ordered on the surface and the strong bonding of the phosphoric acid anchor group to PAA led to the good adhesion and corrosion resistance [41].

4.2.3. Fluoro Alkyl Phosphates (Zonyl UR)

Zonyl UR was deposited on aluminum plate and aluminum oxide powder as a corrosion inhibitor and adhesion promoter. Two different deposition methods were compared: The ultrasonic assisted deposition (USAD) or magnetic stirring. Environmental stability of the coating was observed in a reflux distillation column (100 °C). For aluminum plate, the change in IR spectra was used to evaluate the stability of the coatings, which showed that after 5 h refluxing, barely any changes occurred to the USAD-coated surface. In contrast, clear evidence of corrosion was observed for stirring deposition. For aluminum powder, IR spectra of stirring deposition exhibited broad features, which did not exist in USAD. Solid state nuclear magnetic resonance (NMR) was used to study the interactions between the phosphate group and the aluminum oxide powder. The results confirmed that USAD could provide a more homogeneous film with mostly tridentate interaction and low coverage of mono- and bidentate interaction [12].

4.3. Coatings Dissolved in Paints that Contain Phosphonic Acids or Phosphates

4.3.1. Phenylphosphonic Acid (PPA)

PPA has been used as a self-phosphating agent and an acid catalyst in the formulation of a polyester-melamine coating for aluminum surface [14]. In this work, a chrome-free single-step in situ phosphatizing coating (ISPC) was used on aluminum alloy (3003 or 3105). This work is based on a patent published earlier, where for an ISPC, an optimum amount of an in situ phosphatizing reagent (ISPR) or a mixture of ISPRs is predisposed in the paint system to form a stable and compatible coating formulation [60]. It is believed that when a chrome-free single-step coating of the in situ self-phosphating paint is applied to a bare metal substrate, the phosphatizing reagent chemically and/or physically reacts in situ with the metal surface to produce a metal phosphate layer and simultaneously forms covalent P–O–C (phosphorus-oxygen-carbon) linkages with the polymer resin. Such linkages enhance the adhesion of the coating and suppress substrate corrosion. It was assumed that PPA could also provide phosphate formation on the metal surface in the in-situ phosphatizing treatment [14]. The aluminum panels were spray coated with polyester-melamine paint containing

1 wt % PPA and other formulations containing commercially available acid sources (phosphonosuccinic acid, para toluene sulphonic acid). Coated substrates were analyzed for corrosion behavior using EIS measurements, saltwater immersion, and pencil hardness tests. The paint on cured panels was removed by a razor and analyzed for the glass transition temperature (T_g) using differential scanning calorimetry (DSC). A higher T_g correlates with a coating with a higher cross-linking density in the polymer. T_g of 1 wt % PPA paint formulation was 22 °C. After the paint was exposed to 300 °C, T_g increased to 65 °C indicating an increase in the crosslinking density of the polymer. For formulations containing para toluene sulphonic acid (commercial solution), the crosslinking density dropped as the T_g dropped after severe heating. This resulted from the possible cleavage of polyester-melamine cross-links in favor of the melamine self-condensation in the paint films [61].

After immersion of the coated substrate in 3% NaCl for three days, EIS was used to study the corrosion behavior. It is known from previous work that a surface with an impedance at the low frequency below 10^7 Ω·cm^2 is considered a poor protective barrier [44]. The paint with 1 wt % PPA resulted in the impedance value of 10^{10} Ω·cm^2 at low frequency. This result indicates the superior corrosion protection of PPA-sol-gel coatings. For saltwater immersion test, test panels were scribed with an "X" and then were soaked in a 3% NaCl solution for 66 days. After that, a tape was firmly pressed against the scribed area and pulled to remove. The saltwater corrosion test was evaluated using the ASTM D3359A method, which showed that there was no discoloration of the paint and no paint was removed with the tape. Only very few tiny blisters (Ø < 1 mm) around the scribe were observed, which may result from paint defects. The pencil hardness of the paint (ASTM D3363) was F. The paint was without any discoloration. The authors concluded the paint with 1 wt % of PPA provided excellent corrosion resistance [14].

4.3.2. 2-(Phosphonooxy) Benzoic Acid (Fosfosal)

Fosfosal was used as an acid additive in the polyester-melamine paint and was used in the in situ phosphatizing treatment on aluminum alloy (3003) [14]. Substrates were spray-coated with polyester-melamine paint containing different concentrations (0.5%, 0.75%, and 1%) of Fosfosal. DSC was used to evaluate the cross-linking density of polymer by analyzing the glass transition temperature (T_g). The DSC program involved annealing at 80 °C, then scanning from −50 °C to 300 °C with a heating rate of 10 °C/min. Higher T_g means higher cross-linking density. EIS was used to study the corrosion performance of coated substrates after three days in 3% NaCl immersion. As mentioned earlier, EIS results of lower than 10^7 Ω·cm^2 at low frequency meant poor anti-corrosion properties. Corrosion resistance was further studied after saltwater immersion. Coated substrates were scribed with an "X" and immersed in 3% NaCl solution for 66 days. After drying, a tape was firmly pressed against the "X" area and pulled to remove. The corrosion resistance was evaluated by ASTM method D3359 A The pencil hardness of the painted Al panels was measured using ASTM method D3363. The results of the corrosion test are summarized in Table 7.

Table 7. Measurement results of different Fosfosal concentrations.

Concentration	DSC (T_g)	EIS (Ω·cm^2)	Saltwater Immersion	Pencil Hardness
1%	35 to 59 °C	10^9–10^7	–	4H
0.75%	34 to 55 °C	10^9	Ø 3 mm blister	4H
0.5%	18 to 48 °C	10^9	–	HB

The corrosion resistance (saltwater immersion) for 1% Fosfosal coatings could not be estimated with the ASTM method. For 0.5% Fosfosal, saltwater immersion results were not reproducible; one panel had no paint removal and was evaluated as 5A in the ASTM D3359 A test. One panel had a clump of blisters of size ~Ø 2 mm and the rest had no blistering. DSC results for coatings with 0.5% Fosfosal showed that cross-linking density increased and EIS results showed 0.5% Fosfosal provided sufficient corrosion protection. In case of pencil hardness test, coatings containing 1% and 0.75%

Fosfosal (4H) were harder than for 0.5% Fosfosal (HB). However, EIS results for coatings containing 1% Fosfosal were not reproducible and saltwater immersion results of 0.75% Fosfosal were worse than for 0.5% Fosfosal. Therefore, the authors suggested that 0.5% Fosfosal should be used in the in situ phosphatizing treatment [14].

4.3.3. Phosphonosuccinic Acid (PPSA)

PPSA, as an acid additive, was added to the polyester-melamine paint coated on the aluminum alloy (3003) surface. The EIS accelerated corrosion test was used to study the corrosion performance of these coatings. Polyester-melamine paint with 5% PPSA could not be fully cured. Thus, paint with 1 wt % PPSA was coated on the substrate and cured, but these coatings exhibited poor coating properties. EIS results showed that the impedance at low frequency was in the order of 10^6 $\Omega \cdot cm^2$. After one week immersion in 3% salt solution, many blisters were observed on the paint surface. PPSA performed badly as a catalyst in the polyester-melamine paint system [14].

5. Conclusions

Environmental issues with existing pretreatments for aluminum, together with the need to develop new materials with higher mechanical and chemical performance, is driving the research and industrial communities to develop new pretreatments based on phosphonic acids. The new treatments are shown to have higher mechanical performance and corrosion protection; however, much work in adaptation to industrial processes and their real environmental impact need to be evaluated to demonstrate the commercial feasibility of these treatments. In many cases, the phosphonic acids are incorporated in sol-gel coatings, which look simple in batch-scale application in laboratory research. However, it is not clear if such treatments can be adapted to continuous handling in industrial production. The use of hybrid systems, such as phosphonic acid-modified graphene oxide, is quite promising. However, the economics of such treatments will determine its potential commercial exploitation. We believe that the combination of various organic and inorganic conversion technologies for aluminum is needed to obtain industrially-feasible solutions. The development of novel phosphonic acid chemistry with a focus on self-healing, superior galvanic protection, and higher mechanical performance will have a strong focus in the future. New phosphonic acids should be screened for their toxicological profile and environmental impact before any possible commercial exploitation.

Acknowledgments: We would like to acknowledge the financial support of KTI, Switzerland (TISBA project: 16589.1 PFIW-IW) and Novelis Switzerland.

Author Contributions: Ruohan Zhao as first Author conducted the literature search. All three authors were involved in the writing and editing of the Manuscript .

Conflicts of Interest: The authors declare no conflict of interest.

References

1. Twite, R.L.; Bierwagen, G.P. Review of alternatives to chromate for corrosion protection of aluminum aerospace alloys. *Prog. Org. Coat.* **1998**, *33*, 91–100. [CrossRef]
2. Liu, Y.; Sun, D.; You, H.; Chung, J.S. Corrosion resistance properties of organic–inorganic hybrid coatings on 2024 aluminum alloy. *Appl. Surf. Sci.* **2005**, *246*, 82–89. [CrossRef]
3. Bulusu, A.; Paniagua, S.A.; MacLeod, B.A.; Sigdel, A.K.; Berry, J.J.; Olson, D.C.; Marder, S.R.; Graham, S. Efficient modification of metal oxide surfaces with phosphonic acids by spray coating. *Langmuir* **2013**, *29*, 3935–3942. [CrossRef] [PubMed]
4. Queffélec, C.; Petit, M.; Janvier, P.; Knight, D.A.; Bujoli, B. Surface modification using phosphonic acids and esters. *Chem. Rev.* **2012**, *112*, 3777–3807. [CrossRef] [PubMed]
5. Pellerite, M.J.; Dunbar, T.D.; Boardman, L.D.; Wood, E.J. Effects of fluorination on self-assembled monolayer formation from alkanephosphonic acids on aluminum: Kinetics and structure. *J. Phys. Chem. B* **2003**, *107*, 11726–11736. [CrossRef]

6. Attavar, S.; Diwekar, M.; Linford, M.R.; Davis, M.A.; Blair, S. Passivation of aluminum with alkyl phosphonic acids for biochip applications. *Appl. Surf. Sci.* **2010**, *256*, 7146–7150. [CrossRef]
7. Hoque, E.; DeRose, J.A.; Kulik, G.; Hoffmann, P.; Mathieu, H.J.; Bhushan, B. Alkylphosphonate modified aluminum oxide surfaces. *J. Phys. Chem. B* **2006**, *110*, 10855–10861. [CrossRef] [PubMed]
8. Shida, A.; Sugimura, H.; Futsuhara, M.; Takai, O. Zirconium-phosphate films self-assembled on aluminum substrate toward corrosion protection. *Surf. Coat. Technol.* **2003**, *169–170*, 686–690. [CrossRef]
9. Liu, X.F. Filiform corrosion attack on pretreated aluminum alloy with tailored surface of epoxy coating. *Corros. Sci.* **2007**, *49*, 3494–3513. [CrossRef]
10. Jaehne, E.; Oberoi, S.; Adler, H.-J.P. Ultra thin layers as new concepts for corrosion inhibition and adhesion promotion. *Prog. Org. Coat.* **2008**, *61*, 211–223. [CrossRef]
11. He, L.; Zhao, Y.; Xing, L.; Liu, P.; Wang, Z.; Zhang, Y.; Liu, X. Preparation of phosphonic acid functionalized graphene oxide-modified aluminum powder with enhanced anticorrosive properties. *Appl. Surf. Sci.* **2017**, *411*, 235–239. [CrossRef]
12. Shepard, M.J.; Comer, J.R.; Young, T.L.; McNatt, J.S.; Espe, M.P.; Ramsier, R.D.; Robinson, T.R.; Nelson, L.Y. *Organophosphate Adsorption on Metal Oxide Surfaces*; VSP BV: Oud-Beijerland, The Netherlands, 2004; pp. 225–239.
13. Ulman, A. Formation and structure of self-assembled monolayers. *Chem. Rev.* **1996**, *96*, 1533–1554. [CrossRef] [PubMed]
14. Whitten, M.C.; Burke, V.J.; Neuder, H.A.; Lin, C.-T. Simultaneous acid catalysis and in situ phosphatization using a polyester-melamine paint: A surface phosphatization study. *Ind. Eng. Chem. Res.* **2003**, *42*, 3671–3679. [CrossRef]
15. Dalmoro, V.; Santos, J.H.Z.; Azambuja, D.S. Corrosion behavior of AA2024-T3 alloy treated with phosphonate-containing TEOS. *J. Solid State Electrochem.* **2012**, *16*, 403–414. [CrossRef]
16. Jalili, M.; Rostami, M.; Ramezanzadeh, B. Surface modification of aluminum flakes with amino trimethylene phosphonic acid: Studying the surface characteristics and corrosion behavior of the pigment in the epoxy coating. *Corrosion* **2015**, *71*, 628–640. [CrossRef]
17. Dalmoro, V.; Aleman, C.; Ferreira, C.A.; dos Santos, J.H.Z.; Azambuja, D.S.; Armelin, E. The influence of organophosphonic acid and conducting polymer on the adhesion and protection of epoxy coating on aluminium alloy. *Prog. Org. Coat.* **2015**, *88*, 181–190. [CrossRef]
18. Blajiev, O.L.; Ithurbide, A.; Hubin, A.; Van Haesendonck, C.; Terryn, H. XPS study of the assembling morphology of 3-hydroxy-3-phosphono-butyric acid tert-butyl ester on variously pretreated Al surfaces. *Prog. Org. Coat.* **2008**, *63*, 272–281. [CrossRef]
19. Lee, J.; Bong, J.; Ha, Y.-G.; Park, S.; Ju, S. Durability of self-assembled monolayers on aluminum oxide surface for determining surface wettability. *Appl. Surf. Sci.* **2015**, *330*, 445–448. [CrossRef]
20. Davies, P.R.; Newton, N.G. The chemisorption of organophosphorus compounds at an Al (111) surface. *Appl. Surf. Sci.* **2001**, *181*, 296–306. [CrossRef]
21. Hauffman, T.; Hubin, A.; Terryn, H. Study of the self-assembling of n-octylphosphonic acid layers on aluminum oxide from ethanolic solutions. *Surf. Interface Anal.* **2013**, *45*, 1435–1440. [CrossRef]
22. Sheffer, M.; Groysman, A.; Starosvetsky, D.; Savchenko, N.; Mandler, D. Anion embedded sol-gel films on al for corrosion protection. *Corros. Sci.* **2004**, *46*, 2975–2985. [CrossRef]
23. Iribarren-Mateos, J.I.; Buj-Corral, I.; Vivancos-Calvet, J.; Aleman, C.; Iribarren, J.I.; Armelin, E. Silane and epoxy coatings: A bilayer system to protect aa2024 alloy. *Prog. Org. Coat.* **2015**, *81*, 47–57. [CrossRef]
24. Van den Brand, J.; Van Gils, S.; Beentjes, P.C.J.; Terryn, H.; Sivel, V.; de Wit, J.H.W. Improving the adhesion between epoxy coatings and aluminum substrates. *Prog. Org. Coat.* **2004**, *51*, 339–350. [CrossRef]
25. Essahli, M.; El Asri, M.; Boulahna, A.; Zenkouar, M.; Viguier, M.; Hervaud, Y.; Boutevin, B. New alkylated and perfluoroalkylated phosphonic acids: Synthesis, adhesive and water-repellent properties on aluminum substrates. *J. Fluorine Chem.* **2006**, *127*, 854–860. [CrossRef]
26. Kowalik, T.; Adler, H.J.P.; Plagge, A.; Stratmann, M. Ultrathin layers of phosphorylated cellulose derivatives on aluminium surfaces. *Macromol. Chem. Phys.* **2000**, *201*, 2064–2069. [CrossRef]
27. Adler, H.-J.P.; Jaehne, E.; Stratmann, M.; Grundmeier, G. New concepts for corrosion inhibition and adhesion promotion. *Proc. Annu. Meet. Tech. Program FSCT* **2002**, *80*, 528–545.
28. Hofer, R.; Textor, M.; Spencer, N.D. Alkyl phosphate monolayers, self-assembled from aqueous solution onto metal oxide surfaces. *Langmuir* **2001**, *17*, 4014–4020. [CrossRef]

29. Maxisch, M.; Thissen, P.; Giza, M.; Grundmeier, G. Interface chemistry and molecular interactions of phosphonic acid self-assembled monolayers on oxyhydroxide-covered aluminum in humid environments. *Langmuir* **2011**, *27*, 6042–6048. [CrossRef] [PubMed]
30. Dalmoro, V.; dos Santos, J.H.Z.; Armelin, E.; Alemán, C.; Azambuja, D.S. Phosphonic acid/silica-based films: A potential treatment for corrosion protection. *Corros. Sci.* **2012**, *60*, 173–180. [CrossRef]
31. Guerrero, G.; Chaplais, G.; Mutin, P.H.; Le Bideau, J.; Leclercq, D.; Vioux, A. Grafting of alumina by diphenylphosphinate coupling agents. *Mater. Res. Soc. Symp. Proc.* **2001**, *628*. [CrossRef]
32. Guerrero, G.; Mutin, P.H.; Vioux, A. Organically modified aluminas by grafting and sol-gel processes involving phosphonate derivatives. *J. Mater. Chem.* **2001**, *11*, 3161–3165. [CrossRef]
33. Mutin, P.H.; Guerrero, G.; Vioux, A. Organic-inorganic hybrid materials based on organophosphorus coupling molecules: From metal phosphonates to surface modification of oxides. *C. R. Chim.* **2003**, *6*, 1153–1164. [CrossRef]
34. Textor, M.; Ruiz, L.; Hofer, R.; Rossi, A.; Feldman, K.; Hahner, G.; Spencer, N.D. Structural chemistry of self-assembled monolayers of octadecylphosphoric acid on tantalum oxide surfaces. *Langmuir* **2000**, *16*, 3257–3271. [CrossRef]
35. Alexander, M.R.; Thompson, G.E.; Beamson, G. Characterization of the oxide/hydroxide surface of aluminium using x-ray photoelectron spectroscopy: A procedure for curve fitting the O 1s core level. *Surf. Interface Anal.* **2000**, *29*, 468–477. [CrossRef]
36. Jeurgens, L.P.H.; Sloof, W.G.; Tichelaar, F.D.; Mittemeijer, E.J. Composition and chemical state of the ions of aluminium-oxide films formed by thermal oxidation of aluminium. *Surf. Sci.* **2002**, *506*, 313–332. [CrossRef]
37. Paniagua, S.A.; Hotchkiss, P.J.; Jones, S.C.; Marder, S.R.; Mudalige, A.; Marrikar, F.S.; Pemberton, J.E.; Armstrong, N.R. Phosphonic acid modification of indium-tin oxide electrodes: Combined XPS/UPS/contact angle studies. *J. Phys. Chem. C* **2008**, *112*, 7809–7817. [CrossRef]
38. Van den Brand, J.; Sloof, W.G.; Terryn, H.; de Wit, J.H.W. Correlation between hydroxyl fraction and O/Al atomic ratio as determined from XPS spectra of aluminium oxide layers. *Surf. Interface Anal.* **2004**, *36*, 81–88. [CrossRef]
39. Gouzman, I.; Dubey, M.; Carolus, M.D.; Schwartz, J.; Bernasek, S.L. Monolayer vs. multilayer self-assembled alkylphosphonate films: X-ray photoelectron spectroscopy studies. *Surf. Sci.* **2006**, *600*, 773–781. [CrossRef]
40. Dalmoro, V.; dos Santos, J.H.Z.; Armelin, E.; Alemán, C.; Azambuja, D.S. A synergistic combination of tetraethylorthosilicate and multiphosphonic acid offers excellent corrosion protection to AA1100 aluminum alloy. *Appl. Surf. Sci.* **2013**, *273*, 758–768. [CrossRef]
41. Phung, L.H.; Kleinert, H.; Fuessel, U.; Duc, L.M.; Rammelt, U.; Plieth, W. Influence of self-assembling adhesion promoter on the properties of the epoxy/aluminum interphase. *Int. J. Adhes. Adhes.* **2004**, *25*, 239–245. [CrossRef]
42. Rossi, S.; Fedel, M.; Deflorian, F.; Vadillo, M.D. Localized electrochemical techniques: Theory and practical examples in corrosion studies. *C. R. Chim.* **2008**, *11*, 984–994. [CrossRef]
43. Jabeera, B.; Shibli, S.M.A.; Anirudhan, T.S. Synergistic inhibitive effect of tartarate and tungstate in preventing steel corrosion in aqueous media. *Appl. Surf. Sci.* **2006**, *252*, 3520–3524. [CrossRef]
44. Leidheiser, H., Jr. Towards a better understanding of corrosion beneath organic coatings. *Corrosion* **1983**, *39*, 189–201. [CrossRef]
45. Zhou, L.X.; Chen, J.; Yang, Z.R. Methods for study and evaluation of anticorrosion performance of organic coatings. *Corros. Sci. Prot. Tech.* **2004**, *16*, 375–377.
46. Bertocci, U.; Gabrielli, C.; Huet, F.; Keddam, M. Noise resistance applied to corrosion measurements. I. Theoretical analysis. *J. Electrochem. Soc.* **1997**, *144*, 31–37. [CrossRef]
47. Bacon, R.C.; Smith, J.J.; Rugg, F.M. Electrolytic resistance in evaluating protective merit of coatings on metals. *Ind. Eng. Chem.* **1948**, *40*, 161–167. [CrossRef]
48. Mills, D.; Jamali, S.; Tobiszewski, M.T. Developing electrochemical measurements in order to assess anti-corrosive coatings more effectively. *Prog. Org. Coat.* **2012**, *74*, 385–390. [CrossRef]
49. Ohwaki, T.; Urushihara, W.; Kinugasa, J.; Noishiki, K. Aluminum Alloy Material and Plate Heat Exchanger with Superior Corrosion Resistance. US Patent 20100006277A1, 14 January 2010.
50. Sheffer, M.; Groysman, A.; Mandler, D. Electrodeposition of sol-gel films on al for corrosion protection. *Corros. Sci.* **2003**, *45*, 2893–2904. [CrossRef]

51. Ma, W.; Wu, H.; Higaki, Y.; Otsuka, H.; Takahara, A. A "non-sticky" superhydrophobic surface prepared by self-assembly of fluoroalkyl phosphonic acid on a hierarchically micro/nanostructured alumina gel film. *Chem. Commun.* **2012**, *48*, 6824–6826. [CrossRef] [PubMed]
52. Park, Y.; Han, M.; Ahn, Y. Fabrication of superhydrophobic metal surfaces with self-assembled monolayers of silane derivatives having inter-hydrogen bonding. *Bull. Korean Chem. Soc.* **2011**, *32*, 1091–1094. [CrossRef]
53. Van, T.N.; Lee, Y.K.; Lee, J.; Park, J.Y. Tuning hydrophobicity of TiO_2 layers with silanization and self-assembled nanopatterning. *Langmuir* **2013**, *29*, 3054–3060. [CrossRef] [PubMed]
54. Akporiaye, D.; Stöcker, M. Solid-state n.m.r. and XRD study of the thermal stability of VPI-5: Assignment of 31P and 27Al MAS n.m.r. spectra. *Zeolites* **1992**, *12*, 351–359. [CrossRef]
55. Gao, W.; Dickinson, L.; Grozinger, C.; Morin, F.G.; Reven, L. Self-assembled monolayers of alkylphosphonic acids on metal oxides. *Langmuir* **1996**, *12*, 6429–6435. [CrossRef]
56. Gao, W.; Reven, L. Solid-state nmr studies of self-assembled monolayers. *Langmuir* **1995**, *11*, 1860–1863. [CrossRef]
57. McNatt, J.S.; Morgan, J.M.; Farkas, N.; Ramsier, R.D.; Young, T.L.; Rapp-Cross, J.; Espe, M.P.; Robinson, T.R.; Nelson, L.Y. Sonication assisted growth of fluorophosphate films on alumina surfaces. *Langmuir* **2003**, *19*, 1148–1153. [CrossRef]
58. Neff, G.A.; Page, C.J.; Meintjes, E.; Tsuda, T.; Pilgrim, W.C.; Roberts, N.; Warren, W.W. Hydrolysis of surface-bound phosphonate esters for the self-assembly of multilayer films: Use of solid state magic angle spinning 31p nmr as a probe of reactions on surfaces. *Langmuir* **1996**, *12*, 238–242. [CrossRef]
59. Phung, L.H.; Kleinert, H.; Jansen, I.; Häßler, R.; Jähne, E. Improvement in strength of the aluminium/epoxy bonding joint by modification of the interphase. *Macromol. Symp.* **2004**, *210*, 349–358. [CrossRef]
60. Lin, C.T.; Lin, P.; Hsiao, M.W.; Meldrum, D.A.; Martin, F.L. Chemistry of a single-step phosphate/paint system. *Ind. Eng. Chem. Res.* **1992**, *31*, 424–430. [CrossRef]
61. Gan, S.; Solimeno, R.D.; Jones, F.N.; Hill, L.W. Recent studies of the curing of polyester-melamine enamels. Possible causes of overbake softening. Proceedings of Water-Borne and Higher-Solids Coatings Symposium, Hattiesburg, MS, USA, 1–3 February 1989; pp. 87–108.

© 2017 by the authors. Licensee MDPI, Basel, Switzerland. This article is an open access article distributed under the terms and conditions of the Creative Commons Attribution (CC BY) license (http://creativecommons.org/licenses/by/4.0/).

Article

Easy and Fast Fabrication of Self-Cleaning and Anti-Icing Perfluoroalkyl Silane Film on Aluminium

Peter Rodič [1,*], Barbara Kapun [1], Matjaž Panjan [2] and Ingrid Milošev [1]

[1] Jožef Stefan Institute, Department of Physical and Organic Chemistry, Jamova c. 39, SI-1000 Ljubljana, Slovenia; barbara.kapun@ijs.si (B.K.); ingrid.milosev@ijs.si (I.M.)
[2] Jožef Stefan Institute, Department of Thin Films and Surfaces, Jamova c. 39, SI-1000 Ljubljana, Slovenia; matjaz.panjan@ijs.si
* Correspondence: peter.rodic@ijs.si; Tel.: +386-1-4773-261

Received: 12 February 2020; Accepted: 28 February 2020; Published: 4 March 2020

Abstract: A combination of the chemical etching process in $FeCl_3$ solution and chemical surface grafting by immersion in ethanol solution containing 1H,1H,2H,2H-perfluorodecyltriethoxysilane is a viable route to achieve a hierarchical surface topography and chemical bonding of silane molecules on an aluminium surface leading to (super)hydrophobic characteristics. Characterisation of untreated and treated aluminium surfaces was carried out using contact profilometry, optical tensiometry, scanning electron microscopy coupled with energy-dispersive spectroscopy and X-ray photoelectron spectroscopy to define the surface topography, wettability, morphology and surface composition. Additionally, the dynamic characteristics were evaluated to define bouncing and the self-cleaning effect. A thermal infrared camera was employed to evaluate anti-icing properties. The micro/nano-structured etched aluminium surface grafted with perfluoroalkyl silane film showed excellent superhydrophobicity and bounce dynamics in water droplet tests. The superhydrophobic aluminium surface exhibited the efficient self-cleaning ability of solid pollutants as well as improved anti-icing performance with melting delay.

Keywords: superhydrophobic surface; bounce dynamics; self-cleaning; anti-icing

1. Introduction

High specific strength, low specific weight and good corrosion resistance at atmospheric conditions are some of the main reasons for the extensive use of aluminium in building construction, transportation industries and many other applications in everyday life [1]. In recent years, rapid developments have been made in the area of superhydrophobic anticorrosive coatings. These coating can act as corrosion protection but offer additional functional abilities of an aluminium surface once exposed to the real environment including adsorption of the pollutants and ice formation [2,3]. The definition of superhydrophobicity is based on the water contact angle of the droplet on the surface which must be larger than 150° and have a sliding angle smaller than 10° [4,5]. Thus, naturally hydrophilic aluminium surfaces should be treated to reduce their wettability. Although in recent years, numerous studies related to superhydrophobic surfaces have reported on their excellent performance [2,3,6], there is still a need to develop a convenient, environmentally acceptable and facile method for fabrication of superhydrophobic aluminium surfaces. The latter can trap air in the modified surface topography which prevents aggressive ions from reaching the aluminium surface, consequently offering an efficient mechanism for corrosion protection [2,3,6–9]. Superhydrophobic surfaces have many applications because of their excellent properties such as on anti-icing [8,10–12], anti-fouling [13] and anti-bacterial properties [14]; therefore, there is a high potential for applications in different fields [3].

The wettability of a solid surface is a function of two primary factors: surface roughness and surface chemistry [5]. There are several routes to construct superhydrophobic surfaces such as a sol-gel process [15], anodic oxidation [16,17], chemical vapour deposition [18] and chemical etching [9,19,20]. However, once the practical application of such a coating is considered, a method with simple operation, low-cost and short operation time are required. From this point of view, chemical etching to produce a roughened surface followed by grafting a low-surface energy organic material, such as fatty acid [9,19] or alkyl and perfluoroalkyl silanes [8], are one of the easiest, economical and environmentally acceptable routes [2,20].

Chemical etching of aluminium has been reported in several studies (e.g., [21]); usually, it is performed in strong acid solutions (e.g., HCl, H_2SO_4) in combination with other reagents such as HCl/H_2O_2 [22], HCl/HF-Beck's solution [23–25] or alkaline solutions (e.g., NaOH) [9,26]. The surface etching process can be performed also in solutions of metal chlorides such as NaCl [25], $CuCl_2$ [27,28], and $FeCl_3$ [21,27,29,30]. A comparative study of various etchants confirms that the best performance was obtained with $FeCl_3$ [27]. The latter is also traditionally known as a home or industrially used circuit board etchant and, thus, a multipurpose chemical compound. The process is safe and environmentally acceptable because the ferric ions themselves are not hazardous. The etching process in $FeCl_3$ is usually performed in combination with the electrochemical process [30], and the optimization and novelty of the process would include the fabrication in the absence of electricity [21]. This kind of etching procedure has hitherto been much less investigated, offering room for improvement.

The grafting of aluminium surfaces with various alkyl and fluoroalkyl silanes [8,29] has been often used in the past due to the low surface energy of the CH, CF_2, CF_3 groups in the chain [8,31]. For instance, the treatment of aluminium or its alloys with fluoroalkyl silanes (FAS) in NaOH solution has been utilized to prepare the superhydrophobic coating [8,25,32–34]. The initial silane precursors were hydrolysed, condensed and covalently bonded on activated aluminium surface forming Al–O–Si bonds in an exothermic process [35,36].

Superhydrophobic properties are the basis for several surface functionalities. In that context, the dynamic characterisation of the water droplet on the surface [32] is essential to understand the self-cleaning [22,37] and anti-icing properties [8–12].

In this current work, (super)hydrophobic films on aluminium surfaces were fabricated using a two-step process consisting of an etching procedure in $FeCl_3$ solution (in the absence of electricity) to form a hierarchical micro/nano-structure aluminium surface and grafting at ambient temperature directly in an ethanol solution of 1H,1H,2H,2H-perfluorodecyltriethoxysilane as a low surface energy material. The surface morphology and composition were studied to explore the correlations between the etching process and grafting. Additionally, the bounce dynamics was studied on the superhydrophobic film to better understand the functional properties such as the self-cleaning ability for solid pollutants and melting delay. The latter was evaluated with an innovative approach using a thermal infrared camera.

2. Materials and Methods

2.1. Metal Substrates and Chemicals

A 1 mm thick aluminium (Al > 99.0%) sheet was distributed by GoodFellow, Cambridge Ltd. (Huntingdon, England). It was cut into 20 mm × 20 mm plates. The surface was sequentially ground with Struers LaboSystem LaboPol-20 machine using 1000, 2000 and 4200 SiC abrasive papers (supplied by Struers ApS, Ballerup, Denmark) in the presence of tap water.

A two-step procedure for producing superhydrophobic aluminium surface included etching in a $FeCl_3$ solution for various durations (5, 10, 15, 20, 25 and 30 min) followed by grafting by immersion in an ethanol solution of perfluoroalkyl silane for 30 min. Etching solution was prepared from iron(III) chloride ($FeCl_3 \times 6H_2O$; powder > 97%, CAS no. 10025-77-1, distributed by Sigma–Aldrich) and Milli-Q Direct water with a resistivity of 18.2 MΩ cm at 25 °C (Millipore, Billerica, MA, USA) in a volumetric flask to give a concentration of 1 mol/L. Grafting was performed for 30 min in 1 wt.% 1H,1H,2H,2H-perfluorodecyltriethoxysilane – FAS-10 ($C_{16}H_{19}F_{17}O_3Si$, > 97%, CAS no. 101947-16-4, distributed by Sigma–Aldrich) ethanol solution (C_2H_5OH, absolute, anhydrous > 99.9%, CAS no. 64-17-5, distributed by Sigma–Aldrich).

The samples were positioned at the bottom of the beaker with the ground surface facing up. Both steps were performed at room temperature. After each preparation step (i.e., grinding, etching, grafting), the samples were thoroughly rinsed with distilled water and cleaned by immersion in pure ethanol in an ultrasonic bath to remove all grinding/etching residuals, unreacted FAS-10 and other organic substances present on the surface. Finally, the samples were dried with a stream of compressed nitrogen.

2.2. Surface Characterisation

2.2.1. Weight Loss Test

The weight loss test was performed to evaluate the capability of the etching process. Weight loss (given in %) was determined as a difference in weight of a clean and dried aluminium before and after the etching process using a digital precision laboratory analytical weighing scale (Mettler Toledo AE200 Analytical Balance) with a precision of 0.1 mg. The evaluation was performed on five parallel samples. The data are given as average values ± standard deviations. The obtained data were fitted using linear regression to evaluate etching rate, determined by the slope of the curve.

2.2.2. Surface Topography

Surface 3D topography and linear profiles of the etched aluminium after various etching times were evaluated on three randomly chosen spots, employing a stylus contact profilometer, model Bruker DektakXT, using a 2 µm tip and in a soft-touch mode with force 1 mN. The measured surface was 1 mm × 1 mm, the vertical analysis range 65.5 µm and the vertical resolution 0.167 µm/point. Measured data were analysed using TalyMap Gold 6.2 software. Results are presented as 3D images and line profiles, and their corresponding surface roughness (S_a) are given as average values ± standard deviations.

2.2.3. Wettability

Water, diiodomethane and hexane contact angle measurements were performed at room temperature by the static sessile-drop method on a Krüss FM40 EasyDrop contact-angle measuring system. A small liquid droplet (4 µL) was formed on the end of the syringe which was carefully deposited onto the treated aluminium surface. Digital images of the droplet silhouette were captured with a high-resolution camera, and the contact angle was determined by numerically fitting the droplet image using associated protocol software for drop shape analysis. The values reported herein present the average of at least five measurements on various randomly chosen areas and are reported as average values ± standard deviations.

2.2.4. SEM/EDS Characterisation

A field-emission scanning electron microscope (FE-SEM, JEOL JSM 7600 F, Tokyo, Japan) equipped with an energy-dispersive X-ray spectrometer (EDS, Inca 400, Oxford Instruments, Bucks, England) was used to analyse the morphology and composition of aluminium etched in $FeCl_3$ and grafted with FAS-10. Prior to analysis, samples were coated with a few nanometres thick layer of carbon. The FE-SEM imaging was performed using secondary electron detector (SEI mode), lower secondary electron detector (LEI mode) and backscattering electrons for composition (COMPO mode) of the specimens at 5 and 10 kV energy. The EDS analyses were performed at 10 kV in a point analysis mode. The data were normalised and given as an atomic percentage (at.%). The amount of carbon was excluded from the quantitative analysis.

2.2.5. XPS Characterisation

The chemical compositions of the etched aluminium surface and grafted with FAS-10 were analysed with X-ray photoelectron spectroscopy (XPS) using an PHI TFA XPS spectrometer (Physical Electronics, USA) equipped with aluminium and magnesium monochromatized radiation. The XPS survey and high-energy resolution spectra were collected using Al-K$_\alpha$ radiation (1486.92 eV). The pressure in the chamber was in the range of 10^{-9} mbar. A constant analyser energy mode with 187.9 eV pass energy was used for survey spectra and 39.35 eV pass energy for high-energy resolution spectra. Photoelectrons were collected at a take-off angle 45° relative to the sample surface. The positions of all peaks were normalized with respect to C 1s peak at 284.8 eV. The elemental composition given as an atomic percentage (at.%) was determined from the survey XPS spectra using PHI MultiPak V8.0 software.

2.2.6. Bounce Dynamics

The bounce dynamics of a water droplet were investigated with an ultra-high-speed camera Photron FASTCAM SA-Z (Tokyo, Japan), 2000 frames per second, using 105 mm Nikon F2.8G objective (New York, NY, USA) at ambient temperature. A water droplet (10 µL) was formed on the end of the syringe and released onto the specimen surface from 45 mm height. The bouncing of a water droplet upon the horizontal specimen surface (tilted for 2 degrees) was recorded, and the frames of the movie were analysed in order to evaluate the dynamic properties.

2.2.7. Self-Cleaning Ability

The self-cleaning testing was carried out on ground and treated aluminium samples in size of 50 mm × 40 mm by the following procedures: the aluminium sample on a horizontal stage (tilted for 2 degrees). The solid pollution was simulated by covering the aluminium surface with a layer of graphite multiwalled nanotubes (carbon > 95%, length 1–10 µm, PlasmaChem GmbH, Berlin, Germany). Then, a water droplet of 10 µL was dropped onto the surface from a 2 cm height. The flow of solid pollution along with the water droplet was recorded using a digital camera to evaluate the pollutant removal after the water droplet rolled off from the surface.

2.2.8. Anti-Icing Ability

The anti-icing properties were studied on the ground and treated aluminium samples of a 50 mm × 40 mm size under the overcooled conditions. The test was performed with water droplets of 60 µL which were gently placed on the horizontal substrate. The samples were put in the freezer (−15 °C) for 1 h, then taken out and left at room temperature. Meanwhile, the melting process was evaluated by recording the melting times for droplets on non-treated and treated surfaces using a digital camera and thermal Fluke Ti55FT Infrared Camera (Everett, WA, USA). In parallel, the difference in ice adhesion was also evaluated by tilting the samples vertically (90 degrees) and measuring the time at which the frozen droplets were removed/slid from the surface. Additionally, the anti-icing properties were studied on the ground and treated aluminium under overcooled conditions including the water-dripping test. The test was performed in a freezer at −15 °C. The water droplets of cold water (T = 5 °C) were dropped continuously from a 10 cm height with a dripping rate of ~20 mL/min with the surface tilted at 10°.

3. Results and Discussion

3.1. Surface Characterization

The two-step surface treatment to fabricate the (super)hydrophobic aluminium surface consisted of etching in a FeCl$_3$ solution to obtain the hierarchical micro/nano-structure surface followed by grafting the surface in an ethanol solution of FAS-10 as a low surface energy material. According to the above description, the predicted reaction mechanism during surface treatment (etching + grafting) is schematically presented in Figure 1.

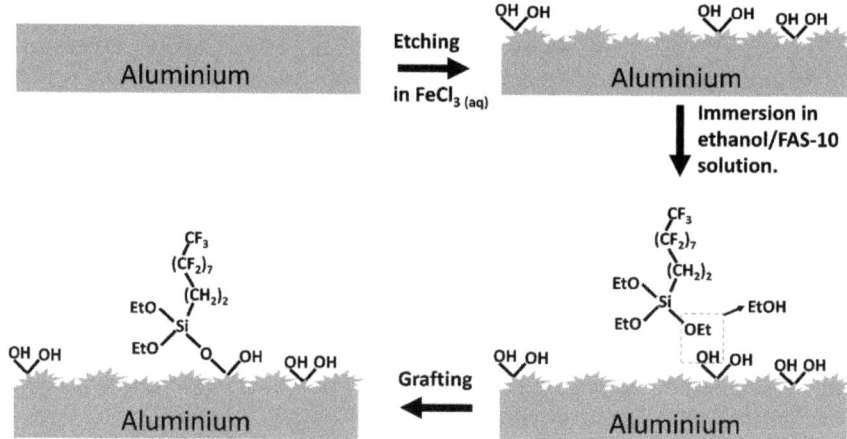

Figure 1. Schematic illustration of the formation of a (super)hydrophobic aluminium surface prepared by the etching process in a FeCl$_3$ solution followed by grafting with perfluoroalkyl silane, FAS-10.

The etching mechanism is based on redox reactions. In the first step, aluminium reacts with ferrous ions in aqueous solution. As a result, aluminium is oxidised and dissolved; at the same time, Fe^{3+} is reduced and transferred into the etchant solution [21,27,38]. The overall oxidation-reduction (redox) reaction of aluminium etching with FeCl$_3$ can be written as follows:

$$Al + Fe^{3+} \rightarrow Al^{3+} + Fe \tag{1}$$

The area of the etched surface is time-dependent and, consequently, also the composition of the aluminium surface due to the progressive removal of the native passive film (Al$_2$O$_3$). After taking the aluminium sample from etching solution, it reacts spontaneously with oxygen/humidity during rinsing with water which causes the passivation of the surface due to the formation of fresh aluminium oxide/hydroxide film containing OH$^-$ groups (Figure 1). In the second step, denoted as grafting, there is a reaction between hydroxylated aluminium surface and (CH$_3$CH$_2$O)$_3$–Si–(CH$_2$)$_2$(CF$_2$)$_7$CF$_3$ (FAS), where –OCH$_2$CH$_3$ (OEt) presents a hydrolysable ethoxy group and –(CH$_2$)$_2$(CF$_2$)$_7$CF$_3$, a non-hydrolysable perfluoroalkyl chain, resulting in the formation of surface film (Figure 1). The interfacial condensation and cross-linking reactions take place between the alkoxy and hydroxy groups of the etched aluminium, leading to robust covalent binding between the FAS molecule and the aluminium surface according to Equation (2):

$$Al(OH)_3 + (CH_3CH_2O)_3\text{-Si-}(CH_2)_2(CF_2)_7CF_3 \rightarrow$$
$$Al(OH)_2\text{-O-Si-}(OCH_2CH_3)_2\text{-}(CH_2)_2(CF_2)_7CF_3 + CH_3CH_2OH \tag{2}$$

The side product of this reaction is ethanol. Such a chemical process is exothermic and, as a result, the monodentate reaction between surface and fluoro-silane is expected [36]. This process allows the integration of (CH$_2$)$_2$(CF$_2$)$_7$CF$_3$ functional groups on the aluminium surface. The perfluoro groups are oriented outward from the surface due to the long perfluoroalkyl chain [9,39].

To sum up, the presence of Al–OH bonding on the freshly FeCl$_3$-etched aluminium surface is necessary to form a covalent bond between the aluminium surface and silicon, –Si–O–Al [35]. The details on the chemical etching and grafting on the surface are discussed below.

3.1.1. Weight Loss Test

The dissolution of aluminium can be quantitively evaluated using weight loss measurements as a function of immersion time (Figure 2a).

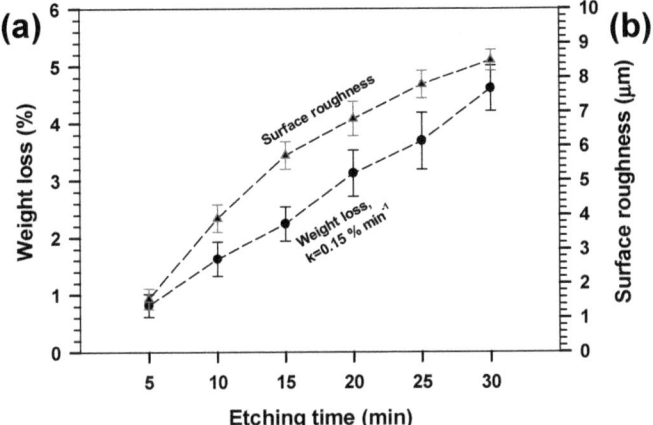

Figure 2. (**a**) Weight loss of aluminium as a function of etching time in 1 M FeCl$_3$ solution, and (**b**) the respective surface roughness determined by contact profilometry. The kinetics of weight loss was estimated from a linear regression of obtained values as a function of etching times.

The intensity of the etching process on the aluminium surface during immersion in FeCl$_3$ solutions could be seen visually due to the aggressive displacement reaction between a freshly ground aluminium surface and Fe ions (seen as bubbles formation and heating the etchant solution). The sample's weight was reduced proportionally with the etching time. At the beginning, the etching was slightly inhibited due to the passive layer of Al$_2$O$_3$, which was later (after 5 min) locally removed. The kinetics of the reaction was related to the diffusion process between the substrate and etchant solution. An increase in weight loss after various etching times confirms a high aggressiveness of FeCl$_3$ solution, where every five minutes of etching induced approximately a 0.75% weight loss (the slope of the curve was 0.15% min^{-1} (Figure 2a).

3.1.2. Surface Topography

The surface roughness (S_a) as a function of the etching time by FeCl$_3$ is quantitatively presented in Figure 2b. The ground aluminium had a small surface roughness, $S_a = 0.12 \pm 0.03$. The surface roughness of the 5 min FeCl$_3$-etched aluminium substrate increased by more than 10 fold, to $S_a = 1.6 \pm 0.3$ µm, and continued to increase linearly up to $S_a = 5.7 \pm 0.4$ µm at 15 min. At a longer etching time, the increase in roughness was slower but after 30 min roughness reached $S_a = 8.5 \pm 0.3$ µm. The increase in surface roughness correlated with that of weight loss; therefore, both parameters can be time-controlled.

The 3D surface profiles show that the dimensions of the roughness features were at different scales and that the topography of the surfaces varied significantly with the etching time (Figure 3).

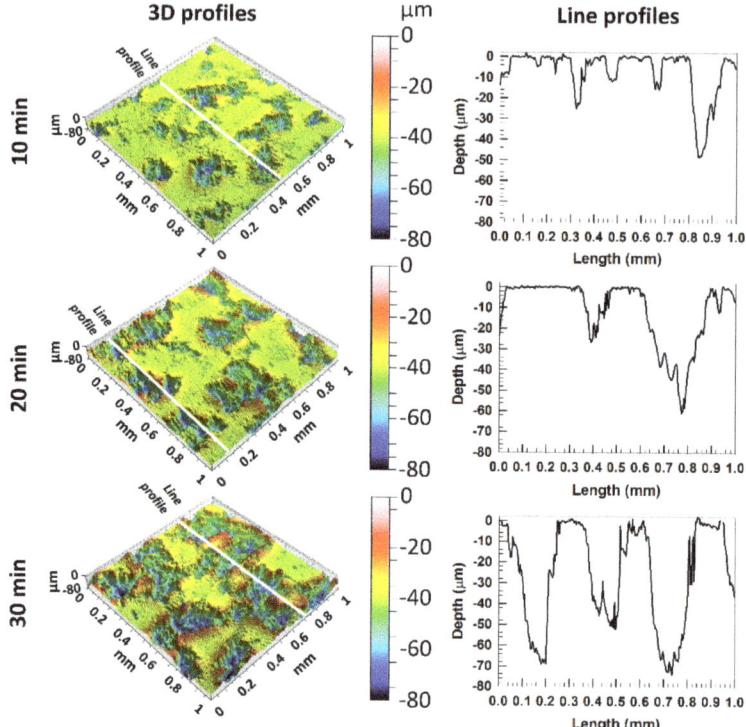

Figure 3. (left panel) 3D surface topography images of aluminium surface etched for 10, 20 and 30 min in 1 M FeCl$_3$ solution. The white lines present the area where single line profiles analyses were performed (right panel). The surface roughnesses determined from 3D profiles are presented in Figure 2.

A macroscale roughness was observed with a variable number of pits. At extended etching times, an increase in the number, size and depth of the pits at the etched aluminium could be observed at both the 3D and line profiles. Moreover, line profiles show that the number of large pits increased three-fold after etching for 30 min compared to that after 10 min; at the same time, their depth almost doubled (from 40 µm to 80 µm). This formation of deep and large pits (Figure 3) explains a slower increase in S_a after longer etching times (Figure 2b).

3.1.3. SEM Characterisation

The topography and morphology of ground and 20 min etched aluminium were characterised using COMPO and LEI modes (Figure 4).

The surface of ground aluminium showed many small pores and some defects that are probably related to the grinding process. Despite the fact that the aluminium used was of 99.0% purity used, it contained impurities that can be seen as bright spots in the SEM image recorded in COMPO mode. The results of the EDS analysis are given in Table 1.

The ground aluminium was mainly composed of Al and O, implying the presence of a thin oxidized layer on the surface (Figure 4a, Table 1, spot 1). It also contained some impurities such as Si and Fe (spots 2, 3). Silicon is related to remnants after the grinding procedure and is seen as dark areas (spot 2), while Fe is an impurity in the metal and forms small Fe-rich intermetallic particles which also contain Si (spot 3).

Table 1. Composition determined by EDS analysis in at.% at enumerated spots on ground and etched aluminium surface using FeCl$_3$ as denoted in Figure 4.

	Numbered Spots	Al	O	Fe	Si
Ground surface	1	40.0	60.0		
	2	39.8	60.0		0.2
	3	29.5	54.6	15.5	0.4
Etched surface	4	82.2	16.5		1.3
	5	98.0	2.0		
	6	56.3	19.9	19.1	4.7

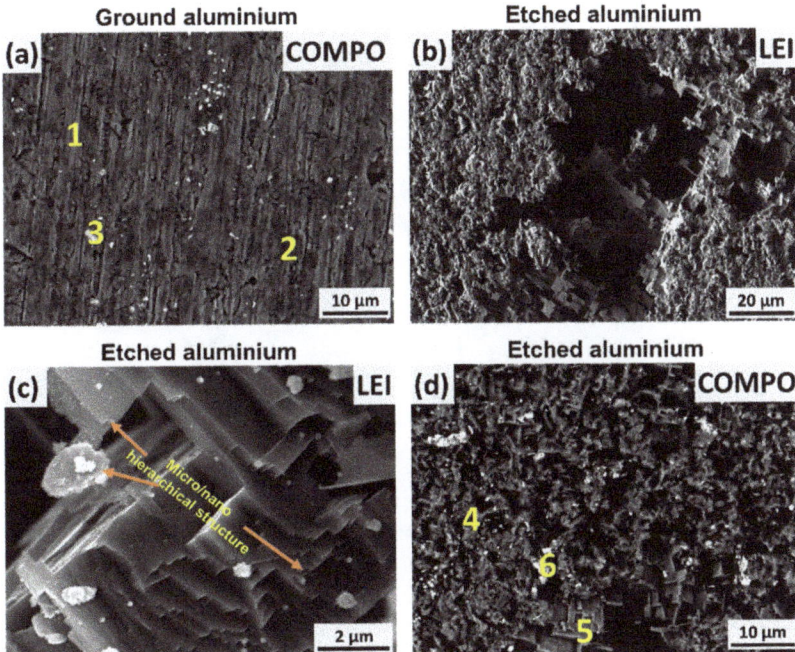

Figure 4. SEM images of (**a**) ground aluminium surface and (**b–d**) ground aluminium etched for 20 min in 1 M FeCl$_3$ recorded in COMPO and LEI modes. The enumerated positions (from 1 to 6) indicate spots where the EDS analysis was performed (Table 1).

The SEM LEI image of the surface after etching for 20 min in FeCl$_3$ solution (Figure 4b) shows many small holes and large deep pits. Moreover, the formation of a micro/nano-structured surface pattern compared to the ground surface is evident (Figure 4c,d). Even after etching, the impurities remained at the surface as confirmed by the SEM COMPO image (Figure 4d, Table 1). It is noteworthy that Fe was not present in the aluminium matrix (spots 4,5) which confirms the efficient etching process between Al and Fe without residual Fe on the surface. Iron was detected at spot 6, but the comparison with ground surface (spot 3) indicates that these are Si-containing Fe impurities which were not removed during the etching process.

The morphology of aluminium after various etching times (10, 20 and 30 min) was recorded by SEM using the SEI mode (Figure 5).

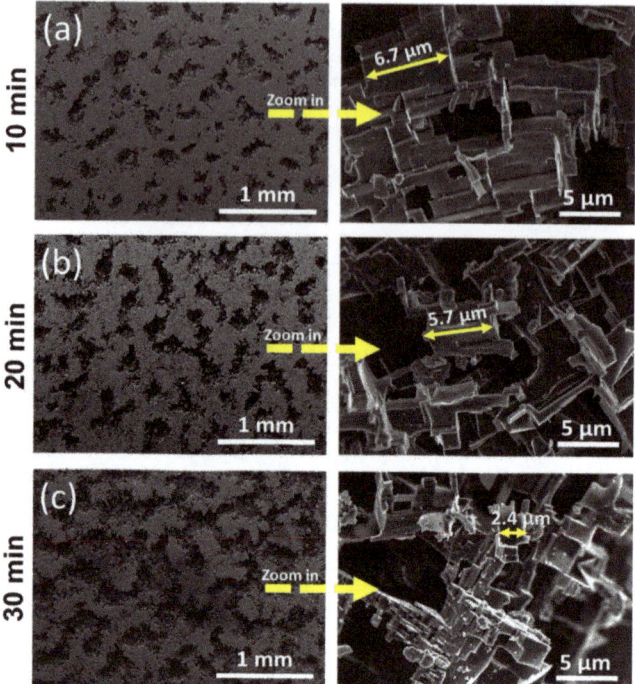

Figure 5. SEM images of surface morphologies of aluminium surface etched for 10, 20 and 30 min in the 1 M FeCl$_3$ solution recorded in SEI mode at lower (left) and higher magnification (right). The arrows present the estimated sizes of the formed hierarchical structures in the formed pits.

Aluminium etched for 10 min shows dispersed rectangular micro-scale pits with a width of about 200 µm distributed throughout the surface (Figure 5a, left). In addition, nano-scale rectangular pits are distributed uniformly across the surface, making a hierarchical structure. Inter-connecting edges around pits are various dimensions; the estimated size is between 6–8 µm (6.7 µm, Figure 5a, right). A binary structure of micro/nano-scale pits is therefore fabricated on Al surface by etching thus enlarging the real surface area and producing a nano-structured roughness.

After longer etching times—approximately 20 and 30 min—the number of pits increased and the size of the micro-scale pits extended due to the connection of small pits to larger holes (Figure 5b,c, left). At higher magnification (Figure 5b,c, right), the reduced size of the interconnected edges of the nano-hierarchical structure can be observed, for example, to 5.7 µm and 2.4 µm.

Such a well-controlled etching process was then further employed to fabricate a superhydrophobic surface on metal during grafting in ethanol solution of FAS-10.

3.2. Surface Wettability

The surface wettability of ground, etched and FAS-10-grafted aluminium samples was studied using water, diiodomethane and hexane (Figure 6).

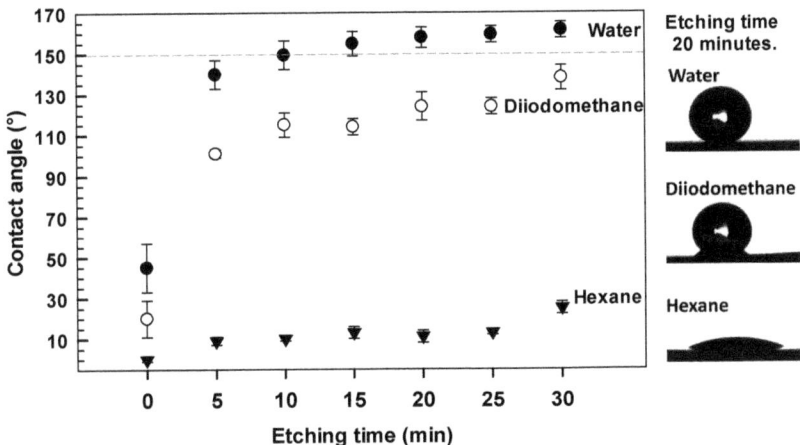

Figure 6. Contact angles measured for water, diiodomethane and hexane at aluminium surface etched for various times in 1 M FeCl$_3$ and grafted for 30 min in 1 wt.% ethanol FAS-10 solutions. The results are presented as mean values ± standard deviations. The dashed line presents the boundary of superhydrophobicity. The images of water, diiodomethane and hexane droplets on aluminium etched for 20 min and grafted with a FAS-10 are presented.

The non-ground aluminium surface was hydrophilic with a water contact angle (WCA) of about 69° due to the presence of a native oxidised layer which spontaneously forms during long exposure to air or moisture. The wettability of Al was enlarged by grinding and etching to produce a rough hydroxylated surface in FeCl$_3$-containing solution. Etched aluminium is superhydrophilic with a WCA of few degrees due to the presence of Al–OH groups formed during the etching process.

The transition from a (super)hydrophilic surface (i.e., the FeCl$_3$-etched aluminium substrate) to a (super)hydrophobic surface occurred after grafting in FAS-10 solution. It is noteworthy that the grafting of as-received or ground aluminium substrate without etching was less efficient. The water contact angle increased only up to 45°. This confirms that surface etching is crucial to fabricate a superhydrophobic surface.

The static water contact angles increase significantly with the etching time (Figure 6), in accordance with previous reports [22]. The two effects were interconnected: an increase in micro/nano roughness by prolonged etching and the presence of low surface energy molecules (FAS). After 5 min etching, the WCA was 142°, and after 20 min etching, all measured water contact angles were above 150°. If the surface was slightly tilted (less than 10°), water drops would slide from the surface confirming that the treated aluminium had a low sliding angle. Further etching does not contribute to an additional WCA increase.

The static contact angles were also measured with less polar solvents: diiodomethane and hexane (Figure 6). The CAs for diiodomethane increased with etching time, but the maximum values were below 138°. Even smaller CAs were obtained for hexane, where a maximum of 25° was reached.

From the obtained wettability data, it can be concluded that etching the surface for 20 min and grafting for 30 min in FAS-10 solution were optimal conditions to obtain superhydrophobic properties with high water repellence. Therefore, further characterisation was carried out on samples fabricated using these optimal parameters.

3.3. Composition, Topography and Morphology of Aluminium Grafted with FAS-10

3.3.1. SEM/EDS Characterisation

The organic film formed by grafting was presumably nanometre-sized and thus cannot be easily analysed by contact profilometry or by the change in sample weight. Therefore, the surface morphology of the etched and FAS-10-grafted surface was investigated by SEM/EDS. Figure 7 shows the typical morphology of aluminium etched for 20 min in $FeCl_3$ and grafted for 30 min using FAS-10.

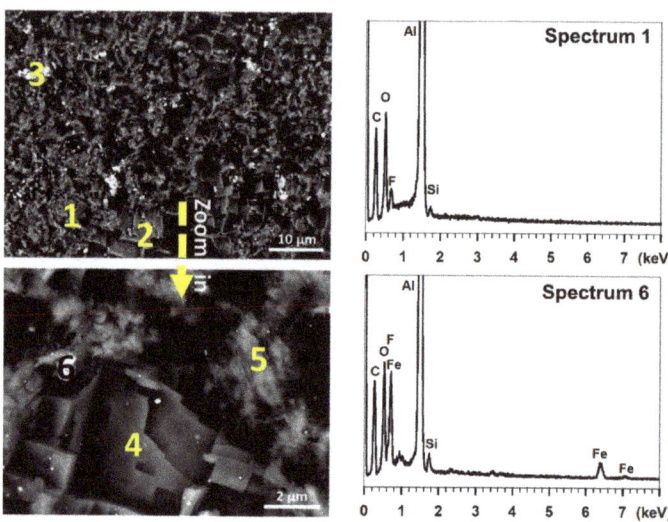

Figure 7. SEM images recorded in COMPO mode of the aluminium surface etched for 20 min in 1 M $FeCl_3$ followed by grafting for 30 min in 1 wt.% ethanol FAS-10 solution. The enumerated positions (from 1 to 6) indicate spots where the EDS analyses were performed (values are given in Table 2). Spectra 1 and 6 are shown as representatives.

Table 2. Composition determined by EDS analysis in at.% at enumerated spots on aluminium treated with FAS-10 as denoted in Figure 7.

Numbered Spots	Al	O	Fe	Si	F
1	81.6	15.2		0.5	2.7
2	61.6	24.6	5.7	1.1	7.0
3	47.6	13.8	33.7	1.4	3.5
4	98.5	1.5			
5	60.6	29.3		3.8	6.3
6	63.6	18.5	12.2	1.1	4.6

After grafting, a film of FAS-10 molecules was formed on the rough surface. Although the formed film was not observed on SEM image recorded in COMPO mode, its presence was confirmed by comparing the chemical compositions of the etched only and superhydrophobic surface using the EDS (Figure 7, Table 2).

Si and F originating from FAS-10 were detected on many analysed spots on the aluminium matrix (Figure 7, spots 1–3, spectrum 1) confirming the efficient grafting process during immersion. At higher magnification in the micro/nano-structured region (Figure 7), it can be noted that the film did not evenly cover the surface. For example, at spot 4, Si and F were not detected. It seems that Si and F were detected only at some spots on aluminium matrix (Figure 7, Spot 5, Table 2) and at spots with higher amounts of Fe (Figure 7, spots 2, 3, 6, Table 2), indicating that these were spots with more preferable bonding due to the presence of Fe-oxide as is shown below.

3.3.2. XPS Characterisation

The XPS analysis was performed to evaluate the chemical composition of aluminium etched for 20 min in FeCl$_3$ and aluminium grafted for 30 min with FAS-10. The survey XPS spectrum of etched aluminium is presented in Appendix A Figure A1a, where the identified elements originated from the substrate (Al), the presence of passive film (Al and O), other impurities present in the aluminium (Fe) or remaining after etching (Fe and Cl) and the effect of the atmosphere (adventitious C). The chemical composition is given in Table 3.

Table 3. Composition determined by XPS analysis in at.% for aluminium surface etched for 20 min in 1 M FeCl$_3$ and aluminium grafted for 30 min in 1 wt.% ethanol FAS-10 solution (Appendix A Figure A1).

Sample	O	C	Al	Cl	Fe	F	Si
Etched aluminium surface	47.5	34.8	13.7	2.8	1.2	-	-
Grafted aluminium surface with FAS-10	26.7	26.8	7.8	-	0.5	36.8	1.4

The etched surface mainly consisted of O and Al and a small amount of Cl and Fe which is in agreement with Equation (1) and proves that Fe was predominantly precipitated into solution, leaving only a residual amount at the surface, probably presented as FeCl$_3$.

The survey XPS spectrum of the grafted aluminium with FAS-10 strongly differed from the etched aluminium, Appendix A Figure A1b. The presence of Al, Si, F, C, O and Fe confirmed the grafting of FAS-10 film on the surface. The F, C and O now became the most abundant elements (Table 3). The Al concentration decreased after grafting but was still visible, indicating that the grafting layer was thinner than 10 nm.

The etched and grafted aluminium surfaces were additionally analysed with high-energy resolution XPS spectra (Figure 8).

The Al 2p and O 1s spectra of etched aluminium confirmed the presence of aluminium oxide/hydroxide (Figure 8a,b). According to the XPS database [40–42], spectra related the presence of AlOOH (74.7 eV, 531.7 eV) and passivation during exposure to air Al$_2$O$_3$ (73.1 eV, 530.7 eV). The peak related to Fe 2p$_{3/2}$ centred at 711 eV has a lot of noise due to the low intensity, confirming the presence of small amount of mixed Fe(II) oxide (multiple splitting at 709.8 eV and 711 eV [43]) and Fe(III) oxide (at 710.9 eV and 712.7 eV [43]) at the surface, Figure 8c. Moreover, XPS spectra confirm that zero-valent iron formed during etching was not present on the surface (usually detected as a peak at 706.8 eV [42–44]). The carbon presented as adventitious carbon at a single sharp peak at 284.8 eV (Figure 8d).

On the other hand, the grafted aluminium showed the broadening of the peaks for Al 2p and O 1s which reflected the presence of the third component related to Al–O–Si (at 75.5 eV and 532.7 eV, respectively) formed during the grafting in the FAS-10 solution. On the other hand, the peak for Fe 2p$_{3/2}$ exhibited a similar shape to a non-grafted sample. Therefore, the XPS data confirmed the formation of the covalent bond between aluminium oxide/hydroxide and (CH$_3$CH$_2$O)$_3$–Si–(CH$_2$)$_2$(CF$_2$)$_7$CF$_3$ presented by Equation (2). This is further corroborated by the C 1s spectrum (Figure 8d). The peak can be resolved into seven components, namely, –CF$_3$ (294.0 eV), –CF$_2$ (291.8 eV), –CF-CF$_x$ (290.8 eV), C–F (289.8 eV), –C–CF$_x$ (286.7 eV), –C–C (284.8 eV) and –C–Si (283.8 eV). The characteristic bands of the fluoroalkyl groups (CF$_2$ and CF$_3$) confirmed the presence of FAS-10 molecules at the surface. The intensity of the CF$_2$ peak is equal to C–C and higher than the CF$_3$ peak, because the FAS-10 contains more CF$_2$ groups than the CF$_3$ and C–C bonds. The Si 2p peaks comprised three components (Figure 8e) with a bending energy at about 101.3 eV for Si–O–C, 102.5 eV for Si–C associated to FAS molecule [40,41] and at 103.7 eV associated with the Si–O–Al species formed during grafting [40,41].

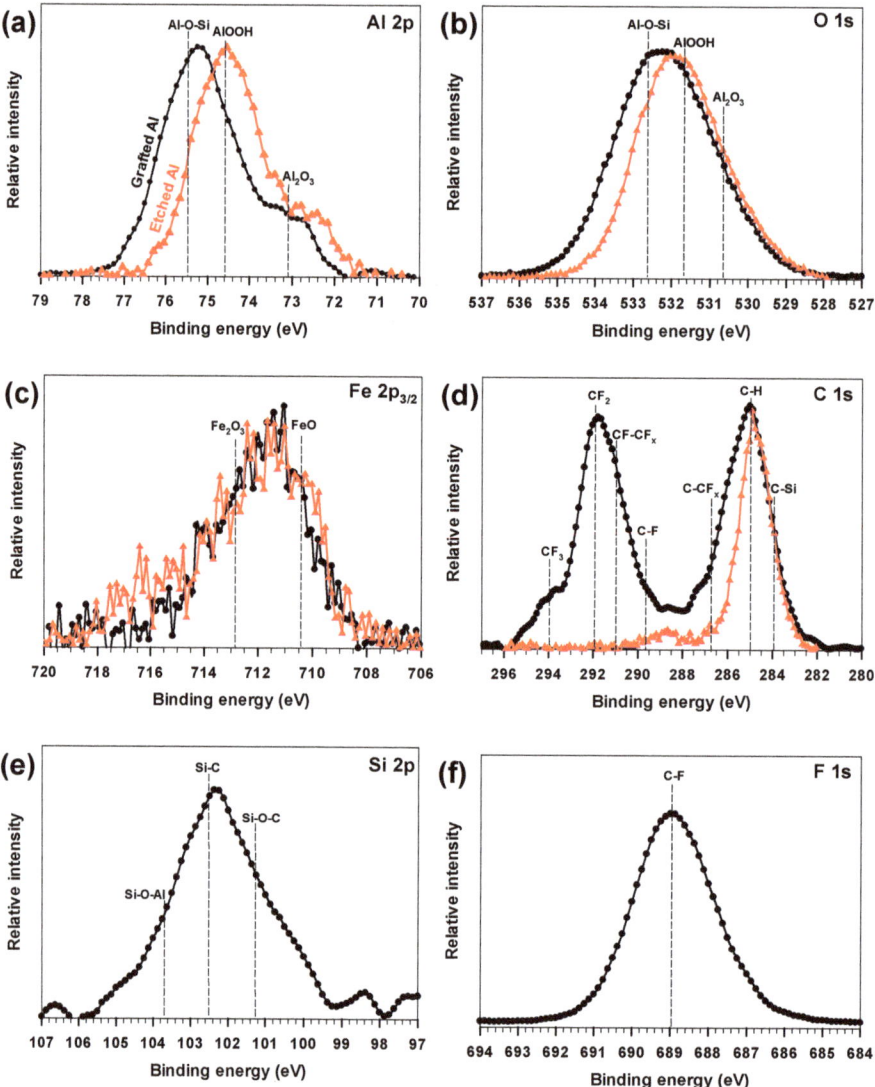

Figure 8. High-energy resolution XPS: (**a**) Al 2p, (**b**) O 1s, (**c**) Fe 2p$_{3/2}$, (**d**) C 1s, (**e**) Si 2p, and (**f**) F 1s. Symbols present experimental spectra, and vertical lines present the position of component peaks. Spectra were recorded for Al etched for 20 min in 1 M FeCl$_3$ (red triangles) and grafted for 30 min in 1 wt.% ethanol FAS-10 solution (black dots).

The presence of the FAS-10 molecule can also be confirmed in F 1s spectra, where a broad peak between 686–691 eV is related to C–F covalent bond in the FAS-10 molecule (Figure 8f). The CF$_3$ and CF$_2$ concentrations present in the C 1s spectrum of the FAS-10-treated Al is in correlation with the molecular structure. The high concentration of CF$_3$ and CF$_2$ on the surface indicates that the molecules were orientated with the Si–O bond to the surface forming Si–O–Al, while the (CH$_2$)$_2$(CF$_2$)$_7$CF$_3$ tail comprised the outermost surface film. Such a surface composition is in correlation with the obtained reduced wettability.

3.4. Bounce Dynamics

The dynamic behaviour of the water drops on the etched and FAS-10-grafted aluminium surface was investigated using water drop impact tests. Treated Al surfaces maintain remarkable non-wetting, superhydrophobic properties, not only in quasi-static conditions (as typically measured by sessile drop method) but also under dynamic conditions leading to rebound of the droplet after impact.

These characteristics were evaluated from the sequential images of one water droplet during the bouncing process on the superhydrophobic surface (Figure 9, Video A1).

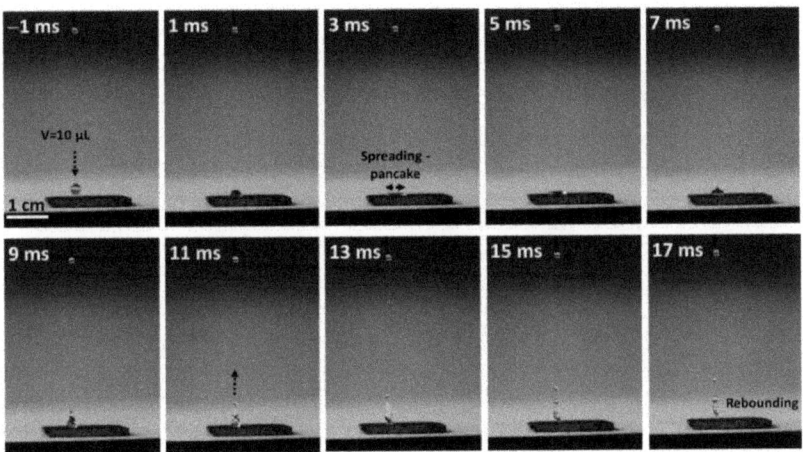

Figure 9. Time-lapsed images of a water droplet (volume 10 µL) bouncing on the horizontal aluminium surface etched for 20 min in 1 M FeCl$_3$ and followed by grafting for 30 min in 1 wt.% ethanol FAS-10 solution. Videos A1 and A2.

The process can be divided into different sequences which include falling toward the surface, impinging the surface, transforming a thin pancake and achieving a maximum contact with the surface, retracting back one water droplet followed by the formation of an inverted baseball bat which rebounds from the surface toward to a maximum height. The water droplet bounded and completely left the coated surface within 14 ms, without leaving any water residuals (Figure 9). The water droplet rebounded a few times on the superhydrophobic surface and, subsequently, left the surface or impinged on the superhydrophobic surface multiple times (Video A1).

The dynamic properties were also evaluated for a water droplet impinged to the superhydrophobic surface, which was tilted for 2° (Video A2). The water droplet rotated and rebounded and then rolled off. This experiment confirmed that even when released from a low height, a water droplet has enough kinetic energy to provide the driving force for bouncing and rotation. Consequently, the main goal to overcome low surface tension and self-gravity of a droplet to achieve the bouncing process was obtained. Moreover, the adhesion of a water droplet was efficiently reduced, and it did not remain on the superhydrophobic surface.

3.5. Self-Cleaning Ability

The self-cleaning test was carried out to demonstrate the removal of solid pollutants (featured by carbon nanotube particles as pollutants) from the treated aluminium surface (Figure 10, Video A3). Pollutants remained on the untreated (ground) aluminium surface which was wetted and contaminated by the particles dispersed in water (Figure 10).

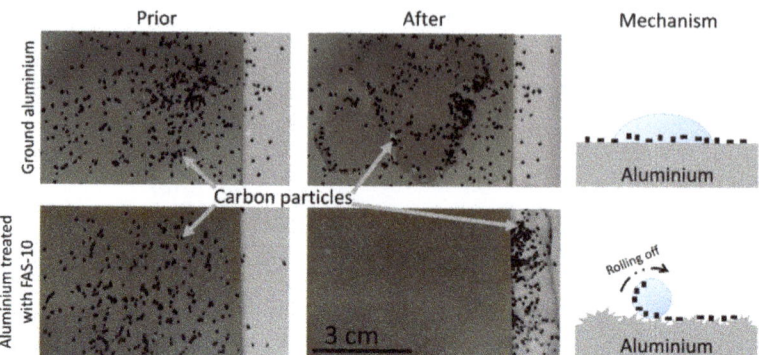

Figure 10. The appearance of ground and treated aluminium surfaces (etched for 20 min in 1 M FeCl$_3$ and followed by grafting for 30 min in 1 wt.% ethanol FAS-10 solution) covered with carbon particles prior and after rinsing with tap water. Video A3. The proposed mechanism of self-cleaning on ground and treated surfaces is presented schematically.

The mechanism is presented schematically in Figure 10. This behaviour indicates that ground aluminium does not have a self-cleaning ability. In contrast, no pollutants or water droplets remained on aluminium treated in FAS-10, because water removes the dirt when water droplets roll (Figure 10, Video A3). This illustrates its excellent water-repelling and self-cleaning ability, as the superhydrophobic aluminium surface does not allow water droplets to penetrate into the grooves, but rather they are suspended on the micro/nano-scale pits. Water droplets cannot stick to the surface and depart easily without a distinct distortion. Low adhesion of the superhydrophobic surface allows droplets to roll off quite easily and get rid of various external contaminants, just like the natural lotus leaf, according to the mechanism schematically presented in Figure 10. This behaviour has been theoretically explained by the Cassie and Baxter equation [5,28,37].

The superhydrophobicity of etched and FAS-10-grafted surfaces results from a hierarchically rough surface covered by an organic film whose low surface energy allows the air to be trapped in "pockets" areas between the water droplets and solid surface thus minimising their contact.

3.6. Anti-Icing Ability

The anti-icing properties of the superhydrophobic surface were characterised using conditions simulating the real environment. To evaluate this effect, two critical aspects have to be considered: icing-delay performance and ice adhesion strength. The icing-delay performance of water droplets deposited on the treated aluminium surface was evaluated in a reverse direction as completely melting delay (i.e., melting-delay time). Figure 11a (Video A4) shows the melting processes of droplets on the treated surfaces frozen at −15 °C. The droplet on the ground surface was quickly melted (120 s); in contrast, the droplets on the treated aluminium surface took longer (260 s) to melt.

The melting process was additionally characterised using the thermal infrared camera (Figure 11b). The droplets on the ground aluminium surface started to melt after 60 s near the edges, some were melted and after 90 s, ice remained only in the centre of larger droplets. Ice was completely melted after 120 s. Frozen droplets on treated aluminium melted much slower (Figure 11a,b, Video A4). After 60 s, the droplets remained frozen, and slow melting can be seen after 120 s up to 260 s.

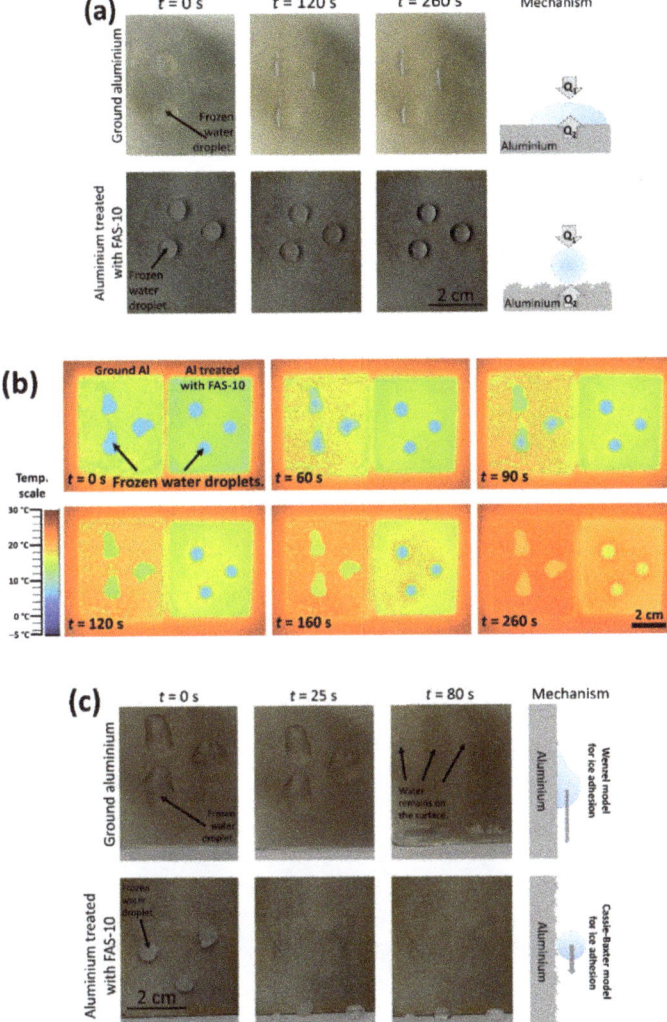

Figure 11. Frames of the de-icing process on ground and treated surfaces after various times (positioned horizontally) after taking the samples from the freezer (−15 °C) to ambient temperature followed using a (**a**) digital camera and (**b**) thermal infrared camera. Figure (**c**) shows the effect on the ice adhesion on the ground and treated aluminium (positioned vertically). Videos A4 and A5. The proposed mechanism of melting-delay and adhesion on ground and treated surfaces are presented schematically in (**a**) and (**c**).

This melting delay is in accordance with thermodynamics. The droplet on the cold surface gains heat from the air in the forms of heat conduction and thermal radiation and absorbs heat by heat conduction, schematically presented in Figure 11a. The difference in the temperature can be expressed according to the absorbed energy of droplet as per Equation (3) [45]:

$$\Delta T = \frac{\rho c_p (T_0 - T_1)}{Q_1 - Q_2} = \frac{\rho c_p (T_0 - T_1)}{\Delta Q} \tag{3}$$

where ΔT is the increased temperature of the droplet; ρ is the water density; c_p is the specific heat capacity of the water; T_0 is the starting temperature of the droplet; T_1 is the sample surface temperature; ΔQ is the net heat increase in unit time; Q_1 is the heat gain from the air; and Q_2 is the heat absorbed from contact with a solid surface. A small Q_2 or a large Q_1 can cause a large ΔQ and results in a small ΔT. Thus, it can explain why droplets suspended on superhydrophobic surfaces (treated aluminium) have a longer melting delay time compared to ground surfaces.

A difference in the heating of the surface area where the droplets were not present was also observed (Figure 11a,b). On the surface of the ground sample, there were small droplets of the water, confirming that the water condensed on the surface during melting. The treated surface was heated more evenly and no water condensed on the surface due to the trapped air in the structure pockets which acts as a thermal insulator.

Figure 11c shows the adhesion of the frozen droplet on the ground and treated aluminium surfaces (Video A5). According to the Wenzle wetting model of the droplet on the aluminium surface, the ice adhesion strength is related to the contact area fractions (f_1) of the liquid droplet on the solid. In the Cassie–Baxter model of the superhydrophobic surface, the actual contact interface is between ice and ice/air pockets and ice/hydrophobic solid, leading to a lower ice adhesion strength. Both mechanisms are presented schematically in Figure 11c. The breakage of the contact between formed ice and superhydrophobic surface practically occurs only along with the real contact interface between ice and solid. Consequently, the measured ice adhesion strength on the superhydrophobic surface is much smaller compared to hydrophilic or hydrophobic aluminium.

The anti-icing properties were additionally tested when placing the ground and treated aluminium samples in the freezer ($T = -15\ °C$) and dropping the cold water onto the surface (Figure 12). The obtained results vividly demonstrate that, on the ground surface, the water droplets were spread on the surface, began to freeze immediately and formed ice film on the entire surface (Figure 12a). In contrast, on the treated aluminium, no film was observed because the droplets bounced or rolled off the surface (Figure 12b); therefore, the aluminium surface remained unsaturated with water despite a continuous water drip.

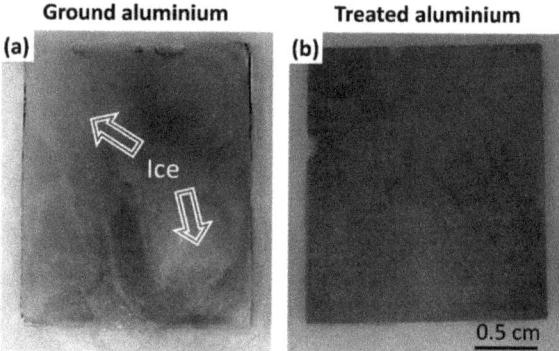

Figure 12. (a) The icing process for ground aluminium and (b) treated aluminium during testing in the freezer at $-15\ °C$ while dropping cold water droplets on the surface for 5 min.

The results confirm that the superhydrophobic aluminium exhibited a high anti-icing potential with the icing-delay and lower ice adhesion and ice film formation on the treated aluminium surface.

4. Conclusions

A (super)hydrophobic aluminium surface was fabricated using a simple, low-cost, two-step process consisting of chemical etching in $FeCl_3$ solution followed by grafting using perfluoroalkyl silane (FAS-10).

The topography and the weight loss of the surface confirm that FeCl$_3$ strongly promotes etching of Al; the effect is time-dependent. Roughening of the aluminium surface by etching in the FeCl$_3$ solution was found to be a crucial parameter to provide a micro/nano-pattern and aluminium oxide/hydroxide structure which then acts as an active surface for further grafting with perfluoroalkyl silane film.

An optimal etching time of 20 min and grafting for 30 min in FAS-10 solution resulted in superhydrophobic aluminium surface with a water contact angle above 150° and a low sliding angle below < 10°.

The XPS data confirmed that the FAS molecules were covalently bonded on the aluminium surface.

The superhydrophobic aluminium surface showed excellent dynamic properties, seen as a bouncing effect without leaving any residual water on the surface. Consequently, such a treated surface showed an excellent self-cleaning ability. Low contact of a water droplet with the treated surface also affected the melting delay, indicating an improved anti-icing effect.

This fabricated aluminium surface is a great candidate for various applications because of the properties gained to increase the durability and functionality of aluminium during exposure to the real environment.

Author Contributions: P.R. prepared the specimens, performed the weight loss, contact angle and profilometer measurements, and evaluated the self-cleaning and anti-icing abilities. The co-authors contributed to the evaluation and discussion on the following techniques employed in this study: B.K., the SEM/EDS results; M.P., the results obtained by the ultra-high-speed camera; I.M., the XPS results. Moreover, P.R. and I.M. wrote the majority of the paper. All authors have read and agreed to the published version of the manuscript.

Funding: This work is a part of M-ERA.NET project entitled "Design of Corrosion Resistant Coatings Targeted for Versatile Applications" (acronym COR_ID). The financial support from the Slovenian Research Agency (research core funding No. P2-0393 and P1-0134) is also acknowledged.

Acknowledgments: The authors thank Primož Fajdiga and Simon Iskra for their valuable technical help.

Conflicts of Interest: The authors declare no conflict of interest.

Appendix A

The following figure of XPS survey spectra complements the high-energy resolution spectra presented in Figure 8.

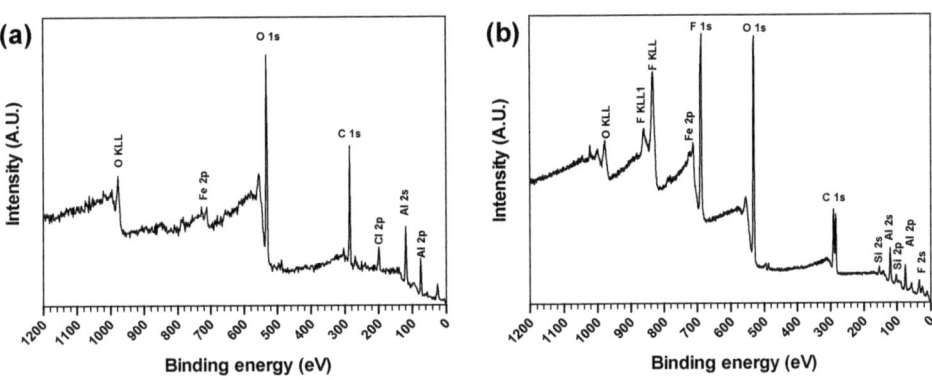

Figure A1. Survey XPS spectra recorded for (**a**) aluminium surface etched for 20 min in 1 M FeCl$_3$ and (**b**) grafted for 30 min in 1 wt.% ethanol FAS-10 solution.

Appendix B

The following videos complement the data and Figures 9–11 in this article. They are available in the online version.

Video A1. Bouncing dynamic characterisation

Video A2. Water rolling off the surface

Video A3. Self-cleaning of solid pollutants
Video A4. Anti-icing properties and melting delay
Video A5. Ice adhesion on ground and treated aluminium

References

1. Hatch, J.E. *Aluminum: Properties and Physical Metallurgy*; ASM International: Cleveland, OH, USA, 1984; ISBN 13 9780871701763.
2. Zhang, D.; Wang, L.; Qian, H.; Li, X. Superhydrophobic surfaces for corrosion protection: a review of recent progresses and future directions. *J. Coat. Technol. Res.* **2016**, *13*, 11–29. [CrossRef]
3. Vazirinasab, E.; Jafari, R.; Momen, G. Application of superhydrophobic coatings as a corrosion barrier: A review. *Surf. Coat. Technol.* **2018**, *341*, 40–56. [CrossRef]
4. Latthe, S.S.; Gurav, A.B.; Maruti, C.S.; Vhatkar, R.S. Recent Progress in Preparation of Superhydrophobic Surfaces: A Review. *J. Surf. Eng. Mater. Adv. Technol.* **2012**, *02*, 76–94.
5. Shirtcliffe, N.J.; McHale, G.; Atherton, S.; Newton, M.I. An introduction to superhydrophobicity. *Adv. Colloid Interface* **2010**, *161*, 124–138. [CrossRef] [PubMed]
6. Barati Darband, G.; Aliofkhazraei, M.; Khorsand, S.; Sokhanvar, S.; Kaboli, A. Science and Engineering of Superhydrophobic Surfaces: Review of Corrosion Resistance, Chemical and Mechanical Stability. *Arab. J. Chem.* **2020**, *13*, 1763–1802. [CrossRef]
7. Mohamed, A.M.A.; Abdullah, A.M.; Younan, N.A. Corrosion behavior of superhydrophobic surfaces: A review. *Arab. J. Chem.* **2015**, *8*, 749–765. [CrossRef]
8. Rodič, P.; Milošev, I. One-step ultrasound fabrication of corrosion resistant, self-cleaning and anti-icing coatings on aluminium. *Surf. Coat. Technol.* **2019**, *369*, 175–185. [CrossRef]
9. Milošev, I.; Bakarič, T.; Zanna, S.; Seyeux, A.; Rodič, P.; Poberžnik, M.; Chiter, F.; Cornette, P.; Costa, D.; Kokalj, A.; et al. Electrochemical, Surface-Analytical, and Computational DFT Study of Alkaline Etched Aluminum Modified by Carboxylic Acids for Corrosion Protection and Hydrophobicity. *J. Electrochem. Soc.* **2019**, *166*, C3131–C3146. [CrossRef]
10. Cao, L.; Jones, A.K.; Sikka, V.K.; Wu, J.; Gao, D. Anti-Icing Superhydrophobic Coatings. *Langmuir* **2009**, *25*, 12444–12448. [CrossRef]
11. Kreder, M.J.; Alvarenga, J.; Kim, P.; Aizenberg, J. Design of anti-icing surfaces: smooth, textured or slippery? *Nat. Rev. Mater.* **2016**, *1*, 1–15. [CrossRef]
12. Wang, L.; Gong, Q.; Zhan, S.; Jiang, L.; Zheng, Y. Robust Anti-Icing Performance of a Flexible Superhydrophobic Surface. *Adv. Mater.* **2016**, *28*, 7729–7735. [CrossRef] [PubMed]
13. Nurioglu, A.G.; Esteves, A.C.C.; With, G. de Non-toxic, non-biocide-release antifouling coatings based on molecular structure design for marine applications. *J. Mater. Chem. B* **2015**, *3*, 6547–6570. [CrossRef]
14. Hizal, F.; Rungraeng, N.; Lee, J.; Jun, S.; Busscher, H.J.; van der Mei, H.C.; Choi, C.-H. Nanoengineered Superhydrophobic Surfaces of Aluminum with Extremely Low Bacterial Adhesivity. *ACS Appl. Mater. Interfaces* **2017**, *9*, 12118–12129. [CrossRef] [PubMed]
15. Lakshmi, R.V.; Bharathidasan, T.; Basu, B.J. Superhydrophobic sol-gel nanocomposite coatings with enhanced hardness. *Appl. Surf. Sci.* **2011**, *257*, 10421–10426. [CrossRef]
16. Zhang, Y.; Wu, J.; Yu, X.; Wu, H. Low-cost one-step fabrication of superhydrophobic surface on Al alloy. *Appl. Surf. Sci.* **2011**, *257*, 7928–7931. [CrossRef]
17. Zhang, B.; Zhu, Q.; Li, Y.; Hou, B. Facile fluorine-free one step fabrication of superhydrophobic aluminum surface towards self-cleaning and marine anticorrosion. *Chem. Eng. J.* **2018**, *352*, 625–633. [CrossRef]
18. Biloiu, C.; Biloiu, I.A.; Sakai, Y.; Suda, Y.; Ohta, A. Amorphous fluorocarbon polymer (a-C:F) films obtained by plasma enhanced chemical vapor deposition from perfluoro-octane (C8F18) vapor I: Deposition, morphology, structural and chemical properties. *J. Vac. Sci. Technol. A* **2003**, *22*, 13–19. [CrossRef]
19. Huang, Y.; Sarkar, D.K.; Grant Chen, X. Superhydrophobic aluminum alloy surfaces prepared by chemical etching process and their corrosion resistance properties. *Appl. Surf. Sci.* **2015**, *356*, 1012–1024. [CrossRef]
20. Varshney, P.; Mohapatra, S.S.; Kumar, A. Superhydrophobic coatings for aluminium surfaces synthesized by chemical etching process. *Int. J. Smart Nano Mater.* **2016**, *7*, 248–264. [CrossRef]
21. Çakır, O. Chemical etching of aluminium. *J. Mater. Process. Technol.* **2008**, *199*, 337–340. [CrossRef]

22. Tong, W.; Xiong, D.; Wang, N.; Yan, C.; Tian, T. Green and timesaving fabrication of a superhydrophobic surface and its application to anti-icing, self-cleaning and oil-water separation. *Surf. Coat. Technol.* **2018**, *352*, 609–618. [CrossRef]
23. Vander Voort, G.F. *Metallography: Principles and Practice*, 6th ed.; ASM International: New York, NY, USA, 1984; ISBN 978-0-87170-672-0.
24. Qian, B.; Shen, Z. Fabrication of Superhydrophobic Surfaces by Dislocation-Selective Chemical Etching on Aluminum, Copper, and Zinc Substrates. *Langmuir* **2005**, *21*, 9007–9009. [CrossRef] [PubMed]
25. Kim, J.-H.; Mirzaei, A.; Kim, H.W.; Kim, S.S. Realization of superhydrophobic aluminum surfaces with novel micro-terrace nano-leaf hierarchical structure. *Appl. Surf. Sci.* **2018**, *451*, 207–217. [CrossRef]
26. Liu, L.; Feng, X.; Guo, M. Eco-Friendly Fabrication of Superhydrophobic Bayerite Array on Al Foil via an Etching and Growth Process. *J. Phys. Chem. C* **2013**, *117*, 25519–25525. [CrossRef]
27. Patil, D.H.; Thorat, S.B.; Khake, R.A.; Mudigonda, S. Comparative Study of $FeCl_3$ and $CuCl_2$ on Geometrical Features Using Photochemical Machining of Monel 400. *Procedia CIRP* **2018**, *68*, 144–149. [CrossRef]
28. Zhan, Z.; Li, Z.; Yu, Z.; Singh, S.; Guo, C. Superhydrophobic Al Surfaces with Properties of Anticorrosion and Reparability. *ACS Omega* **2018**, *3*, 17425–17429. [CrossRef]
29. Parin, R.; Col, D.D.; Bortolin, S.; Martucci, A. Dropwise condensation over superhydrophobic aluminium surfaces. *J. Phys. Conf. Ser.* **2016**, *745*, 1–8. [CrossRef]
30. Ruan, M.; Li, W.; Wang, B.; Deng, B.; Ma, F.; Yu, Z. Preparation and anti-icing behavior of superhydrophobic surfaces on aluminum alloy substrates. *Langmuir* **2013**, *29*, 8482–8491. [CrossRef]
31. Nishino, T.; Meguro, M.; Nakamae, K.; Matsushita, M.; Ueda, Y. The Lowest Surface Free Energy Based on $-CF_3$ Alignment. *Langmuir* **1999**, *15*, 4321–4323. [CrossRef]
32. Bernagozzi, I.; Antonini, C.; Villa, F.; Marengo, M. Fabricating superhydrophobic aluminum: An optimized one-step wet synthesis using fluoroalkyl silane. *Colloids Surf, A Physicochem Eng Asp.* **2014**, *441*, 919–924. [CrossRef]
33. Brassard, J.-D.; Sarkar, D.K.; Perron, J. Fluorine Based Superhydrophobic Coatings. *Appl. Sci.* **2012**, *2*, 453–464. [CrossRef]
34. Saleema, N.; Sarkar, D.K.; Gallant, D.; Paynter, R.W.; Chen, X.-G. Chemical Nature of Superhydrophobic Aluminum Alloy Surfaces Produced via a One-Step Process Using Fluoroalkyl-Silane in a Base Medium. *ACS Appl. Mater. Interfaces* **2011**, *3*, 4775–4781. [CrossRef] [PubMed]
35. Poberžnik, M.; Costa, D.; Hemeryck, A.; Kokalj, A. Insight into the Bonding of Silanols to Oxidized Aluminum Surfaces. *J. Phys. Chem. C* **2018**, *122*, 9417–9431. [CrossRef]
36. Poberžnik, M.; Kokalj, A. Implausibility of bidentate bonding of the silanol headgroup to oxidized aluminum surfaces. *Appl. Surf. Sci.* **2019**, *492*, 909–918. [CrossRef]
37. Kumar, A.; Gogoi, B. Development of durable self-cleaning superhydrophobic coatings for aluminium surfaces via chemical etching method. *Tribol. Int.* **2018**, *122*, 114–118. [CrossRef]
38. Rao, P.N.; Kunzru, D. Fabrication of microchannels on stainless steel by wet chemical etching. *J. Micromech. Microeng.* **2007**, *17*, N99–N106.
39. Poberžnik, M. Quantum mechanical modeling of the oxidation of aluminum surfaces and their interactions with corrosion inhibitors. Ph.D. Thesis, University of Ljubljana, Ljubljana, Slovenia, 2019; pp. 75–83.
40. Milošev, I.; Jovanović, Ž.; Bajat, J.B.; Jančić-Heinemann, R.; Mišković-Stanković, V.B. Surface Analysis and Electrochemical Behavior of Aluminum Pretreated by Vinyltriethoxysilane Films in Mild NaCl Solution. *J. Electrochem. Soc.* **2012**, *159*, C303–C311. [CrossRef]
41. Bajat, J.B.; Milošev, I.; Jovanović, Ž.; Jančić-Heinemann, R.M.; Dimitrijević, M.; Mišković-Stanković, V.B. Corrosion protection of aluminium pretreated by vinyltriethoxysilane in sodium chloride solution. *Corros. Sci.* **2010**, *52*, 1060–1069. [CrossRef]
42. NIST X-ray Photoelectron Spectroscopy Database. *NIST Standard Reference Database Number 20*; National Institute of Standards and Technology: Gaithersburg, MD, USA, 2000; Volume 2000. [CrossRef]
43. Milošev, I.; Strehblow, H.-H. The behavior of stainless steels in physiological solution containing complexing agent studied by X-ray photoelectron spectroscopy. *J. Biomed. Mater. Res.* **2000**, *52*, 404–412. [CrossRef]

44. Santana, I.; Pepe, A.; Jimenez-Pique, E.; Pellice, S.; Milošev, I.; Ceré, S. Corrosion protection of carbon steel by silica-based hybrid coatings containing cerium salts: Effect of silica nanoparticle content. *Surf. Coat. Tech.* **2015**, *265*, 106–116. [CrossRef]
45. Shen, Y.; Tao, J.; Tao, H.; Chen, S.; Pan, L.; Wang, T. Superhydrophobic Ti_6Al_4V surfaces with regular array patterns for anti-icing applications. *RSC Adv.* **2015**, *5*, 32813–32818. [CrossRef]

 © 2020 by the authors. Licensee MDPI, Basel, Switzerland. This article is an open access article distributed under the terms and conditions of the Creative Commons Attribution (CC BY) license (http://creativecommons.org/licenses/by/4.0/).

Article

Optical Evidence for the Assembly of Sensors Based on Reduced Graphene Oxide and Polydiphenylamine for the Detection of Epidermal Growth Factor Receptor

Mihaela Baibarac [1,*], Monica Daescu [1] and Szilard N. Fejer [2]

[1] National Institute of Materials Physics, Laboratory of Optical Processes in Nanostructured Materials, Atomistilor Street 405A, P.O. Box MG-7, Magurele, 077125 Bucharest, Romania; monica.daescu@infim.ro
[2] Pro-Vitam Ltd., Muncitorilor Street 16, 520032 Sfantu Gheorghe, Romania; fejersz@gmail.com
* Correspondence: barac@infim.ro; Tel.: +40-213-690-170

Abstract: Using Raman scattering and FTIR spectroscopy, new optical evidence for the assembly of sensors based on reduced graphene oxide (RGO) and polydiphenylamine (PDPA) for the electrochemical detection of the epidermal growth factor receptor (EGFR) are reported. The assembly process of the RGO sheets electrochemical functionalized with PDPA involves the chemical adsorption of 1,4-phenylene diisothiocyanate (PDITC), followed by an incubation with protein G in phosphate buffer (PB) solution and after that the interaction with EGFR antibodies solution. Taking into account the changes reported by Raman scattering and FTIR spectroscopy, a chemical mechanism of the assembling process for this sensor is proposed. The preliminary testing of the electrochemical activity of the sensors based on RGO and PDPA was reported by cyclic voltammetry.

Keywords: reduced graphene oxide; polymer; coatings; epidermal growth factor receptor; Raman scattering; FTIR spectroscopy

1. Introduction

The epidermal growth factor receptor (EGFR) is a biomarker often used for many tumors in various diseases such as breast cancer, gliomas, laryngeal cancers, carcinoma, and so on [1,2]. Different methods were developed for the detection of EGFR, the most used being immunohistochemistry [3,4], enzyme-linked immunosorbent assay (ELISA), [5] and Western blotting [6]. The main platforms used until now for the EGFR detection were lab-on-chip sensors [7], biochips-based on microfluids [8], Au nanoparticles which show surface plasmon resonance [9] and label-free electrochemical immunosensors [10]. In comparison with this progress, a new platform based on reduced graphene oxide (RGO) electrochemical functionalized with polydiphenylamine (PDPA) is proposed to be used for the EGFR detection in this work. These platforms are considered more attractive in comparison with the Au nanoparticles or Au plate, as a consequence of the fact that it is no longer necessary to interact with cysteamine in order to generate new amine-type bonds that would later allow interaction with 1,4-phenylene diisothiocyanate (PDITC). An example which supports this is the case of the RGO sheets electrochemically functionalized with poly(5-amino-1-naphtol) [11]. Despite the greater progress made concerning the assembly of immunosensors for EGFR detection, the information concerning the optical evidence of the stages of assembling these platforms is missing. In order to overcome this limitation, in this work, we report several optical studies carried out using Raman scattering and FTIR spectroscopy concerning the assembly of RGO sheets electrochemically functionalized with PDPA in order to be used in the future for the electrochemical detection of EGFR. In this work, a short characterization of these platforms is shown by cyclic voltammetry. Our results open up new perspectives for highly reproducible platforms for clinical screening of cancer tumors.

2. Materials and Methods

The main chemical compounds used in this study were: diphenylamine (DPA), $H_4SiW_{12}O_{40}$ xH_2O, HCl, graphite, dimethylformamide (DMF), ethanol, protein G, PDITC, EGFR antibody, EGFR antigen, ethanolamine, Tween 80, Na_2HPO_4, and NaH_2PO_4, all purchased from Sigma-Aldrich (St. Louis, MO, USA). Screen-printed carbon electrodes (SPCE) modified with RGO (SPCE-RGO) were purchased from Methrohm DropSens (Herisau, Switzerland). The configuration of SPCE-RGO consists of: (i) a working electrode, which in our case corresponds to the RGO sheets deposited onto carbon electrode; (ii) a counter electrode from carbon; and (iii) reference electrode from Ag, which in fact is a pseudoreference electrode that shows a shift in potential of -131 mV in comparison with the classical reference electrode Ag/AgCl. Electrochemical functionalization of SPCE-RGO was carried out according to Ref. [12]. Briefly, this involved using a semi-aqueous solution of 2×10^{-2} M DPA and 1 M HCl in $DMF:H_2O = 1:1$ (volumetric ratio) in the presence of 5×10^{-3} M $H_4SiW_{12}O_{40}$. The electrochemical functionalization of SPCE-RGO with PDPA doped with the $H_4SiW_{12}O_{40}$ heteropolyanions was carried out by cyclic voltammetry, in the potential range of +100 to +960 mV vs. SCE. After recording 20 cyclic voltammograms onto the SPCE-RGO surface, a platform of the type SPCE-RGO covalently functionalized with PDPA doped with the $H_4SiW_{12}O_{40}$ heteropolyanions (SPCE-RGO/PDPA) was produced. The cyclic voltammograms were stopped at +960 mV vs. SCE.

The assembling process of the RGO sheets electrochemically functionalized with PDPA for the electrochemical detection of EGFR was carried out in four steps. In the first step, the interaction of the SPCE-RGO/PDPA platform with 0.5, 1, 2, and 4 mg mL^{-1} PDITC in the ethanol for 20 min was carried out in order to obtain SPCE-RGO/PDPA modified with PDITC (SPCE-RGO/PDPA-PDITC). To eliminate the excess PDITC, the SPCE-RGO/PDPA-PDITC platform was washed four times with 5 mL PB solution with a pH equal to 7.4. In the second step, the SPCE-RGO/PDPA-PDITC platform was incubated with 15 μg mL^{-1} protein G in PB solution with a pH equal to 8.5, for 1 h. In order to remove the unbound protein G on the SPCE-RGO/PDPA-PDITC-G platform, a rinse with 0.1% Tween in PB solution with the same pH (10 mL) was performed. In the third step, the deactivation of the thiocyanate groups of the SPCE-RGO/PDPA-PDITC-G platform was carried out by their immersion in a solution of ethanolamine (0.1 M), for 30 min, and then washed with the PB solution with a pH of 8.5. In the fourth step, the SPCE-RGO/PDPA-PDITC-G platform interacted with a solution of EGFR antibodies (15 μg mL^{-1}) in the PB solution with a pH equal to 8.5. After 45 min., the SPCE-RGO/PDPA-PDITC-G-EGFR antibody platform was stored at a temperature of 4 °C and then incubated with 1 μg mL^{-1} EGFR antigen solution with a pH equal to 7.4 for 1 h. Before testing the SPCE-RGO/PDPA-PDITC-G-EGFR antibodies/EGFR platforms in the presence of the $[Fe(CN)_6]^{3-}/[Fe(CN)_6]^{4-}$ solution, a rinse of these with PB solution (10 mL, pH 7.4) was performed.

The electrochemical functionalization of SPCE-RGO with PDPA, as well as the testing of the SPCE-RGO/PDPA-PDITC-G-EGFR antibodies/EGFR platforms in the presence of the $[Fe(CN)_6]^{3-}/[Fe(CN)_6]^{4-}$ sample, were recorded with a potentiostat/galvanostat Voltalab 80 model, purchased from Radiometer Analytical (Lyon, France).

The Raman spectra of the SPCE-RGO functionalized with PDPA platform and its evolution during the assembling process were recorded with a Raman spectrophotometer, T64000 model, from Horiba Jobin Yvon (Edison, NJ, USA), which was endowed with an Ar laser (excitation wavelength of 514 nm). Complementary studies were performed with a FT Raman spectrophotometer, MultiRam model, from Bruker (Billerica, MA, USA), which was endowed with a YAG:Nd laser (excitation wavelength of 1064 nm). In the case of all Raman spectra, a baseline operation was applied.

The infrared (IR) spectra of the SPCE-RGO functionalized with PDPA platform and its evolution during the assembling process were recorded with a FTIR spectrophotometer, Carry 600 series, from Agilent (Santa Clara, CA, USA).

3. Results and Discussion

3.1. Optical Evidences by Raman Scattering and FTIR Spectroscopy Studies Concerning the Assembling of the Sensorial Platforms for EGFR Detection

Figure 1a shows the Raman spectrum of the SPCE-RGO/PDPA platform, which is characterized by two intense bands with the maximum at 1592 and 1350 cm^{-1} that are accompanied of other two Raman lines of low intensity at 1176, 1133, and 996 cm^{-1}.

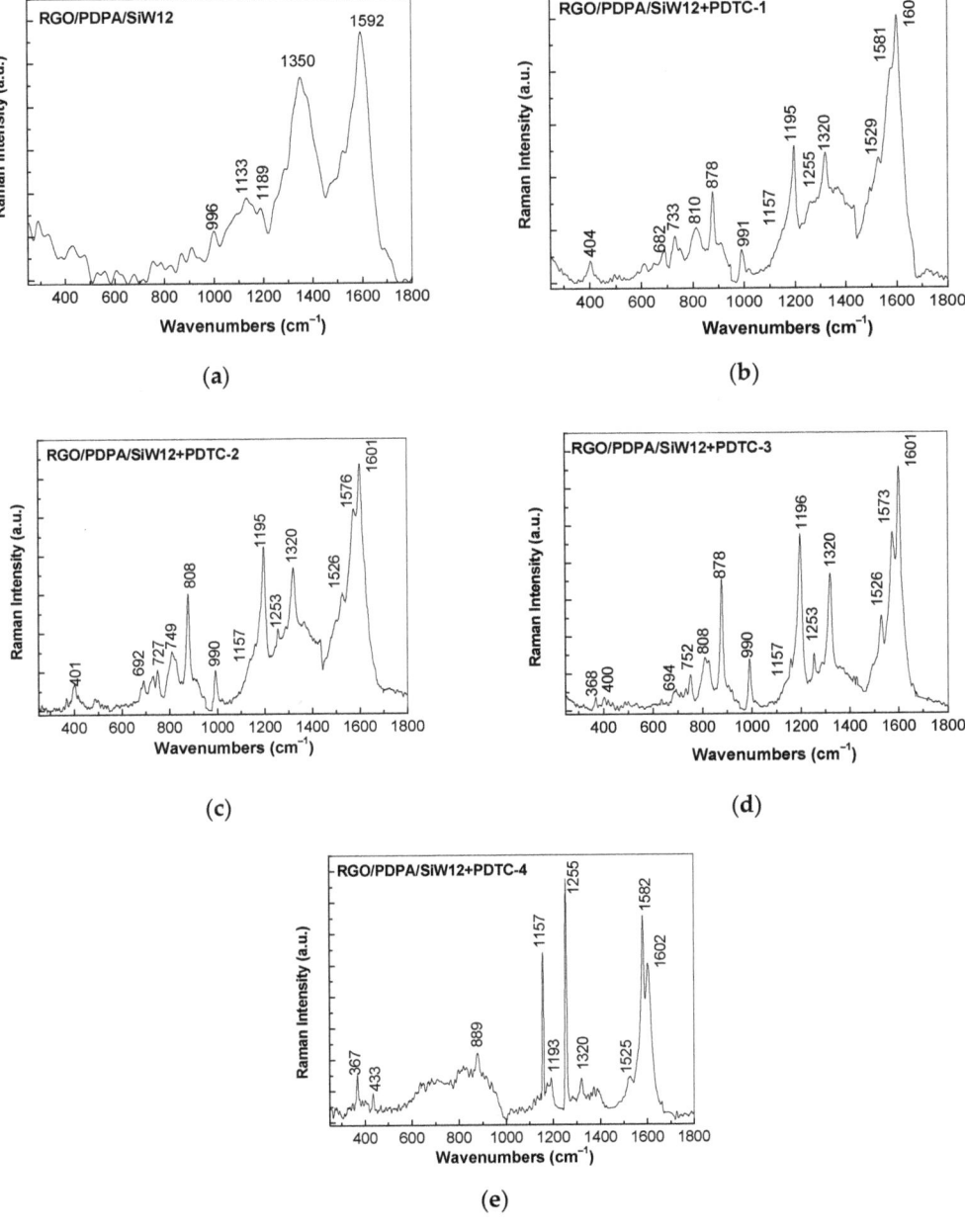

Figure 1. Cont.

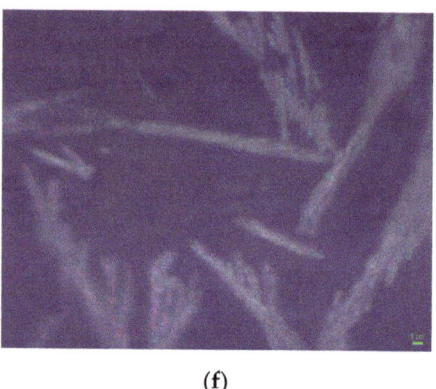

(f) (g)

Figure 1. The Raman spectrum of the screen-printed carbon electrodes modified with reduced graphene oxide and polydiphenylamine (SPCE-RGO/PDPA) platform before (**a**) and after the interaction with the 1,4-phenylene diisothiocyanate (PDITC) solution in ethanol (1 mL) with a concentration equal to 0.5 (**b**), 1 (**c**), 2 (**d**), and 4 mg mL^{-1} (**e**). (**f,g**) show the optical images of the platform SPCE-RGO/PDPA after the interaction with the PDITC solution having the concentration equal to 0.5 and 4 mg mL^{-1}. All spectra are recorded at the excitation wavelength of 514 nm.

As shown in our previous article [13], the RGO Raman spectrum at an excitation wavelength of 514 nm is characterized by two lines at 1349 and 1573 cm^{-1}, these being assigned to the graphitic lattice defects and E$_{2g}$ in-plane phonon in the Brillouin zone G point [14]. According to our previous studies, the main Raman lines of PDPA in a doped state were reported to be localized at 1176, 1342, 1367, 1492, 1585, and 1613 cm^{-1}, these being assigned to the vibrational modes C–H bending in the benzene ring, C–N in the N,N'-diphenyl benzidine radical cation, C=N stretching, C–C stretching in the quinoid ring + C–C stretching in the benzene ring, and C–C stretching in benzene ring + C–H bending in benzene ring, respectively [15,16]. The Raman line at 1592 cm^{-1} in Figure 1a, confirms the presence of PDPA on the RGO sheet's surface. The Raman line at 996 cm^{-1} (Figure 1a) belongs to the vibrational modes of W=O in the H$_4$SiW$_{12}$O$_{40}$ [17]. The up-shift of the Raman line assigned to the C–H bending vibrational mode in the benzene rings of the polymer from 1176 to 1189 cm^{-1} can be explained if we accept that a covalent functionalization process of the RGO sheets with conjugated polymer takes place, when the steric hindrance effects are induced in the PDPA macromolecular chain. A puzzling fact is the presence of the Raman line at 1133 cm^{-1} (Figure 1a), which is located not far from the Raman line from 1150 cm^{-1} assigned to the vibrational mode of C–H in-plane bending of the polymers having triphenylamine as repeating units [18]. The Raman spectrum of PDITC, at the excitation wavelength of 514 nm, shows lines with peaks at 1157, 1257, 1583, and 1603 cm^{-1}, attributed to the vibrational modes of the C–S bending, C–H in the benzene ring + C–C stretching + C–N stretching, C=C+C–C stretching in the benzene ring, and C–C stretching + C–H bending in the benzene ring, respectively [19–23]. The interaction of the SPCE-RGO/PDPA with the PDITC solutions with increasing concentration leads to the following changes in the Raman spectra of Figure 1: (i) an up-shift of the Raman lines from 1592 and 1189 cm^{-1} to 1601 and 1196 cm^{-1}, respectively; (ii) a down-shift of the Raman line from 1350 to 1320 cm^{-1}; (iii) as the PDITC concentration increases from 0.5 to 4 mg mL^{-1}, a change in the values between the Raman lines peaked at 1157 and 1195–1196 cm^{-1}, respectively, ($I_{1157}/I_{1193-1196}$) as well as those at 1255 and 1320 cm^{-1} (I_{1255}/I_{1320}) vary from 0.26 and 0.41 (Figure 1b) to 4.27 and 5 (Figure 1e), respectively. These changes indicate a chemical interaction of SPCE-RGO/PDPA with PDITC, which, according to Figure 1f,g, involve the appearance of one-dimensional structures.

Similar vibrational changes are seen in the Raman spectra recorded at the excitation wavelength of 1064 nm (Figure 2).

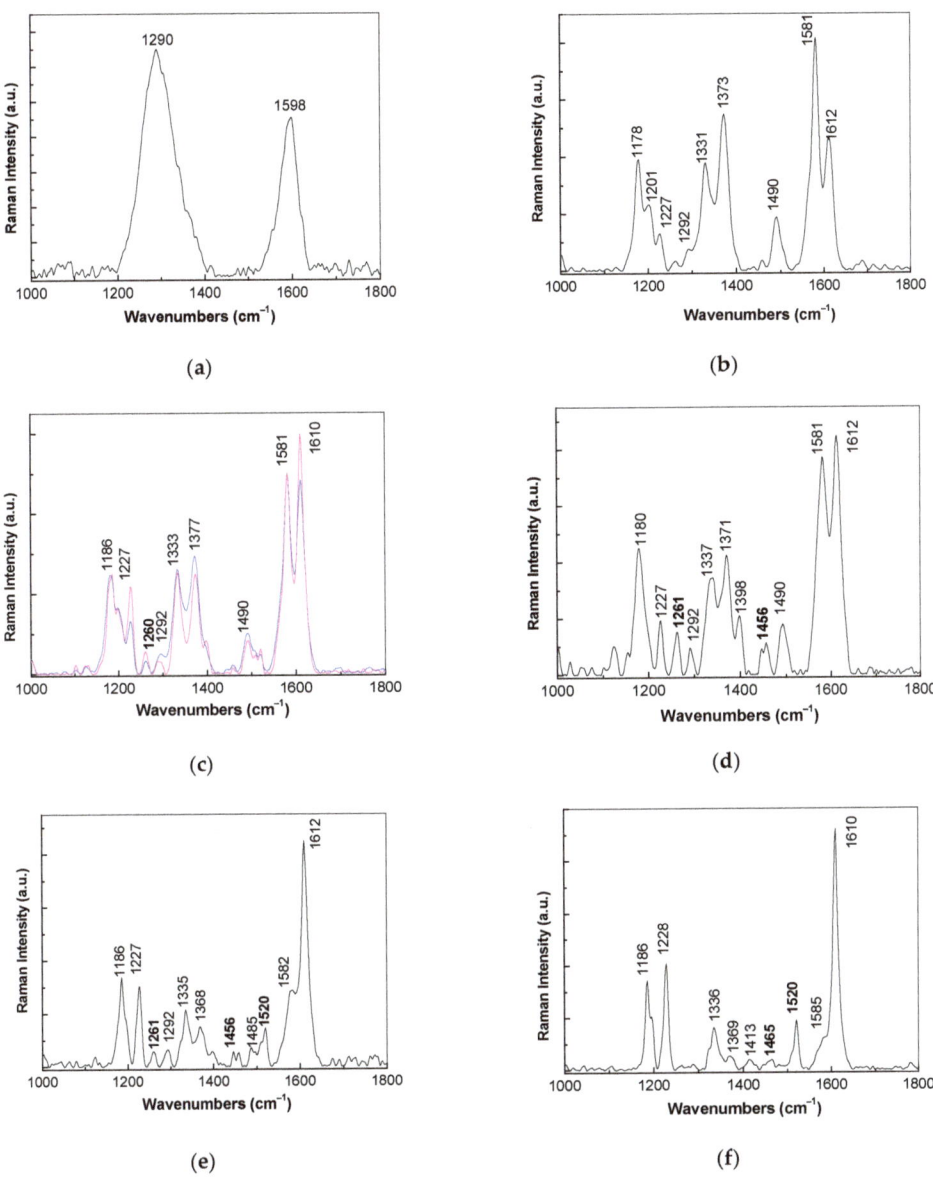

Figure 2. The Raman spectra of SPCE-RGO (**a**), SPCE-RGO/PDPA (**b**), SPCE-RGO/PDPA interacted with 2 (blue curve) and 4 mg mL^{-1} (magenta curve) PDITC in ethanol (**c**), and after the interaction of SPCE-RGO/PDPA/PDITC with protein G (**d**), anti-epidermal growth factor receptor (EGFR) (**e**) and EGFR (**f**).

In this last case, we observe that:

(a) the Raman lines of the RGO sheets peaked at 1290 and 1598 cm^{-1}, with the ratio between the intensities of the two bands being equal to 1.42 (Figure 2a);

(b) after the recording of 20 cyclic voltammograms onto the SPCE-RGO surface, the SPCE-RGO/PDPA platform is characterized by the intense Raman lines peaking at 1178, 1331, 1373, 1490, 1581, and 1612 cm^{-1} belonging to PDPA doped with the H$_4$SiW$_{12}$O$_{40}$ heteropolyanions and two Raman lines of low intensity at 1227 and 1292 cm^{-1} belonging to the vibrational mode of the C–N stretching of PDPA and the D band of the RGO sheets (Figure 2b);

(c) the interaction of the SPCE-RGO/PDPA platform with PDITC leads to: (c_1) a gradual increase in the intensity of the Raman lines peaked at 1227 and 1610 cm^{-1} simultaneously with an up-shift of the Raman line from 1178 to 1186 cm^{-1}, (c_2) the change in the ratio between the intensities of the Raman lines at 1178–1186 and 1227 cm^{-1} ($I_{1178-1186}/I_{1227}$) from 3.19 (Figure 2b) to 1.14 (magenta curve in Figure 2c) as well as those at 1581 and 1612–1610 cm^{-1} ($I_{1581}/I_{1612-1610}$) from 1.75 (Figure 2b) to 0.84 (magenta curve in Figure 2c), and (c_3) the appearance of the Raman line at 1260 cm^{-1}; all these changes indicate a chemical adsorption of PDITC on the surface of the SPCE-RGO/PDPA platform, which will be labeled in the following as the SPCE-RGO/PDPA-PDITC platform;

(d) the interaction of the SPCE-RGO/PDPA-PDITC platform with protein G induces in the Raman spectrum shown in Figure 2d the appearance of a new line at 1456 cm^{-1} simultaneously with an up-shift of the Raman line assigned to the vibrational mode of C–N in the N,N′-diphenyl benzidine radical cation, from 1331 to 1337 cm^{-1};

(e) the interaction of the SPCE-RGO/PDPA-PDITC-G platform with the EGFR antibodies highlights an increase in the intensity of the Raman line at 1227 cm^{-1} and a decrease in the intensity of the Raman lines situated in the spectral range 1300–1400 cm^{-1}, and a change in the ratio between the Raman lines at 1582 and 1612 cm^{-1} (I_{1582}/I_{1612}) of 0.34 (Figure 2e); in addition, these interactions induce an increase in the relative intensity of the Raman line localized in the spectral range 1000–1200 cm^{-1} from 0.51 (Figure 2c) to 0.6 (Figure 2d) and 1.17 (Figure 2e) and a down-shift of the Raman line from 1490 cm^{-1} (Figure 2b,c) to 1485 cm^{-1} (Figure 2e); and

(f) after the incubation of the SPCE-RGO/PDPA-PDITC-G-EGFR antibody platform with EGFR antigen, an additional decrease in the intensity of the Raman lines localized in the spectral range 1300–1400 cm^{-1} accompanied by an increase in the intensity of the Raman line at 1520 cm^{-1}, as well as a change in the I_{1582}/I_{1612} ratio at 2.74 and the appearance of a new Raman line at 1465 cm^{-1}, occur (Figure 2f).

The new Raman lines reported in Figure 2 peaking at 1260, 1456, 1465, and 1520 cm^{-1} belong to PDITC, protein G, EGFR antibodies, and EGFR, as shown in the Raman spectra recorded at the excitation wavelength of 1064 (Figure 3).

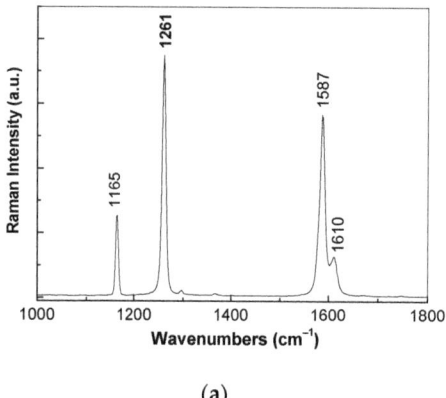

(a)

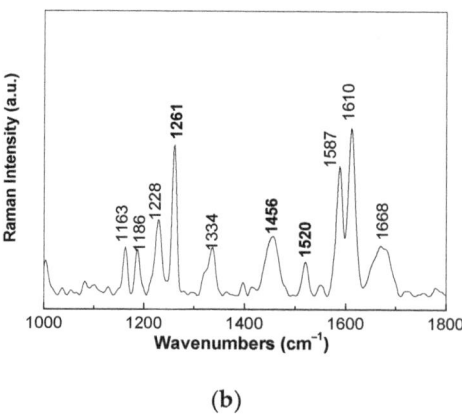

(b)

Figure 3. *Cont.*

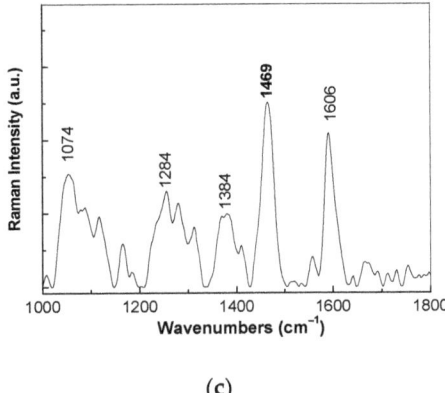
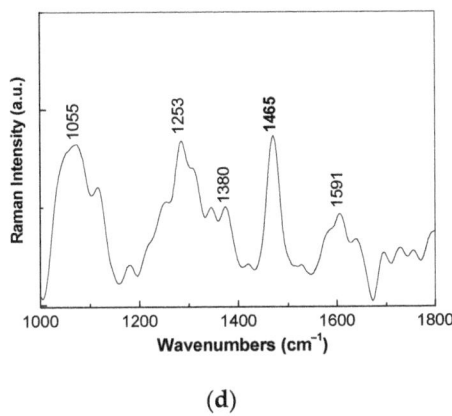

Figure 3. The Raman spectra of PDITC (**a**), protein G (**b**), anti-EGFR (**c**), and EGFR (**d**).

In this context, we note that: (i) down-shift of the Raman line from 1490 cm^{-1} (Figure 2b,c) to 1485 cm^{-1} (Figure 2e) can be explained, taking into account both the presence of Raman line of protein G at 1456 cm^{-1} (Figure 3b) and its chemical interaction with the SPCE-RGO/PDPA-PDITC; and (ii) new Raman line of EGFR at 1465 cm^{-1}, assigned to the vibrational mode of the CH$_2$ scissoring [24] (Figure 2f), confirms a chemical adsorption of a part the EGFR antibodies and EGFR antigen onto the SPCE-RGO/PDPA-PDITC-G platform surface. This can be explained by an incomplete deactivation of the thiocyanate groups due to steric effects induced by the presence of protein G on the surface. The remaining thiocyanate groups can anchor to the surface the anti-EGFR antibodies that get non-covalently attached to protein G, and similarly the EGFR antigen that gets caught by the anti-EGFR antibodies.

Additional information is obtained by FTIR spectroscopy, as shown in Figure 4. The main IR bands of the SPCE-RGO/PDPA platform peak at 694, 750, 779, 881, 916, 970, 1014, 1164, 1251, 1315, 1493, 1593, and 1651 cm^{-1}, being assigned to the following vibrational modes: inter-ring deformation, ring deformation, W–O$_c$–W (octahedral edge-sharing), C–H in-plane bending of the quinoid ring (Q), W–O$_b$–W (octahedral corner-sharing), W–O$_d$ (terminal), Si–O$_a$, C–H bending in the benzene ring (B) + quinoid ring (Q), radical cation structure, C$_{aromatic}$–N stretching, C–C stretching + C–H bending, C–C stretching, and –NH$^+$=Q=Q=NH$^+$–, respectively [19,25–27]. The interaction of the SPCE-RGO/PDPA platform with PDITC, protein G, EGFR antibodies, and EGFR antigen induces in Figure 4 the following changes: (i) a decrease in the absorbance of the IR bands at 694 and 750 cm^{-1}; (ii) an up-shift of IR band from 1164 to 1184 cm^{-1}; (iii) a gradual increase in the absorbance of the IR bands at 1251, 1315, 1593, and 1651 cm^{-1}; and (iv) the appearance of two IR bands with maxima at 1699 and 1780 cm^{-1}, both assigned to the C=O vibrational mode whose absorbance gradually increases as the platform interacts with protein G, EGFR antibodies, and EGFR antigen.

According to Figure 5, the IR spectrum of protein G is dominated by two IR bands at 1518 and 1634 cm^{-1}, which were assigned to the vibrational modes C–C + C–H and C=C, respectively [28]. Other IR bands of low absorbance are remarked in Figure 5a, at 1084, 1234, and 1393 cm^{-1} that were attributed to the vibrational modes of bonds CH, COH and COO, respectively [28]. In the case of the EGFR antibodies, the IR bands localized at 1038–1109 and 1649 cm^{-1} (Figure 5b) were assigned to the vibrational mode C–O + C–H and C=O [28]. All these vibrational changes can be explained by taking into account the chemical mechanism of the assembling of these platforms shown in Scheme 1.

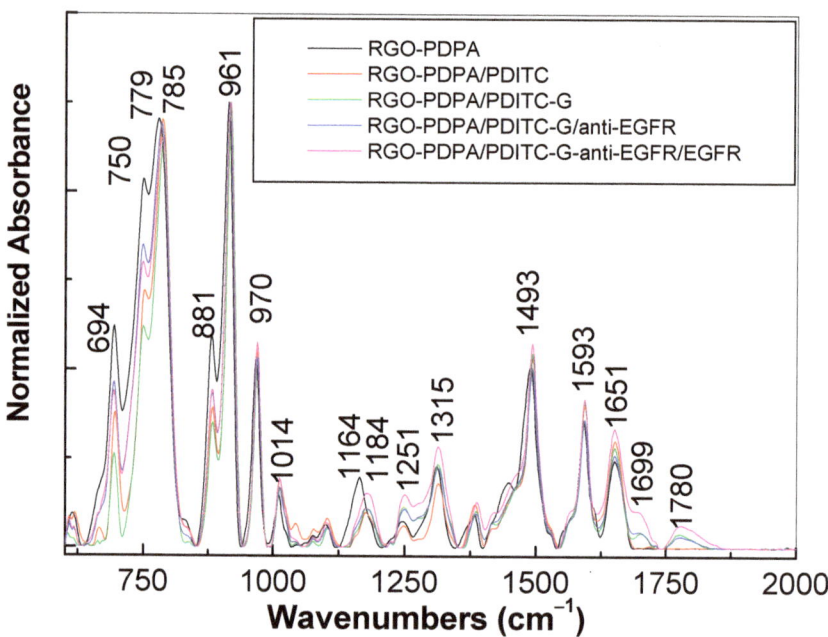

Figure 4. The IR spectrum of the SPCE-RGO/PDPA platform before (black curve) and after the successive interaction with the PDITC solution in C_2H_5OH (1 mL) with a concentration equal to 4 mg mL^{-1} (red curve), protein G (green curve), anti-EGFR (blue curve), and EGR (magenta curve).

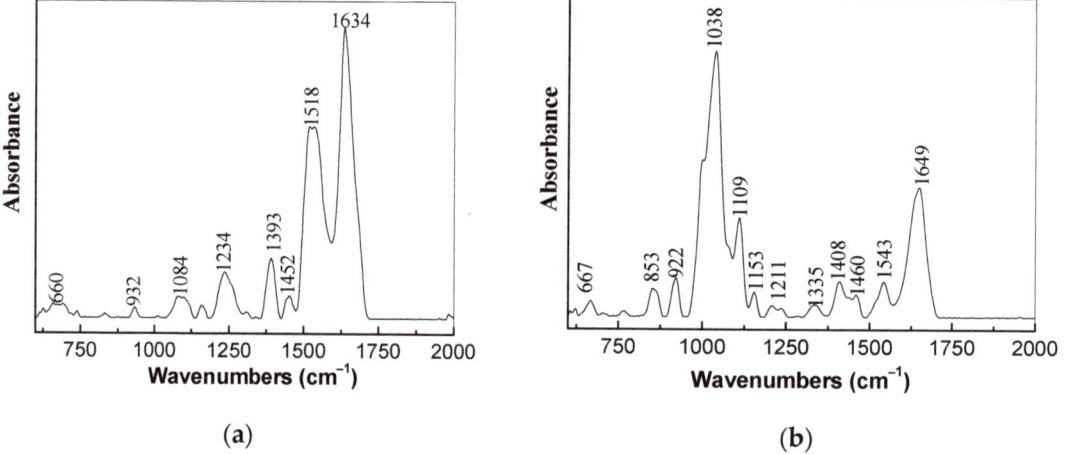

Figure 5. The IR spectrum of the protein G (**a**) and EGFR antibody (**b**).

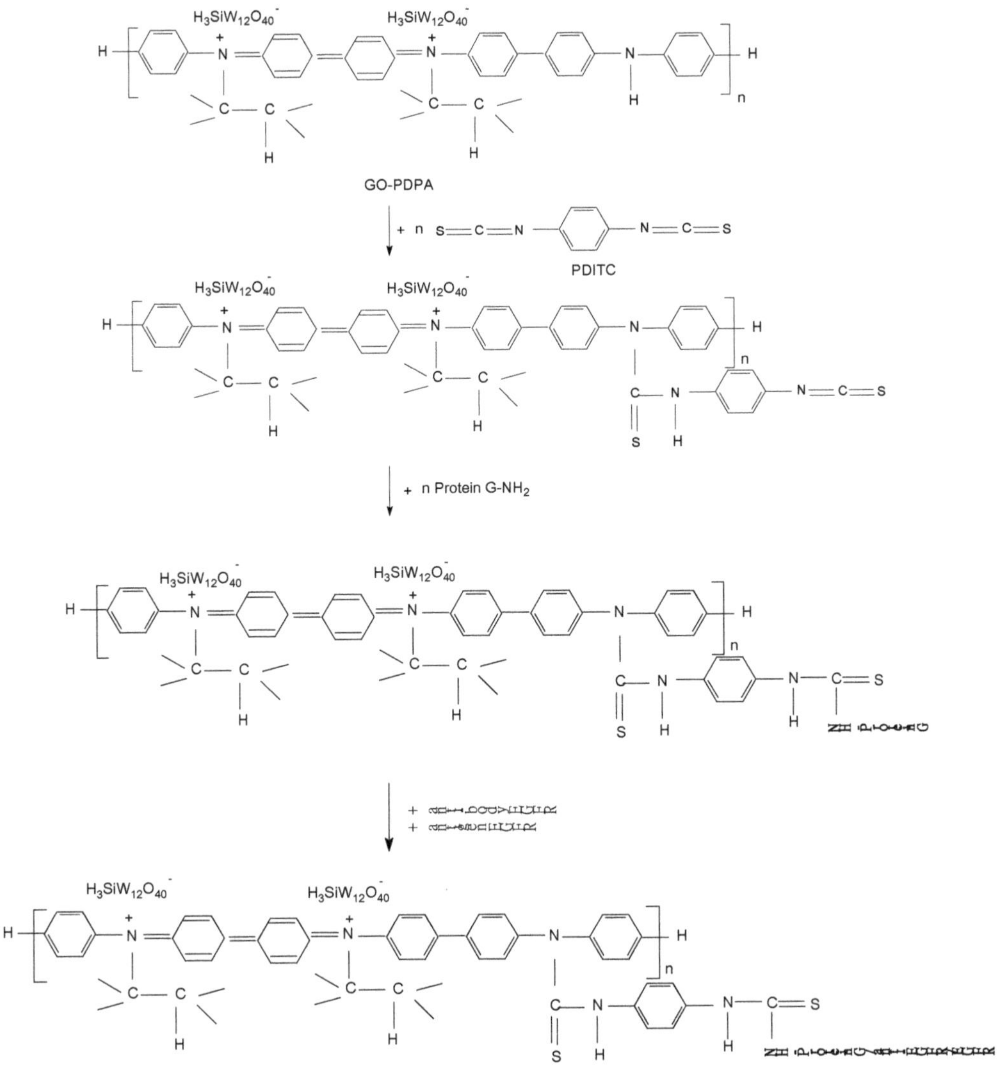

Scheme 1. The interaction of SPCE-RGO/PDPA with PDITC followed of the chemical interaction with protein G, EGFR antibodies, and EGFR.

3.2. The Electrochemical Properties of the Platforms SPCE-RGO/PDPA-PDITC-G-EGFR Antibodies-EGFR

Figure 6a–c show the fifth cyclic voltammogram of the Au electrode, SPCE-RGO and SPCE-RGO/PDPA-PDITC-G-anti-EGFR/EGFR in 5 mM $K_3[Fe(CN)_6]/K_4[Fe(CN)_6]$ solution, depending on the scan rate. The main changes in the potential of the anodic and cathodic peaks as well as their current densities are summarized in Table 1.

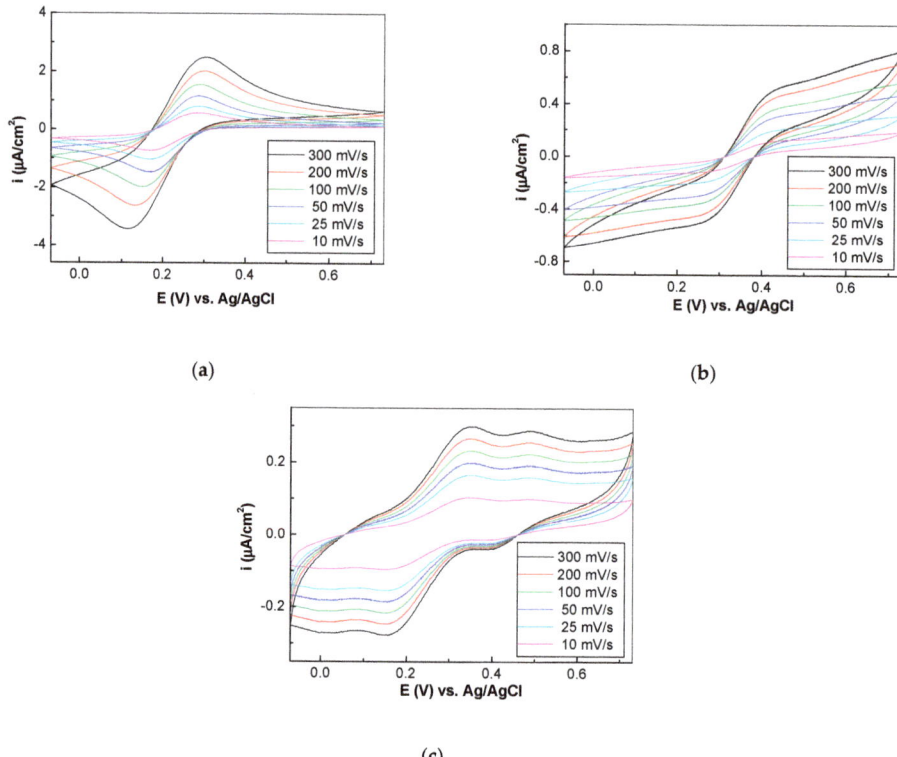

Figure 6. Cyclic voltammograms of the electrodes of Au (**a**), SPCE-RGO (**b**), SPCE-RGO/PDPA-PDITC-G-anti-EGFR/EGFR (**c**), with 5 mM K$_3$[Fe(CN)$_6$]/K$_4$[Fe(CN)$_6$] solution in 0.1 M PB with pH = 7.4. Cyclic voltammograms were recorded with scan rates equal to 300 (black curve), 200 (red curve), 100 (green curve), 50 (blue curve), 25 (cyan curve) and 10 mV s^{-1} (magenta curve).

Table 1. The potential of the anodic and cathodic peaks (E_{pa}, E_{pc}) as well as current densities (i_{pa}, i_{pc}) of the cyclic voltammograms as depending on the scan rate (v) of the following electrodes: Au, SPCE-RGO, SPCE-RGO/PDPA-PDITC-G-anti-EGFR/EGFR. $\Delta E = E_{pa} - E_{pc}$ corresponds to the potential of separation of the anodic and cathodic peaks.

Electrode	v (mV s^{-1})	E_{pa} (mV)	E_{pc} (mV)	ΔE (mV)	i_{pa} (μA cm^{-2})	i_{pc} (μA cm^{-2})	i_{pa}/i_{pc}
0	300	300	114	186	2.49	3.43	0.72
	200	295	131	164	2.01	2.63	0.76
	100	284	144	140	1.55	1.97	0.78
	50	283	167	116	1.16	1.47	0.79
	25	281	166	115	0.81	1.03	0.79
	10	283	168	115	0.58	0.74	0.78
SPCE-RGO	300	435	258	177	0.52	0.50	1.04
	200	428	254	174	0.45	0.45	1
	100	421	258	163	0.35	0.44	1.03
	50	417	264	153	0.28	0.29	0.96
	25	416	270	146	0.19	0.19	1
	10	410	279	131	0.11	0.11	1
SPCE-RGO/PDPA-PDITC-G-anti-EGFR/EGFR	300	345	147	198	0.3	0.28	1.07
	200	343	148	195	0.26	0.25	1.04
	100	341	150	191	0.23	0.21	1.09
	50	338	152	186	0.19	0.18	1.06
	25	334	155	179	0.16	0.15	1.07
	10	330	160	170	0.10	0.09	1.11

In the case of Au electrode, the ratio between current densities of the anodic and cathodic peaks is different from one. In the case of the electrodes SPCE-RGO, SPCE-RGO/PDPA-PDITC-G-anti-EGFR/EGFR, the values of the ratio between current densities of the anodic and cathodic peaks is ~1. Regardless of the electrode type, i.e., Au, SPCE-RGO, and SPCE-RGO/PDPA-PDITC-G-anti-EGFR/EGFR, the potential of separation of the anodic and cathodic peaks has a difference of 56.5/n, where n corresponds to the number of electrodes involved in the electrochemical process. These results indicate that at the electrode–electrolyte interface, an irreversible process occurs.

This process must to be understood by the electrostatic interaction between the positively charged amine entities of the SPCE-RGO/PDPA-PDITC-G-EGFR antibodies/EGFR platform and negative charges of $[Fe(CN)_6]^{3-}/^{4-}$.

4. Conclusions

In this work, new optical evidence of the assembly process of sensors based on RGO sheets functionalized with PDPA in a doped state are reported by Raman scattering and FTIR spectroscopy. Our results allow us to conclude that:

i. the interaction of the SPCE-RGO /PDPA platform with PDITC leads to a covalent functionalization of this platform, evidenced by an up-shift of the Raman line, from 1176 to 1189 cm^{-1}, and the appearance of a new Raman line at 1133 cm^{-1};

ii. the successive interactions of the SPCE-RGO/PDPA-PDITC with protein G, EGFR antibodies, and EGFR were highlighted by (a) new Raman lines at 1260, 1456, 1465, and 1520 cm^{-1} belonging to PDITC, protein G, EGFR antibodies, and EGFR, respectively, and (b) the IR spectra by the appearance of new IR bands at 1699 and 1780 cm^{-1}; and

iii. at the interface of the SPCE-RGO/PDPA-PDITC-G-EGFR antibodies/EGFR platform with $K_3[Fe(CN)_6]/K_4[Fe(CN)_6]$ solution in PB with a pH equal to 7.4, the irreversible processes were reported.

Author Contributions: Conceptualization, M.B. and S.N.F.; investigation, M.B. and M.D.; writing—original draft preparation, M.B. and S.N.F.; writing—review and editing, M.B.; supervision, M.B. All authors have read and agreed to the published version of the manuscript.

Funding: This work was funded in the framework of a project co-funded by the European Regional Development Fund under the Competitiveness Operational Program 2014-2020 entitled "Physicochemical analysis, nanostructured materials and devices for applications in the pharmaceutical field and medical in Romania", financing contract no. 58/05.09.2016 signed between the National Institute of Materials Physics and the National Authority for Scientific Research and Innovation as an Intermediate Body, on behalf of the Ministry of European Funds as Managing Authority for Operational Program Competitiveness (POC), sub-contract of type D, no. 2570/29.11.2017, between the National Institute of Materials Physics and Pro-Vitam Ltd.

Institutional Review Board Statement: Not applicable.

Informed Consent Statement: Not applicable.

Data Availability Statement: Samples of the SPCE-RGO /PDPA platforms are available from the authors.

Conflicts of Interest: The authors declare no conflict of interest. The funders had no role in the design of the study; in the collection, analyses, or interpretation of data; in the writing of the manuscript, or in the decision to publish the results.

References

1. Nicholson, R.; Gee, J.; Harper, M. EGFR and cancer prognosis. *Eur. J. Cancer* **2001**, *37*, 9–15. [CrossRef]
2. Arteaga, C.L. Epidermal growth factor receptor dependence in human tumors: More than just expression? *Oncologist* **2002**, *7*, 31–39. [CrossRef] [PubMed]

3. Hirsch, F.R.; Dziadziuszko, R.; Thatcher, N.; Mann, H.; Wartkins, C.; Parums, D.V.; Speake, G.; Holloway, B.; Bunn, P.A.; Franklin, W.A. Epidermal growth factor receptor immunohistochemistry: Comparison of antibodies and cutoff points to predict benefit from gefitinib in a phase 3 placebo-controlled study in advanced non-small-cell lung cancer. *Cancer* **2008**, *112*, 1114–1121. [CrossRef]
4. Buffet, W.; Geboes, K.P.; Dehertogh, G. EGFR-immunohistochemistry in colorectal cancer and non-small cell lung cancer: Comparison of 3 commercially available EGFR-antibodies. *Acta Gastroenterol. Belg.* **2008**, *71*, 213–218. [PubMed]
5. Thariat, J.; Etienne-Grimaldi, M.C.; Grall, D.; Bensadoun, R.J.; Cayre, A.; Penault-Llorca, F.; Verracini, L.; Francoual, M.; Formento, J.L.; Dassonville, O.; et al. Epidermal growth factor receptor protein detection in head and neck cancer patients: A many-faceted picture. *Clin. Cancer Res.* **2012**, *18*, 1313–1322. [CrossRef]
6. Pfeiffer, P.; Nexo, E.; Bentzen, S.M.; Clausen, P.P.; Andersen, K.W.; Rose, C.D. Enzyme-linked immunosorbent assay of epidermal growth factor receptor in lung cancer: Comparisons with immunohistochemistry, clinicopathological features and prognosis. *Br. J. Cancer* **1998**, *78*, 96–99. [CrossRef] [PubMed]
7. Weigum, S.E.; Floriano, P.N.; Redding, P.N.; Yeh, S.W.; Westbrook, S.D.; McGuff, S.; Lin, H.; Miller, A.; Vllarreal, F.R.; Rowan, F.; et al. Nano-bio-chip sensor platform for examination of oral exfoliative cytology. *Cancer Prev. Res.* **2010**, *3*, 518–528. [CrossRef] [PubMed]
8. Weigum, S.E.; Floriano, P.N.; Christodoulides, N.; McDevitt, J.T. Cell-based sensor for analysis of EGFR biomarker expression in oral cancer. *Lab Chip* **2007**, *7*, 995–1003. [CrossRef]
9. El-Sayed, I.H.; Huang, X.; El-Sayed, M.A. Surface plasmon resonance scattering and absorption of anti-egfr antibody conjugated gold nanoparticles in cancer diagnostics: Applications in oral cancer. *Nano Lett.* **2005**, *5*, 829–834. [CrossRef]
10. Vasudev, A.; Kaushik, A.; Bhansali, S. Electrochemical immunosensor for label free epidermal growth factor receptor (EGFR) detection. *Biosens. Bioelectron.* **2013**, *39*, 300–305. [CrossRef]
11. Baibarac, M.; Daescu, M.; Socol, M.; Bartha, C.; Negrila, C.; Fejer, S.N. Influence of reduced graphene oxide on the electropolymerization of 5-amino-1-naphthol and the interaction of 1,4-phenylene diisothiocyanate with the poly(5-amino-1-naphtol)/reduced graphene oxide composite. *Polymers* **2020**, *12*, 1299. [CrossRef]
12. Baibarac, M.; Stroe, M.; Fejer, S.N. Vibrational and photoluminescence properties of polydiphenylamine doped with silicotungstic acid heteropolyanions and their composites with reduced graphene oxide. *J. Mol. Struct.* **2019**, *1184*, 25–35. [CrossRef]
13. Smaranda, I.; Benito, A.M.; Maser, W.K.; Baltog, I.; Baibarac, M. Electrochemical grafting of reduced graphene oxide with polydiphenylamine doped with heteropolyanions and its optical properties. *J. Phys. Chem. C* **2014**, *118*, 25704–25717. [CrossRef]
14. Malard, L.M.; Pimenta, M.A.; Dresselhaus, G.; Dresselhaus, M.S. Raman spectroscopy in graphene. *Phys. Rep.* **2009**, *473*, 51–87. [CrossRef]
15. Guo, Y.; Li, K.; Yu, X.; Clark, J.H. Mesoporous $H_3PW_{12}O_{40}$-silica composite: Efficient and reusable solid acid catalyst for the synthesis of diphenolic acid from levulinic acid. *Appl. Catal. B Environ.* **2008**, *81*, 182–191. [CrossRef]
16. Baibarac, M.; Baltog, I.; Lefrant, S.; Gómez-Romero, P. Polydiphenylamine/carbon nanotube composites for applications in rechargeable lithium batteries. *Mater. Sci. Eng. B* **2011**, *176*, 110–120. [CrossRef]
17. Legagneux, N.; Basset, J.M.; Thomas, A.; Lefebvre, F.; Goguet, A.; Sa, J.; Hardacre, C. Characterization of silica-supported dodecatungstic heteropolyacids as a function of their dihydroxylation temperature. *Dalton Trans.* **2009**, *12*, 2235–2240.
18. Kvarnström, C.; Petr, A.; Damlin, P.; Lindfors, T.; Ivaska, A.; Dunsch, L. Raman and FTIR spectroscopic characterization of electrochemically synthesized poly(triphenylamine), PTPA. *J. Solid State Electrochem.* **2002**, *6*, 505–512. [CrossRef]
19. Quillard, S.; Louarn, G.; Lefrant, S.; MacDiarmid, A.G. Vibrational analysis of polyaniline—a comparative study of leucoemeraldine, emeraldine and pernigranilien bases. *Phys. Rev. B* **1994**, *50*, 12496–12508. [CrossRef] [PubMed]
20. Janik, I.; Carmichael, I.; Tripathi, G. Transient Raman spectra, structure and thermochemistry of the thiocyanate dimer radical anion in water. *J. Chem. Phys.* **2017**, *146*, 214305. [CrossRef]
21. Moritz, A. Infra-red and Raman spectra of methyl thiocyanate and methyl-d3 thiocyanate. *Spectrochim. Acta* **1966**, *22*, 1021–1028. [CrossRef]
22. Hill, W.; Wehling, B. Potential- and pH-dependent surface-enhanced Raman scattering of p-mercapto aniline on silver and gold substrates. *J. Phys. Chem.* **1993**, *97*, 9451–9455. [CrossRef]
23. Wang, P.; Li, H.; Cui, C.; Jiang, J. In situ surface-enhanced Raman spectroscopy study of thiocyanate ions adsorbed on silver nanoparticles under high pressure. *Chem. Phys.* **2019**, *516*, 1–5. [CrossRef]
24. Zhang, T.; Qin, Y.; Tan, T.; Lv, Y. Targeted live cell raman imaging and visualization of cancer biomarkers with thermal-stimuli responsive imprinted nanoprobes. *Part. Part. Syst. Charact.* **2018**, *35*, 1800390. [CrossRef]
25. Sui, C.; Li, C.; Guo, X.; Cheng, T.; Gao, Y.; Zhou, G.; Gong, J.; Du, J. Facile synthesis of silver nanoparticles-modified $PVA/H_4SiW_{12}O_{40}$ nanofibers-based electrospinning to enhance photocatalytic activity. *Appl. Surf. Sci.* **2012**, *258*, 7105–7111. [CrossRef]
26. Raj, M.R.; Anandan, S.; Zhou, M.; AshokKumar, M. A facile one-step synthesis of hollow polydiphenylamine. *Int. J. Polym. Mater.* **2013**, *62*, 23–27. [CrossRef]
27. De Santana, H.; Dias, F. Characterization and properties of polydiphenylamine electrochemically modified by iodide species. *Mater. Chem. Phys.* **2003**, *82*, 882–886. [CrossRef]
28. Barth, A. Infrared spectroscopy of proteins. *Biochim. et Biophys. Acta (BBA) Bioenerg.* **2007**, *1767*, 1073–1101. [CrossRef] [PubMed]

Article

Fabrication of Microalloy Nitrided Layer on Low Carbon Steel by Nitriding Combined with Surface Nano-Alloying Pretreatment

Jian Sun [1],* and Quantong Yao [2]

[1] Department of Materials Science and Engineering, Hefei University of Technology, Hefei 230009, China
[2] Key Laboratory of Electromagnetic Processing of Materials, Ministry of Education, Northeastern University, Shenyang 110819, China; tongtong282@126.com
* Correspondence: sunjian@hfut.edu.cn; Tel.: +86-551-6290-4557

Academic Editor: Yilei Zhang
Received: 14 October 2016; Accepted: 8 November 2016; Published: 17 November 2016

Abstract: Surface mechanical attrition treatment (SMAT) is an effective method to accelerate the nitriding process of metallic materials. In this work, a novel technique named surface nano-alloying (SNA) was developed on the basis of surface mechanical attrition treatment, which was employed as a pretreatment for the nitriding of low carbon steel materials. The microstructure and surface properties of treated samples were investigated by SEM, XRD, TEM and the Vickers hardness test. Experimental results showed that a surface alloying layer (Cr element) of about 10–20 μm in thickness was formed on the low carbon steel sample after the surface nano-alloying treatment. After nitriding for the SNA sample, a complex compound layer composed of $Fe_{2-3}N$, FeCr and Cr_2N phases was fabricated. Moreover, the thickness of this compound layer was about 50 μm. Meanwhile, both the surface hardness and wear resistance of the SNA nitrided sample are better that those of the SMAT nitrided sample. This work offers a new approach for improving the nitriding process of steel materials.

Keywords: surface mechanical attrition treatment; surface nano-alloying; nitriding; diffusion; surface properties

---

1. Introduction

Surface mechanical attrition treatment (SMAT) is an effective surface modification technique to improve the surface properties of metallic materials including hardness, wear and corrosion resistance by server plastic deformation [1–8]. Using SMAT, different microstructures can be obtained within the deformed surface layer through the depth from the treated surface to the strain-free matrix, i.e., from nano-sized grains to sub-micro-sized and micro-sized crystalline on a bulk material [1,3–6]. Previous studies showed that the atom diffusion ability on SMATed materials was significantly enhanced due to the formation of a large number of grain boundaries or other defects (vacancy, dislocation and interfaces) by SMAT [9–14]. For example, Tong and Zhang showed that the gaseous nitriding of a SMATed pure iron plate and the plasma nitriding of a SMATed AISI 304 stainless steel specimen were also achieved recently at 300 °C and 400 °C, respectively. These temperatures are evidently lower than those in conventional nitriding processes (500–550 °C) [9,10]. In addition, the experiment of chromizing behaviors of the SMATed steel sample showed that the formation temperature of chromium compounds was found to be much lower and the amount of chromium carbides was higher than those in the coarse-grained counterpart [11]. Moreover, the significantly enhanced aluminizing behaviors of a low carbon steel at temperatures far below the austenitizing temperature, with a nanostructured surface layer produced by surface mechanical attrition treatment, was found to show us that SMAT seems to be appropriate for enhancing the atom diffusion ability efficiently [12].

As we know, mechanical alloying (MA) is a well-established method to prepare metastable phases (such as supersaturated solid solutions) by energetic collisions in a ball mill [15]. Additionally, surface mechanical alloying (SMA) was developed based on MA to fabricate coatings on metallic materials via ball milling [16]. Recently, a novel surface technology named surface nano-alloying (SNA) was developed on the basis of SMAT to fabricate a nanocrystalline composite surface on metallic materials [17–21]. Commonly, a small amount of alloy powder was added to the chamber before the SMAT process. Subsequently, the impacts of the milling balls delivered particles from the powder charge and attached them to the surface. Furthermore, due to the formation of a large number of deformation-induced non-equilibrium grain boundaries, non-equilibrium excessive vacancies and other defects, such as dislocations, twins and lattice strains induced by SMAT, the surface becomes chemically active, which may promote interdiffusion between the alloy element and substrate. Ultimately, a composite surface including the supersaturated solid solution, intermetallic compound and amorphous phases can be formed by this method [22]. At present, this method has become a hot topic in the surface engineering field. For example, Du et al. showed that Ni powders were welded into the surface of iron plates by SMAT [22]. Révész et al. illustrated that Cu plates were coated mechanically with a mixture of Ti and Zr powders via a combination of deposition, wear, deformation and defect-enhanced diffusion processes [23].

In this paper, surface nano-alloying was employed as a pretreatment for a low carbon steel sample. Subsequently, the nitriding behaviors of the surface nano-alloying sample were investigated in comparison with those of original and SMATed samples.

2. Materials and Methods

A 20 steel plate (4 × 100 × 100 mm^3 in size) with chemical composition of (in wt.%): 0.2 C, 0.17 Si, 0.5 Mn, 0.25 Cr, 0.035 P (max.), 0.035 S (max.) and balance Fe, was used in the present study. The as-received sample was annealed at 950 °C for 1 h in vacuum to eliminate the effect of mechanical deformation and to obtain homogeneous coarse grains. Prior to surface nano-alloying, the specimens were machined and surface-polished with silicon carbide paper to grade 1200.

SNA was performed in a vacuum chamber as the schematic diagram of the experimental apparatus shown in Figure 1. GCr15 steel balls with a diameter of 4 mm and about 1 g Cr powders were placed at the bottom of the chamber. This chamber was vibrated by a generator with a frequency of 50 Hz. The balls generated mechanical attrition on the surface of iron plate, producing plastic deformation in the surface layer, and welding of the Ni powders onto the surface of sample. In this work, the sample was treated for 60 min. Subsequently, the samples were treated by plasma nitriding at 550 °C for 4 h. The processing of plasma nitriding was described in our previous work [24].

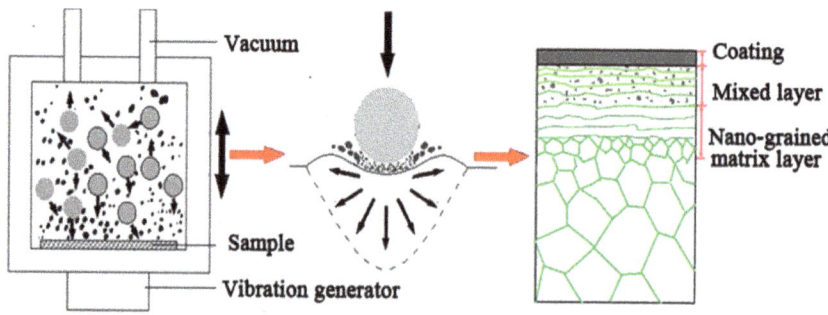

Figure 1. The schematic diagram of SNA processing.

The phase information of treated samples were analyzed by a X Pert Pro PW3040/60 X-ray diffractometer (XRD) (Panalytical, Almelo, The Netherlands) using Cu Kα radiation (40 kV, 40 mA).

Cross-sectional morphologies of various samples were observed by using Zeiss Ultra 55 scanning electron microscope (SEM) (Zeiss, Oberkochen, Germany). The element concentrations through the cross-section were analyzed by EDS (Zeiss, Oberkochen, Germany). Microstructure characterizations were also examined by using a Jeol-4000FX transmission electron microscope (TEM) (Jeol, Tokyo, Japan) with an operating voltage of 200 kV. The samples for TEM observation were prepared by grinding and mechanical polishing followed by ion-thinning at lower temperature.

Microhardness (HV) tests on the experimental samples were performed on a L101MVD model Vickers microhardness tester (Mitutoyo, Tokyo, Japan). The testing was carried out at a load of 25 g with duration of 10 s. Wear resistance tests were carried out on Optimol SRV®tribometer (Optimol, Berlin, Germany) with WC-6% Co ball. The duration of test is 30 min and the loads are 5 N, 15 N, 30 N, and 50 N, respectively. The wear volume loss was calculated by Archard formula.

3. Results and Discussion

3.1. Microstructure of SNA Sample

After SNA, the visual color of the sample surface changed from silver-white to gray-black, indicating the introduction of impurities into the surface layer during SNA. Figure 2 shows the cross-sectional SEM image and the corresponding EDS analysis of the SNA sample. It can be seen that the microstructure of the SNA sample were taken from different regions. Evidently, severe plastic deformation during the SNA process results in an inhomogeneous and non-uniform surface alloying layer (Cr element). The maximum thickness of the alloying layer can exceed 20 μm in some regions. In addition, a deformed layer (substrate) of about 100 μm in thickness can be found below the alloying layer.

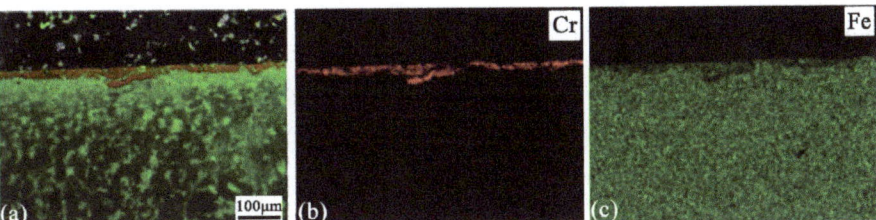

Figure 2. Cross-sectional morphology of SNA sample (**a**) and the corresponding EDS element mapping images: (**b**) Cr, (**c**) Fe.

3.2. Microstructure of SNA Sample after Nitriding

The image shown in Figure 3 is the cross-sectional SEM microstructure of the SNA sample after nitriding at 550 °C for 4 h, in comparison with that of the SMAT sample after nitriding at the same conditions. For the SMAT nitrided sample, the nitrided layer includes the compound layer, transition zone and diffusion layer from the surface to the inside, respectively, as shown in Figure 3a. The compound layer thickness in the SMAT nitrided sample is 30 μm. However, an obviously different microstructure was detected on the SNA nitrided sample. As shown in Figure 3b, it can be observed that a complex compound layer about 50 μm thick was formed on the SNA nitrided sample. Especially, the overall compound layer can be categorized into three parts from the top surface into the interior. The outermost layer is very dense, the thickness of which is about 20 μm. The middle layer displays a needle-like shape and the inner layer is not uniform. The formation of the complex compound layer may originate from the interdiffusion between nitrogen, chromium and ferrite atoms.

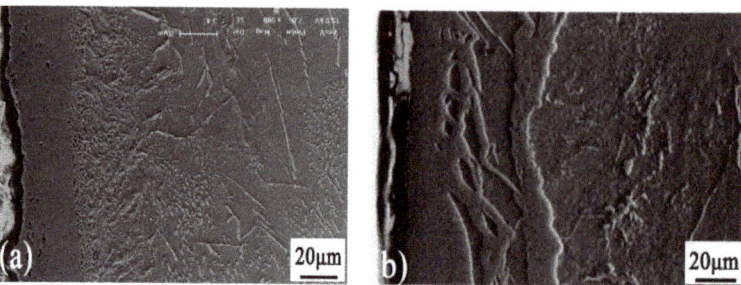

Figure 3. Cross-sectional SEM images of SMAT nitrided sample (**a**) and SNA nitrided sample (**b**).

To further investigate the phase information of the SNA nitrided sample, the XRD analysis was employed and its results are shown in Figure 4. It can be observed that the surface layer of the SMAT nitrided sample consists of $Fe_{2-3}N$ and Fe_4N phases. However, in the SNA sample nitrided under the same conditions, strong FeCr and Cr_2N phase diffraction peaks can be observed, indicating that interdiffusion may happen during the nitriding process. These results can be confirmed by the TEM observation and the corresponding selected area electron diffraction (SAED, shown in Figure 5). From the TEM results, it can be seen that the composition of the surface layer of the SNA nitrided sample was found to be composed of an ultrafine polycrystalline compound. The phases include $Fe_{2-3}N$, FeCr and Cr_2N, and these results are in agreement with the XRD analysis. As we know, SMAT can enhance the nitrogen diffusion ability in pure iron materials and result in significantly improved surface properties relative to those of the coarse-grain nitrided sample due to the "nano-effect". In this paper, the SNA method can not only offer the "nano-effect", but it also brings the alloy element into the nitrides, which may have a significant impact on the hardness and wear resistance, as will be shown later.

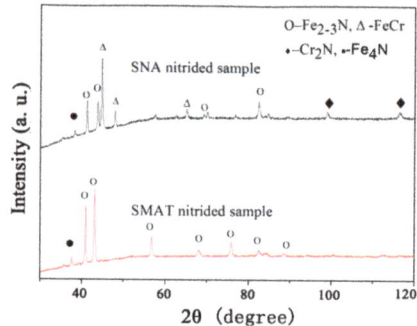

Figure 4. XRD pattern of two kinds of samples.

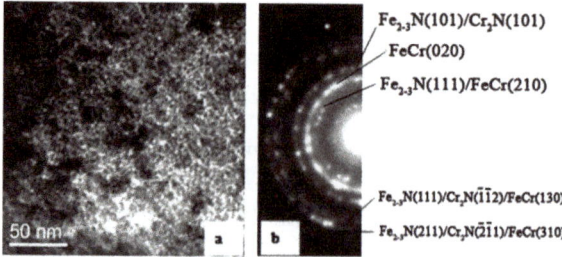

Figure 5. TEM observation (**a**) and corresponding SAED (**b**) of SNA nitrided sample.

3.3. Surface Properties of SNA Nitrided Sample

The variation of the hardness along the depth in the SNA sample nitrided at 550 °C is shown in Figure 6, in comparison with that of the SMAT sample nitrided in the same conditions. It should be noted that a hardness value of 1168 HV was obtained on the surface layer of the SNA nitrided sample, which is a significantly higher value than that recorded for the SMAT nitrided sample (846 HV). The variation of the hardness along the depth revealed the presence of a hardened surface layer, which was defined by the distance from the surface to the point where the microhardness was 50 HV higher than that in the matrix. This hardened surface layer is more than 350 μm in thickness for the SNA nitrided sample. In comparison, the hardness was approximately 846 HV at the surface and decreased sharply to the substrate value within 280 μm of the surface of the SMAT nitrided sample.

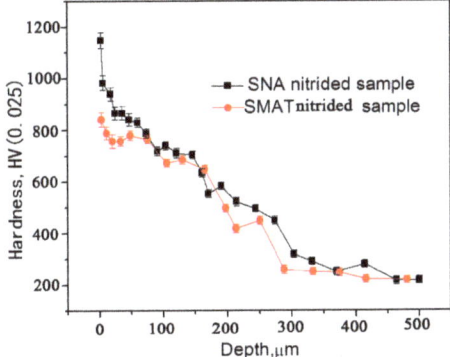

Figure 6. Hardness variations along the depth of both the SNA nitrided sample and the SMAT nitrided sample.

Measurements of volume losses of wear scars at various loads ranging from 5 to 50 N were performed on the SNA nitrided sample and the SMAT nitrided sample, respectively, as depicted in Figure 7. The wear volume loss values of both samples increased with the longer testing time, but the wear volume losses of the SNA nitrided sample were remarkably smaller than those of the SMAT nitrided sample for each sliding time. For example, the wear loss of the SNA nitrided sample was less than one half of that of the SMAT nitrided sample at load of 50 N. This observation indicates that the wear resistance of the SNA nitrided sample is better than that of the SMAT nitrided sample.

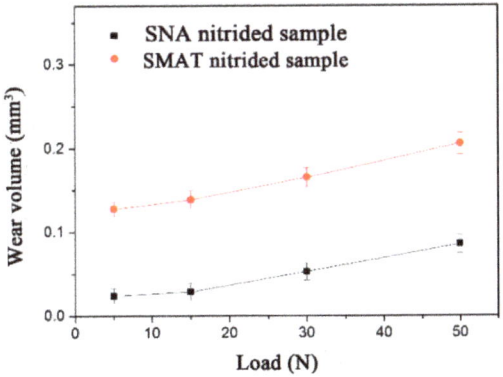

Figure 7. Variations in the wear volume losses with load of both the SNA nitrided sample and the SMAT nitrided sample.

4. Conclusions

The nitriding behavior of low carbon steel with a nano-alloying surface layer induced by SMAT with Cr powders was investigated in comparison with that of the SMAT sample after nitriding at the same conditions. The following conclusions were reached:

- After the surface nano-alloying treatment, a surface alloying layer (Cr element) of about 10–20 μm in thickness was formed on the low carbon steel sample. Additionally, a deformed substrate layer of about 100 μm can be detected at the subsurface.
- After nitriding for the SNA sample, a complex compound layer composed of $Fe_{2-3}N$, FeCr and Cr_2N phases was fabricated. Moreover, the thickness of this compound layer is about 50 μm.
- Both the surface hardness and wear resistance of the SNA nitrided sample are better than those of the SMAT nitrided sample.

Acknowledgments: This work was supported by the Fundamental Research Funds for the Central Universities (JZ2014HGBZ0042, JZ2016HGTA0702 and 2014HGQC0005), and the Anhui Provincial Natural Science Foundation (JZ2015AKZR0655).

Author Contributions: Jian Sun conceived and designed the experiments; Jian Sun performed the experiments; Jian Sun and Quantong Yao analyzed the data; Jian Sun wrote the paper.

Conflicts of Interest: The authors declare no conflict of interest.

References

1. Lu, K.; Lu, J. Surface nanocrystallization (SNC) of metallic materials: Presentation of the concept behind a new approach. *J. Mater. Sci. Technol.* **1993**, *15*, 193–197.
2. Tao, N.R.; Wang, Z.B.; Tong, W.P.; Sui, M.L.; Lu, J.; Lu, K. An investigation of surface nanocrystallization mechanism in Fe induced by surface mechanical attrition treatment. *Acta Mater.* **2002**, *50*, 4603–4616. [CrossRef]
3. Wu, X.; Tao, N.R.; Hong, Y.; Xu, B.; Lu, J.; Lu, K. Microstructure and evolution of mechanically-induced ultrafine grain in surface layer of Al-alloy subjected to USSP. *Acta Mater.* **2002**, *50*, 2075–2084. [CrossRef]
4. Yong, X.P.; Liu, G.; Lu, K.; Lu, J. Characterization and properties of nanostructured surface layer in low carbon steel subjected to surface mechanical attrition. *J. Mater. Sci. Technol.* **2003**, *19*, 1–4.
5. Roland, T.; Retraint, D.; Lu, K.; Lu, J. Enhanced mechanical behavior of a nanocrystallised stainless steel and its thermal stability. *Mater. Sci. Eng. A* **2007**, *445–456*, 281–288. [CrossRef]
6. Wang, Y.M.; Pan, D.; Lu, K.; Hemkeri, K.J. Microsample tensile testing of nanocrystalline copper. *Scr. Mater.* **2003**, *48*, 1581–1586. [CrossRef]
7. Amanov, A.; Cho, I.S.; Kim, D.E.; Pyun, Y.S. Fretting wear and friction reduction of CP titanium and Ti-6Al-4V alloy by ultrasonic nanocrystalline surface modification. *Surf. Coat. Technol.* **2012**, *27*, 135–142. [CrossRef]
8. Chang, H.W.; Kelly, P.M.; Shi, Y.N.; Zhang, M.X. Thermal stability of nanocrystallized surface produced by surface mechanical attrition treatment in aluminum alloys. *Surf. Coat. Technol.* **2012**, *26*, 3970–3980. [CrossRef]
9. Tong, W.P.; Tao, N.R.; Wang, Z.B.; Lu, J.; Lu, K. Nitriding iron at lower temperatures. *Science* **2003**, *299*, 686–688. [CrossRef] [PubMed]
10. Zhang, H.W.; Wang, L.; Hei, Z.K.; Liu, G.; Lu, J.; Lu, K. Low-temperature plasma nitriding of AISI 304 stainless steel with nano-structured surface layer. *Z. Metallkunde* **2003**, *94*, 1143–1147. [CrossRef]
11. Wang, Z.B.; Lu, J.; Lu, K. Chromizing behaviours of a low carbon steel processed by means of surface mechanical attrition treatment. *Acta Mater.* **2005**, *53*, 2081–2089. [CrossRef]
12. Si, X.; Lu, B.; Wang, Z.B. Aluminizing low carbon steel at lower temperatures. *J. Mater. Sci. Technol.* **2009**, *25*, 433–436.
13. Wang, Z.B.; Tao, N.R.; Tong, W.P.; Lu, J.; Lu, K. Diffusion of chromium in nanocrystalline iron produced by means of surface mechanical attrition treatment. *Acta Mater.* **2003**, *51*, 4319–4329. [CrossRef]
14. Tong, W.P.; Liu, C.Z.; Wang, W.; Tao, N.R.; Wang, Z.B.; Zuo, L.; He, J.C. Gaseous nitriding of iron with a nanostructured surface layer. *Scr. Mater.* **2007**, *57*, 533–536. [CrossRef]

15. Takacs, L.; Torosyan, A.R. Surface mechanical alloying of an aluminum plate. *J. Alloys Compd.* **2007**, *434*, 686–688. [CrossRef]
16. Baláž, P.; Takacs, L.; Ohtani, T.; Mack, D.E.; Boldižárová, E. Properties of a new nanosized tin sulphide phase obtained by mechanochemical route. *J. Alloys Compd.* **2002**, *337*, 76–82. [CrossRef]
17. Li, W.; Wang, X.D.; Meng, Q.P.; Rong, Y.H. Interdiffusion of alloying elements in nanocrystalline Fe-30 wt.% Ni alloy during surface mechanical attrition treatment and its effect on α→γ transformation. *Scr. Mater.* **2008**, *59*, 344–347. [CrossRef]
18. Meng, Y.F.; Shen, Y.F.; Chen, C.; Li, Y.C.; Feng, X.M. Effects of Cu content and mechanical alloying parameters on the preparation of W-Cu composite coatings on copper substrate. *J. Alloys Compd.* **2014**, *585*, 368–375. [CrossRef]
19. Wen, L.; Wang, Y.M.; Zhou, Y.; Guo, L.X.; Qu, J.H. Iron-rich layer introduced by SMAT and its effect on corrosion resistance and wear behavior of 2024 Al alloy. *Mater. Chem. Phys.* **2011**, *126*, 301–309. [CrossRef]
20. Romankov, S.; Hayasaka, Y.; Kasai, E.; Yoon, J.M. Fabrication of nanostructured Mo coatings on Al and Ti substrates by ball impact cladding. *Surf. Coat. Technol.* **2010**, *205*, 2313–2321. [CrossRef]
21. Romankov, S.; Sha, W.; Kaloshkin, S.D.; Kaevitser, K. Fabrication of Ti-Al coatings by mechanical alloying method. *Surf. Coat. Technol.* **2006**, *201*, 3235–3245. [CrossRef]
22. An, Y.L.; Du, H.Y.; Wei, Y.H.; Wang, N.; Hou, L.F.; Lin, W.M. Interfacial structure and mechanical properties of surface iron–nickel alloying layer in pure iron fabricated by surface mechanical attrition alloy treatment. *Mater. Des.* **2013**, *46*, 427–433. [CrossRef]
23. Revesz, A.; Takacs, L. Coating a Cu plate with a Zr-Ti powder mixture using surface mechanical attrition treatment. *Surf. Coat. Technol.* **2009**, *203*, 3026–3031. [CrossRef]
24. Sun, J.; Tong, W.P.; Zhang, H.; Zuo, L.; Wang, Z.B. Low-temperature plasma nitriding of titanium layer on Ti/Al clad sheet. *Mater. Des.* **2013**, *47*, 408–415. [CrossRef]

© 2016 by the authors; licensee MDPI, Basel, Switzerland. This article is an open access article distributed under the terms and conditions of the Creative Commons Attribution (CC-BY) license (http://creativecommons.org/licenses/by/4.0/).

Article

Surface Segregation of Amphiphilic PDMS-Based Films Containing Terpolymers with Siloxane, Fluorinated and Ethoxylated Side Chains

Elisa Martinelli [1], Elisa Guazzelli [1], Antonella Glisenti [2] and Giancarlo Galli [1,*]

[1] Dipartimento di Chimica e Chimica Industriale and UdR Pisa INSTM, Università di Pisa, 56124 Pisa, Italy; elisa.martinelli@unipi.it (E.M.); elisa.guazzelli@hotmail.it (E.G.)
[2] Dipartimento di Scienze Chimiche, Università di Padova, 35131 Padova, Italy; antonella.glisenti@unipd.it
[*] Correspondence: giancarlo.galli@unipi.it; Tel.: +39-050-2219-272

Received: 29 January 2019; Accepted: 25 February 2019; Published: 26 February 2019

Abstract: (Meth)acrylic terpolymers carrying siloxane (Si), fluoroalkyl (F) and ethoxylated (EG) side chains were synthesized with comparable molar compositions and different lengths of the Si and EG side chains, while the length of the fluorinated side chain was kept constant. Such terpolymers were used as surface-active modifiers of polydimethylsiloxane (PDMS)-based films with a loading of 4 wt%. The surface chemical compositions of both the films and the pristine terpolymers were determined by angle-resolved X-ray photoelectron spectroscopy (AR-XPS) at different photoemission angles. The terpolymer was effectively segregated to the polymer−air interface of the films independent of the length of the constituent side chains. However, the specific details of the film surface modification depended upon the chemical structure of the terpolymer itself. The exceptionally high enrichment in F chains at the surface caused the accumulation of EG chains at the surface as well. The response of the films to the water environment was also proven to strictly depend on the type of terpolymer contained. While terpolymers with shorter EG chains appeared not to be affected by immersion in water for seven days, those containing longer EG chains underwent a massive surface reconstruction.

Keywords: surface-active polymer; surface segregation; surface modification; amphiphilic polymer; polysiloxane; fluoropolymer; PEGylated polymer; X-ray photoelectron spectroscopy

1. Introduction

The dispersion of non-reactive surface-active additives is generally regarded as a facile and straightforward method to modify the surface properties of a polymer film without affecting its bulk properties to a significant extent [1–4]. The surface segregation process and, therefore, the selective accumulation of the additive at the interface with the external environment (air, water, organic vapors) is a complex phenomenon that depends on several factors that may add to each other, including the additive molecular structure and composition, molecular weight, surface tension, chemical compatibility with the host matrix and chemical affinity with the external environment [5–10]. A special class of surface-active additives consists of amphiphilic copolymers [11–13]. Random copolymers in particular have gained a great deal of interest in the last decade being their synthesis easier and more suitable for an industrial scale production than the block copolymer counterparts [14]. Notably, random copolymers have been proven to generate nanostructured materials as a result of their self-assembling in solution [15–20], bulk and at the surface of polymer thin films [4,21]. In the field of coatings, surface-active additives have been utilized for several purposes, especially for the development of antifouling (AF)/fouling-release (FR) coatings to combat marine biofouling [12]. In this regard, reactive [22–25] and non-reactive amphiphilic copolymers [21,26–29] generally composed of poly(ethylene glycol) (PEG) chains, as the hydrophilic component, and polysiloxane and/or fluoropolymer chains, as the hydrophobic component, have been investigated. The hydrophilic component of election is PEG on the basis of its known high resistance to the adhesion of proteins, bacteria, cells and marine organisms [30,31]. Moreover, fluorinated and siloxane chains, besides performing a FR action attributed to their hydrophobicity [32,33], can also play other distinct roles. In particular, fluorinated chains are anticipated to promote the diffusion and accumulation of the entire copolymer to the coating surface, as a result of their ability to self-segregate and self-organize at the outermost surface layers of polymer films [34–36]. Polysiloxane chains are also known to have a low surface tension behavior, albeit this character is not so distinct as for fluorinated chains [37–39], and have been introduced into several types and architectures of polymers to modify the surface properties [40,41]. However, in this case, the main role of polysiloxane chains is to act as a compatibilizer with the host elastomeric polydimethylsiloxane (PDMS) matrix to prevent macrophase separation. Moreover, low modulus PDMS-based films are generally thought to favor removal of marine organisms by a peeling-like, lower-energy mechanism [33,42].

In this work, we prepared novel amphiphilic films composed of a condensation-cured PDMS matrix, in which an opportunely designed surface-active amphiphilic terpolymer was physically dispersed in order to modify the surface structure and property of the films derived therefrom. The synthesized terpolymers, consisting of a (meth)acrylic backbone carrying fluorinated (F), ethoxylated (EG) and siloxane (Si) side chains, possessed comparable mole percentages of each type of side chains, but different lengths of the EG and Si and chains, while the length of the F chain was fixed. In particular, we studied the surface nanoscale composition and surface segregation of the terpolymer by means of angle-resolved X-ray photoelectron spectroscopy (AR-XPS). Interestingly, we found that the chemical composition of the PDMS-based films within the first few nanometers (~3 nm) of the surface was almost equal to that of the corresponding pristine terpolymer included in the formulation, independent of its chemical structure. However, the amount of each element at the surface as well as the effectiveness of surface segregation were strongly affected by the structure of the terpolymer itself. A peculiar reconstruction of the film surface was found when the PDMS-based films were immersed in water, which unexpectedly did not result in an increased concentration of the hydrophilic EG chains at the outer surface to contact water.

2. Materials and Methods

2.1. Materials

1H,1H,2H,2H-perfluorooctyl acrylate (F) (Fluorochem, 97%), polyethyleneglycol methyl ether methacrylate (EGa (M_n = 300 g mol^{-1}) and EGb (M_n = 1100 g mol^{-1})), bismuth neodecanoate (BiND) (all from Aldrich), monomethacryloxypropyl-terminated poly(dimethyl siloxane) (Sia (M_n = 1000 g mol^{-1}) and Sib (M_n = 5000 g mol^{-1})), bis(silanol)-terminated poly(dimethyl siloxane) (HO-PDMS-OH (M_n = 26,000 g mol^{-1}, 0.1% OH)), poly(diethoxy siloxane) (ES40) (all from ABCR) were used as received. 2,2'-azobis-isobutyronitrile (AIBN) (from Fluka) was recrystallized from methanol. Diethylene glycol dimethyl ether (diglyme) was kept at 100 °C over sodium for 4 h and then distilled under reduced pressure.

2.2. General Procedure for the Preparation of Terpolymers

In a typical preparation of a terpolymer p(Sib-F-EGb), monomers Sib (3.400 g, 0.68 mmol), F (0.860 g, 2.05 mmol) and EGb (0,780 g, 0.71 mmol), free-radical initiator AIBN (53,7 mg) and anhydrous solvent diglyme (20 mL) were introduced into a Carius tube. The solution was outgassed by four freeze-pump-thaw cycles. The polymerization reaction was let to proceed under stirring at 65 °C for 72 h. The crude product was purified by several precipitations from chloroform solutions into methanol (yield 48%). The obtained terpolymer p(Sib-F-EGb) contained 26 mol% Sib, 45 mol% F and 29 mol% EGb (M_n = 21000 g mol^{-1}, M_w/M_n = 2.2).

^{1}H-nuclear magnetic resonance (NMR) (CDCl$_3$): δ (ppm) = 4.6–3.8 (COOCH$_2$), 3.8–3.5 (CH$_2$O), 3.4 (OCH$_3$), 2.5 (CF$_2$CH$_2$), 2.1–0.7 (CH$_2$CCH$_3$, CH$_2$CH, COOCH$_2$CH$_2$CH$_2$Si, SiCH$_2$CH$_2$CH$_2$CH$_3$), 0.5 (SiCH$_2$), 0.1 (SiCH$_3$).

^{19}F-NMR (CDCl$_3$/CF$_3$COOH): δ (ppm) = −5.5 (CF$_3$), −38.5 (CF$_2$CH$_2$), −46 to −48 (CF$_2$), −51 (CF$_2$CF$_3$).

Fourier transform-infrared (FT–IR) (film): (cm^{-1}) = 2963–2906 (ν C–H aliphatic), 1738 (ν C=O ester), 1260 and 799 (ν Si–CH$_3$), 1207–1020 (ν C–F, ν C–O, ν Si–O), 662 (ω CF$_2$).

2.3. Preparation of Films

Glass slides (76 × 26 mm^2) were rinsed with acetone and dried in an oven for 30 min. A solution of HO-PDMS-OH (5.0 g), ES40 (0.125 g) and BiND (50 mg) in ethyl acetate (25 mL) was spray-coated onto the glass slides using a Badger model 250 airbrush (50 psi air pressure). The films were dried at room temperature for a day and annealed at 120 °C for 12 h to form a thin bottom layer (thickness ~2 µm). On top of it, a solution of the same amounts of HO-PDMS-OH, ES40 and BiND with a terpolymer p(Si-F-EG) (200 mg) in ethyl acetate (20 mL) was cast and cured at room temperature for a day and later at 120 °C for 12 h to give a thicker top layer (overall thickness ~200 µm). The blend films containing 4 wt% terpolymer (with respect to the PDMS matrix) in the top layer were named p(Si-F-EG)4. A film of PDMS alone was also prepared in the same way as a standard film.

Films of the pristine terpolymers were prepared by spin-coating (5000 rpm for 20 s) a filtered 3 wt% solution in chloroform and dried at room temperature for 12 h and then at 120 °C for 12 h (thickness ~200 nm).

2.4. Characterization

^{1}H-NMR and ^{19}F-NMR spectra were recorded with a Varian Gemini VRX300 spectrometer (Palo Alto, CA, USA) on CDCl$_3$ and CDCl$_3$/CF$_3$COOH solutions, respectively. Gel permeation chromatography (GPC) analyses were carried out using a Jasco PU–1580 liquid refractive index detector (Hachioji-shi, Tokyo, Japan). CHCl$_3$ was used as an eluent with a flow rate of 1 mL min^{-1} and poly(methyl methacrylate) standards were used for calibration.

Differential scanning calorimetry (DSC) analysis was performed with a Mettler DSC-30 instrument (Columbus, OH, USA) from −150 to 80 °C at heating/cooling rate of 10 °C min^{-1} under a dry nitrogen flow. The glass transition temperature (T_g) was taken as the inflection temperature in the second heating cycle.

Contact angles were measured by the sessile droplet (10 µL) method with a FTA200 Camtel goniometer (Portsmouth, VA, USA), using water (θ_w) (J. T. Baker, HPLC grade) as wetting liquid after 10 s from deposition.

Atomic force microscopy (AFM) experiments under ambient conditions were carried out in intermittent contact (tapping) mode with a Multimode system equipped with a Nanoscope IIIa controller (Veeco Instruments, New York, USA) using silicon cantilevers with a nominal force constant of 42 N m^{-1} from Olympus type OMCL-AC160TS (Tokyo, Japan) at a resonance frequency of about 320 kHz. The scan rate was kept at 1 Hz, while the tip–sample forces were carefully minimized to avoid artifacts. Tip radius of less than 7 nm (manufacturer's information) was used. To quantify the variation in the microscale structure of the coatings, the root-mean-square roughness (RMS) was determined over regions of 1 × 1 µm^2 and 10 × 10 µm^2:

$$\text{RMS} = \sqrt{\frac{1}{mn}\sum_{j=1}^{n}\sum_{i=1}^{m} Z^2(x_i, y_j)} \quad (1)$$

with Z the height and x, y the in-plane coordinates stored by the AFM software (version 6.13).

X-ray photoelectron spectroscopy (XPS) spectra were recorded by using a Perkin-Elmer PHI 5600 spectrometer (Chanhassen, MN, USA) with a standard Al-Kα source (1486.6 eV) operating at

300 W. The working pressure was less than 10^{-8} Pa. The spectrometer was calibrated by assuming the binding energy (BE) of the Au $4f_{7/2}$ line to be 84.0 eV with respect to the Fermi level. Extended (survey) spectra were collected in the range 0–1350 eV (187.85 eV pass energy, 0.4 eV step, 0.05 s step^{-1}). Detailed spectra were recorded for the following regions: C(1s), O(1s), Si(2p) and F(1s) (11.75 eV pass energy, 0.1 eV step, 0.1 s step^{-1}). The standard deviation (SD) in the BE values of the XPS line was 0.1 eV. The spectra were recorded at two photoemission angles φ (between the surface normal and the path taken by the photoelectrons) of 70° and 20°, corresponding to sampling depths d of ~3 nm and ~9 nm, respectively ($d = d_0 \cos\varphi$, where d_0 is the maximum information depth (d_0 ~10 nm for the C(1s) line)). The software used for background subtraction (Shirley type) [43] and quantitative analysis was the PHI software (version 5.2) for data collection in PHI 5600ci Multitechnique (Chanhassen, MN, USA). The atomic percentage was evaluated using the PHI sensitivity factors (considering both the cross-section and the beam out depth) with triplicate measurements on different film spots and the estimated experimental error was ±0.5% [44]. To take into account charging problems, the C(1s) peak was considered at 284.5 eV and the peak BE differences were evaluated. The XPS peak fitting procedure was carried out by means of Voigt functions and the results evaluated through the χ^2 function [45]. The deconvolution of XPS signals was performed with the software Igor Pro.

3. Results and Discussion

3.1. Synthesis of Terpolymers

Amphiphilic terpolymers p(Si-F-EG) with siloxane, fluorinated and ethoxylated side chains were prepared by free-radical polymerization of monomethacryloxypropyl-terminated poly(dimethyl siloxane) (Si), 1H,1H,2H,2H-perfluorooctyl acrylate (F) and polyethyleneglycol methyl ether methacrylate (EG). The terpolymers were characterized by a similar molar content of the three co-units, but differed in the average lengths, i.e. the average number degrees of polymerization of the EG and Si side chains, m ~4 (EGa) and m ~22 (EGb) and n ~11 (Sia) and n ~65 (Sib), while that of the F side chain was fixed (6 CF_2 groups) (Scheme 1). The formation of terpolymers was confirmed by ^{1}H-NMR, ^{19}F-NMR and FT-IR analyses. Their chemical composition was evaluated from the integrated areas of the signals at 0.5 ppm ($SiCH_2$ of Sia and Sib), 2.5 ppm (CH_2CF_2 of F) and 3.3 ppm (OCH_3 of EGa and EGb). Therefore, by alternatively changing the length of the siloxane and ethoxylated chains it was possible to modify the content of hydrophobic/hydrophilic co-units, that is the amphiphilic character, of the surface-active terpolymer and the eventual ability of the film surface to interact/reconstruct after immersion in water.

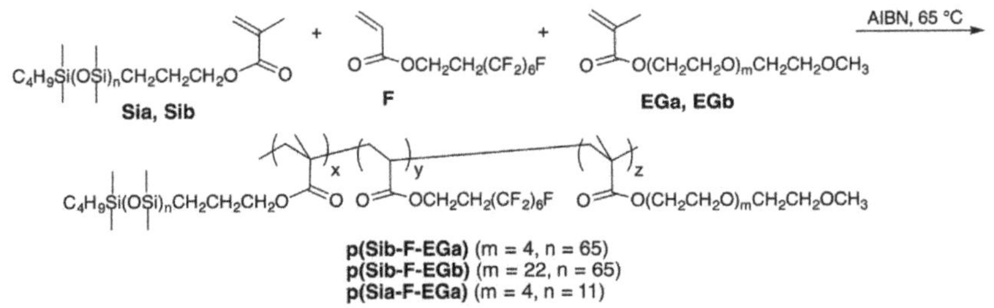

Scheme 1. Synthesis of amphiphilic terpolymers p(Si-F-EG) consisting of x, y, and z mol% co-units carrying Si, F, and EG side chains, respectively.

The thermal behavior of the terpolymers strictly depended on the type of the constituent co-units (Table 1). In particular, p(Sia-F-EGa) was completely amorphous and showed two glass transition temperatures (T_g) at −124 °C and −52 °C similar to those of the corresponding homopolymers p(Sia) and p(EGa), respectively. Differently, p(Sib-F-EGb) showed two melting transitions at −53 °C and 21 °C due to the longer Sib and EGb side chains, respectively, in addition to a T_g at −129 °C typical of the siloxane chain. The T_g of the EGb, expected at ca. −60 °C, was not detected because it was superimposed to the melting peak of the Sib (Figure S1). p(Sib-F-EGa) displayed an intermediate behavior showing only the two thermal transitions of the longer polysiloxane Sib, while the T_g of the shorter EGa was hidden by the melting peak of the Sib. These results indicate that all the terpolymers were microphase separated in the bulk in EG-rich and Si-rich domains, each of which displayed the thermal behavior of the respective homopolymer.

Table 1. Physical-chemical properties of terpolymers p(Si-F-EG).

Terpolymer	Composition [a] (mol%)	M_n [b] (g/mol)	M_w/M_n [b]	$T_{g,Si}$ [c] (°C)	$T_{g,EG}$ [c] (°C)	$T_{m,Si}$ [c] (°C)	$T_{m,EG}$ [c] (°C)
p(Sib-F-EGb)	26/45/29	21000	2.22	−129	-	−53	21
p(Sib-F-EGa)	22/52/26	21000	2.06	−129	-	−54	-
p(Sia-F-EGa)	19/58/23	9000	2.94	−124	−52	-	-

[a] Mole percentage of siloxane (Si), fluorinated (F) and ethoxylated (EG) side chains in the terpolymer. [b] By gel permeation chromatography (GPC). [c] Glass transition temperature and melting temperature of Si and EG.

The amphiphilic terpolymers were used as non-reactive, physically dispersed surface-active additives in condensation cured PDMS matrices. The terpolymers were dissolved in ethyl acetate together with the bis(silanol)-terminated PDMS matrix and the ES40 cross-linker, in the presence of bismuth neodecanoate as catalyst. The solution was cast on glass slides, previously modified by deposition of a thin layer (~2 μm) of cross-linked PDMS. This bottom thinner layer acted as a primer to improve adhesion between the glass substrate and the film formulation (overall thickness ~200 μm) to avoid delamination of the film during the subsequent tests upon immersion in water.

Each terpolymers were dispersed in a 4 wt% content with respect to the PDMS matrix in the top layer and the corresponding films were named p(Si-F-EG)4.

3.2. Surface Segregation of the Amphiphilic Surface-Active Terpolymer

Surface segregation of the terpolymer in the condensation-cured matrix films was investigated by AR-XPS at photoemission angles φ of 70° and 20°. The survey spectra of the samples did not show the presence of elements other than Si, C, O, and F (Figure 1). The XPS atomic surface compositions of the films p(Si-F-EG)4 are reported in Table 2 together with those of the respective terpolymers. For comparison, the calculated theoretical values for ideal homogeneous samples are also added in Table 2.

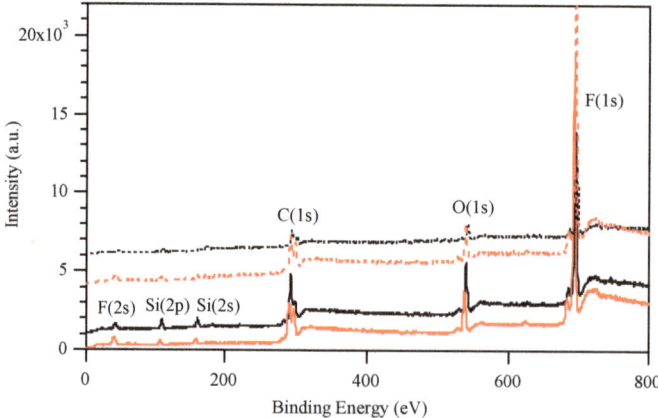

Figure 1. X-ray photoelectron spectroscopy (XPS) survey spectra for the amphiphilic polydimethylsiloxane (PDMS)-based film p(Sia-F-EGa)4 (dashed line) and the respective terpolymer p(Sia-F-EGa) (continuous line) at φ of 70° (red) and 20° (black).

Table 2. XPS atomic surface composition of amphiphilic PDMS-based films p(Si-F-EG)4 and respective terpolymers p(Si-F-EG). The errors on the values of composition were estimated to be ±0.5%.

Film	φ (°)	C (%)	O (%)	F (%)	Si (%)	C_{exp}/C_{theor}	O_{exp}/O_{theor}	F_{exp}/F_{theor}	Si_{exp}/Si_{theor}
	theor.	51.4	22.0	8.4	18.2				
p(Sib-F-EGa)	70	42.2	13.4	37.7	6.7	0.8	0.6	4.5	0.4
	20	45.2	20.3	20.5	14.0	0.9	0.9	2.4	0.8
	theor.	54.0	24.2	5.5	16.3				
p(Sib-F-EGb)	70	44.7	17.1	29.9	8.3	0.8	0.7	5.4	0.5
	20	47.0	22.3	17.3	13.4	0.9	0.9	3.1	0.8
	theor.	52.3	16.6	23.9	7.2				
p(Sia-F-EGa)	70	43.5	10.0	44.1	2.4	0.8	0.6	1.8	0.3
	20	45.7	14.3	34.8	5.2	0.9	0.9	1.5	0.7
	theor.	50.0	24.9	~0.4	24.7				
p(Sib-F-EGa)4	70	41.1	11.2	43.0	4.7	0.8	0.4	107.5	0.2
	20	45.9	18.8	23.3	12.0	0.9	0.8	58.3	0.5
	theor.	50.2	25.0	~0.2	24.6				
p(Sib-F-EGb)4	70	42.8	15.8	34.2	7.2	0.9	0.6	171.0	0.3
	20	48.0	22.7	16.8	12.5	1.0	0.9	84.0	0.5
	theor.	50.1	24.6	1.1	24.2				
p(Sia-F-EGa)4	70	42.6	10.1	45.4	1.9	0.8	0.4	41.2	0.1
	20	46.3	12.3	37.9	3.5	0.9	0.5	34.4	0.1

All the pristine terpolymer films showed a high enrichment in fluorine at the surface, the fluorine atomic percentage (F_{exp}) being higher than the theoretical value (F_{theor}) for all the samples. However, while F_{exp} was higher for p(Sia-F-EGa) with a larger F_{theor}, the F_{exp}/F_{theor} ratio was found to decrease from 5.4 and 4.5 down to 1.8 in going from p(Sib-F-EGb) and p(Sib-F-EGa) to p(Sia-F-EGa) ($\varphi = 70°$). Thus, a longer siloxane chain (Sib, n ~65) promoted a more effective surface segregation of the terpolymer than a shorter one (Sia, n ~11). Consistently, the Si_{exp}/Si_{theor} ratio followed the same trend, passing from 0.5 and 0.4 down to 0.3. Because of the great enrichment in fluorine in the top few nanometers, the atomic percentages of all the other elements C, O and Si were significantly lower than the theoretical ones. Moreover, the atomic percentages varied with φ. In particular, the F percentage markedly increased, whereas the C, O and Si percentages decreased with increasing φ. Thus, there was a composition gradient along the normal to the film surface into the bulk.

It was surprising that the surface chemical composition of the PDMS-based films containing 4 wt% terpolymer did not differ significantly from that of the respective pristine terpolymer and the F_{exp} was even slightly higher for all the PDMS-based films, despite the fact that PDMS was the largely major component (96 wt%) in all formulations. Thus, neither the thickness of the film nor the presence of the PDMS matrix affected the surface migration of the fluorinated side chains to a significant extent. Similar findings were reported for hydrophobic, i.e. not amphiphilic, (meth)acrylic copolymers containing siloxane and fluorinated side chains, for which a very effective surface segregation was found [9,46]. The observed surface enhancement in fluorine content resulted in an exceptionally high value of F_{exp}/F_{theor}, varying from 41.2 for p(Sia-F-EGa), to 107.5 for p(Sib-F-EGa) up to 171.0 for p(Sib-F-EGb). Accordingly, a strong decrease in Si_{exp} was also observed for all the films, and in particular for p(Sia-F-EGa)4 with the higher amount of fluorine at the surface and the shorter siloxane chains.

The C(1s) high resolution spectra for the PDMS-based films and the terpolymers are shown in Figure 2. In all cases the C(1s) signal was resolved in five contributions, as is shown for p(Sib-F-EGa)4, as a representative illustration, in Figure 3: (i) C−C and C−Si at 284.5 eV, (ii) C−O at 286.1 eV, (iii) C(=O)O at 289.1 eV, (iv) CF_2 at 291.7 eV and (v) CF_3 at 294.1 eV. From a qualitative point of view, the most striking finding was that the individual components of the C(1s) spectra of the blend films almost fully overlapped those of the respective terpolymers. This finding demonstrates that the amphiphilic terpolymer was completely located to the polymer−air interface of the PDMS-based films, independent of the length of the EG or Si chains, and the structural arrangement of each constituent at the molecular level was basically the same for the terpolymer whether alone or in the blend. Although the peaks (ii) to (v) had almost the same intensities in the blends and respective terpolymers, a decrease in the intensity of peak (i) was observed, especially for the formulations containing p(Sib-F-EGa) and p(Sib-F-EGb), which is in agreement with a better segregation of the F side chains to the topmost surface layers for these terpolymers, as already suggested by the F_{exp}/F_{theor} ratio (Table 3). Finally, one notes that the contribution due to C—O groups of the EG side chains was also remarkably higher for all the films than the theoretical value calculated for the respective terpolymers (Table 3). Thus, the EG side chains, in spite of their high surface energy, were pulled to the outer surface by the lowest surface energy F side chains. In particular, films with the shorter Ea chains displayed a larger surface enrichment in C—O groups (peak (ii), $C-O_{exp}/C-O_{theo}$ ~5 and ~3 for p(Sib-F-EGa) and p(Sia-F-EGa), respectively) with respect to films containing EGb ($C-O_{exp}/C-O_{theo}$ ~1.3), owing to the higher mobility of the shorter EGa chains. The dependence of the surface segregation of the EG chains on their chain length was also confirmed by the fact that the C—O percentage at $\varphi = 70°$ was higher than that at $\varphi = 20°$ for the films containing EGa and lower for those containing EGb (Table 3). Therefore, for copolymers containing EGa side chains the concentration of the EG chains within the outer ~3 nm of the polymer−air interface was maximized, in spite of their high surface energy (~43 mN m^{-1}). On the other hand, the intensity of the peak associated with the Si—C contribution markedly increased, while that of the CF_2 signal significantly decreased and that of the CF_3 signal completely disappeared in some cases (Figure S2).

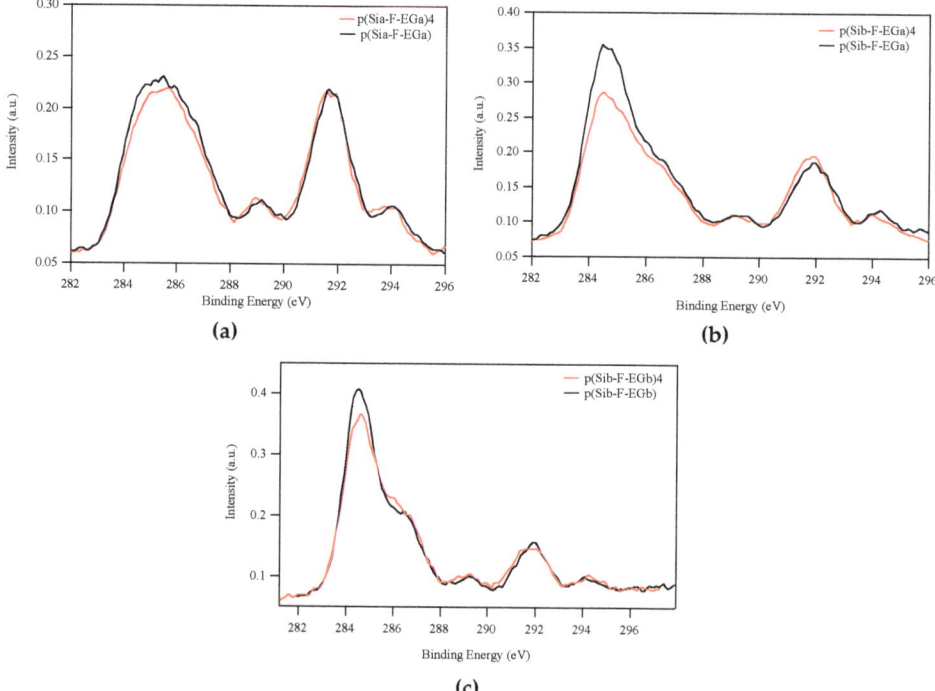

Figure 2. Area-normalized XPS C(1s) spectra ($\varphi = 70°$) for the amphiphilic PDMS-based films p(Si-F-EG)4 (red) and the respective terpolymers p(Si-F-EG) (black). (**a**) p(Sia-F-EGa)4 and p(Sia-F-EGa); (**b**) p(Sib-F-EGa)4 and p(Sib-F-EGa); (**c**) p(Sib-F-EGb)4 and p(Sib-F-EGb).

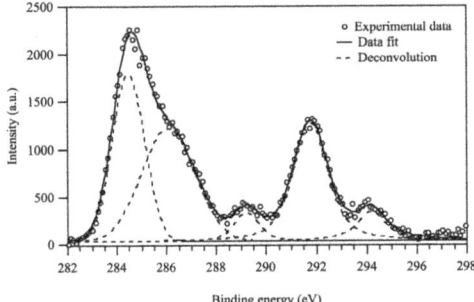

Figure 3. Deconvolution of the C(1s) XPS spectrum of p(Sib-F-EGa)4 (φ = 70°).

Table 3. Percent contributions of peaks (i)–(v) to the XPS C(1s) signal for the amphiphilic PDMS-based films p(Si-F-EG)4.

Film	φ (°)	Peak (i) (%)	Peak (ii) (%)	Peak (iii) (%)	Peak (iv) (%)	Peak (v) (%)
p(Sib-F-EGa)4[a,b]	70	29.1	32.3	4.8	27.2	6.6
	20	61.0	24.3	3.5	11.2	-
p(Sib-F-EGb)4[a,b]	70	44.5	32.0	4.4	14.4	4.7
	20	50.6	38.3	2.6	8.5	-
p(Sia-F-EGa)4[a,b]	70	14.6	38.0	6.0	34.7	6.7
	20	35.0	27.6	9.1	23.3	5.0

[a] Experimental percent contributions of peaks (i)–(v) for terpolymers p(Sib-F-EGa): 37.5%, 35.2%, 4.2%, 17.6%, 5.5%; p(Sib-F-EGb): 49.8%, 31.4%, 3.2%, 12.5%, 3.1%; p(Sia-F-EGa): 16.2%, 40.1%, 5.3%, 32.1%, 6.3%. [b] Theoretical percent contributions of peaks (i)–(v) for terpolymers p(Sib-F-EGa): 80.0%, 10.0%, 2.4%, 6.3%, 1.3%; p(Sib-F-EGb): 67.5% 26.0%, 1.8%, 3.9%, 0.8%; p(Sia-F-EGa): 49.3%, 23.5%, 6.1%, 17.6%, 3.5%.

3.3. Surface Composition after Immersion in Water

In order to evaluate the response of the amphiphilic PDMS-based films to the water environment, an AR-XPS analysis was also carried out on the films after immersion in water for 7 days. The chemical composition of these film surfaces is however indicative of the actual composition as it represents a kinetically trapped condition and not the equilibrium state reached by the polymer surface upon immersion in water. The atomic compositions of the surfaces after water immersion are collected in Table 4. Generally, the film surfaces were highly enriched in fluorine with respect to the theoretical amount and its percent content decreased with decreasing φ, thus showing that a composition gradient was maintained upon contact with water. On the other hand, the atomic percentages of all the other elements were lower than the theoretical ones and increased with decreasing φ. However, the chemical composition of the film surface as well as its modification upon immersion in water strictly depended on the type of terpolymer introduced in the formulation. In particular, the films containing p(Sib-F-EGb) underwent a surface reconstruction after immersion which resulted in a significant surface depletion in F moieties and enrichment in Si and O. A similar trend, although less marked, was observed for the film p(Sib-F-EGa)4, while the film p(Sia-F-EGa)4 did not display a significant variation in the atomic percentages and only showed a slight increase in F upon immersion in water. Consistently, the C(1s) signal of p(Sia-F-EGa)4 after immersion almost exactly overlapped the corresponding signal before immersion, with a slight increase in the intensity of the peaks (iv) and (v) due to the CF_2 and CF_3 groups (Figure 4c). For p(Sib-F-EGa)4 the C(1s) signals before and after immersion were very similar, with a slight increase in the intensity of peak (i) associated with the C−Si contribution (Figure 4a). However, in neither case was an increase in the C−O contribution (peak (ii)) detected, indicating that the amount of EG as well as F chains remained practically unchanged after immersion in water.

Table 4. XPS atomic surface composition of the amphiphilic PDMS-based films p(Si-F-EG)4 after 7 days of immersion in water. The errors on the values of composition were estimated to be ±0.5%.

Film	φ (°)	C (%)	O (%)	F (%)	Si (%)	C_{exp}/C_{theor}	O_{exp}/O_{theor}	F_{exp}/F_{theor}	Si_{exp}/Si_{theor}
	theor.	50.0	24.9	~0.4	24.7				
p(Sib-F-EGa)4	70	43.4	14.7	34.3	7.6	0.9	0.6	85.7	0.3
	20	47.8	20.0	19.9	12.3	1.0	0.8	49.7	0.5
	theor.	50.2	25.0	~0.2	24.6				
p(Sib-F-EGb)4	70	45.0	21.7	21.6	11.7	0.9	0.9	108.0	0.5
	20	48.9	24.6	11.2	15.3	1.0	1.0	56.0	0.6
	theor.	50.1	24.6	1.1	24.2				
p(Sia-F-EGa)4	70	40.3	10.2	47.3	2.2	0.8	0.4	43.0	0.1
	20	43.5	12.6	40.5	3.4	0.9	0.5	36.8	0.1

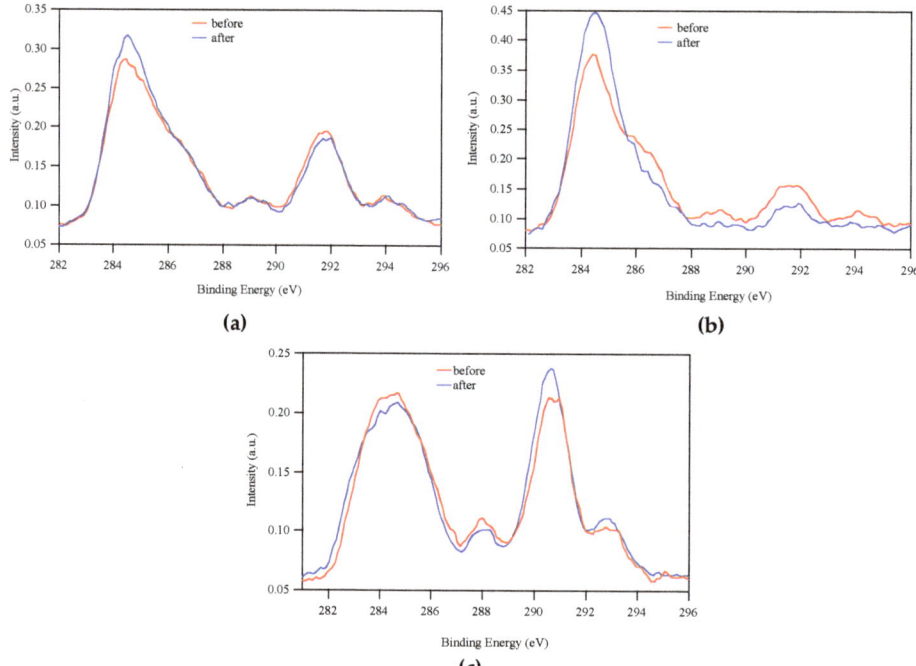

Figure 4. Area-normalized C(1s) XPS spectra ($\varphi = 70°$) for p(Sib-F-EGa)4 (**a**), p(Sib-F-EGb)4 (**b**) and p(Sia-F-EGa)4 (**c**) before (red) and after (blue) immersion in water.

The comparison of C(1s) signals before and after immersion for p(Sib-F-EGb)4 revealed the presence of significant differences, indicating that these films were subjected to a massive surface reconstruction as a result of the combination of Sib, EGb and F chains in the same chemical structure. The intensities of the peaks due to CF_3, CF_2 and $C(=O)O$ moieties markedly decreased, while that associated with C−Si increased (Figure 4b). Unexpectedly, the intensity of the signal due to the C−O groups of the longer EGb chains decreased upon immersion in water (Figure 4b). However, the C−O/($CF_2 + CF_3$) ratio increased (Table 5), indicating that the hydrophilic/hydrophobic balance, i.e. the amphiphilic character, of the film surface was larger after immersion.

Table 5. Percent contributions of peaks (i)–(v) to the XPS C(1s) signal for the amphiphilic PDMS-based films ($\varphi = 70°$) after immersion in water for 7 days.

Film	Peak (i) (%)	Peak (ii) (%)	Peak (iii) (%)	Peak (iv) (%)	Peak (v) (%)
p(Sib-F-EGa)4	33.7	30.4	4.6	24.6	6.7
p(Sib-F-EGb)4	66.7	27.1	-	6.2	-
p(Sia-F-EGa)4	13.2	38.3	4.2	37.3	7.0

Overall, it appears that, when F and EGa were combined in the same macromolecular structure, the maximum concentration of the EG chains at the outermost surface (cf. C(1s) signal percentages at $\varphi = 70°$ and $20°$ for p(Sib-F-EGa)4 and p(Sia-F-EGa)4) did not provide the necessary driving force for reconstruction, resulting in a chemical stability of the surface upon contact with water at least for the investigated time of 7 days. A very different process occurred when F and EGb were combined instead. A greater accumulation of the longer EGb chains in a region immediately below the outer surface (cf. C(1s) signal percentages at $\varphi = 70°$ and $20°$ for p(Sib-F-EGb)4) provided the right drive for surface reconstruction, resulting from the tendency of the EG chains to migrate to contact water. Actually, a reconstruction mechanism appears mainly to involve the migration of the hydrophobic fluorinated chains away from the polymer–water interface, rather than the migration of the hydrophilic ethoxylated chains toward the surface, as suggested by the reduction in the amount of C–O groups after immersion in water. This was possibly due to the effect of the concomitant presence of F chains that mechanically dragged part of the EG chains into the film bulk, thus preventing the effective migration of the latter to the polymer–water interface. However, the decrease in the F chains was much more marked than that in the EG ones, resulting in an increased C–O/(CF$_2$ + CF$_3$) ratio and consistently in a higher hydrophilicity of the whole system. This peculiar reconstruction mechanism differs from that generally reported for other amphiphilic copolymers containing ethoxylated and fluorinated components, which led to a substantial increase in surface concentration of the ethoxylated chains as a result of their major exposure to water [47–50].

In agreement with these last conclusions, measurements of static contact angle with water (θ_w) at different immersion times in water up to 6 days showed that θ_w decreased from $113° \pm 2$, $111° \pm 2$ and $106° \pm 2$ to $106° \pm 2$, $106° \pm 2$ and $84° \pm 2$ for p(Sib-F-EGa)4, p(Sia-F-EGa)4 and p(Sib-F-EGb)4, respectively. The last film surfaces underwent a more pronounced reconstruction becoming more hydrophilic after contact with water. Atomic force microscopy (AFM) measurements showed that all film surfaces were featureless and very smooth (RMS ~7 nm (1×1 µm^2)). Therefore, the effect of surface roughness on θ_w could be neglected.

4. Concluding Remarks

Novel surface-active amphiphilic terpolymers composed of a (meth)acrylic backbone with fluorinated (F), ethoxylated (EG) and siloxane (Si) side chains were engineered with variable lengths of EG and Si chains for one given length of F chains to create chemically modified PDMS-based film surfaces within the outermost few nanometers. AR-XPS analysis proved that all the PDMS-based films containing the terpolymer displayed a surface chemical composition close to that of the respective parent terpolymer, indicating that the presence of the PDMS matrix as the major component in the formulation did not inhibit the strong surface segregation of the terpolymer, which was responsible for the exceptionally high enrichment in fluorine of the outermost surface layers with respect to the bulk of the film. Even though the experimental amount of fluorine at the surface was lower for the sample p(Sib-F-EGb)4 consisting of both longer siloxane and ethoxylated side chains, the F_{exp}/F_{theor} ratio was the highest, indicating that the surface segregation process was the most effective for this terpolymer. The substantial surface segregation of the fluorinated chains produced an accumulation of the high surface energy ethoxylated chains at the polymer–air interface, thus resulting in an amphiphilic, chemically heterogeneous surface structure at the nanoscale level of the otherwise hydrophobic siloxane surface. In particular, the concentration of CH_2CH_2O groups within the first ~3 nm of the polymer surface was maximized for terpolymers containing the shorter and more mobile EGa side chains.

The sensitivity of the polymer films to the water environment depended on the structure of the terpolymer and in particular on the length of the EG side chains and their content at the polymer−air interface. In fact, for films p(Sia-F-EGa)4 and p(Sib-F-EGa)4, with the maximum percentage of CH_2CH_2O groups at the outermost surface layers ($\varphi = 70°$), the $C-O/(CF_3 + CF_2)$ ratio, namely an estimation of the hydrophilic/hydrophobic balance of the system, did not change upon immersion in water, since the chemistry of the film surface was stable upon immersion. On the other hand, for films of p(Sib-F-EGb)4, with a higher content of CH_2CH_2O groups in the bulk ($\varphi = 20°$), the $C-O/(CF_3 + CF_2)$ significantly increased showing that the film surface became more hydrophilic upon immersion in water. However, in no case was an enhancement detected in ethoxylated chain concentration at the surface after immersion. The reconstruction process involved a massive migration of the fluorinated tails away from the surface, which led to a marked increased in the amphiphilicity degree, although part of the ethoxylated chains were also concurrently dragged into the bulk.

The features of surface structure and reconstruction of amphiphilic polymer coatings are highly relevant to potential application, notably when distinct interactions with the water environment are involved. Simultaneous incorporation of opposite, hydrophobic and hydrophilic, functions into a surface-active terpolymer-based coating leads to a chemically heterogeneous and dynamically rearranging coating surface. Such coating is conceivably able to better resist adhesion from diverse fouling agents, such as bacteria, cells and organisms, especially those that exhibit contrasting preferences for hydrophilic, or otherwise hydrophobic, surface characters. One prime example is in the field of marine antifouling/fouling-release coating application.

Supplementary Materials: The following are available online at http://www.mdpi.com/2079-6412/9/3/153/s1, Figure S1: Differential scanning calorimetry (DSC) heating curves of the amphiphilic terpolymers p(Sib-F-EGa) and p(Sib-F-EGb). Figure S2: Area-normalized C(1s) XPS spectra of p(Sib-F-EGa)4 (a), p(Sib-F-EGb)4 (b) and p(Sia-F-EGa)4 (c).

Author Contributions: Conceptualization, G.G., E.G. and E.M.; methodology, G.G.; software, E.G. and A.G.; validation, E.M., E.G. and A.G.; formal analysis, E.M.; investigation, E.M., E.G. and A.G.; resources, G.G.; data curation, E.M. and E.G.; writing—original draft preparation, E.M.; writing—review and editing, G.G.; visualization, E.G.; supervision, G.G.; project administration, G.G. and E.M.; funding acquisition, G.G.

Funding: This research was funded by support from the University of Pisa (Progetti di Ricerca di Ateneo, PRA_2017_17).

Conflicts of Interest: The authors declare no conflict of interest.

References

1. Narrainen, A.P.; Hutchings, L.R.; Ansari, I.; Thompson, R.L.; Clarke, N. Multi-End-Functionalized polymers: Additives to modify polymer properties at surfaces and interfaces. *Macromolecules* **2007**, *40*, 1969–1980. [CrossRef]
2. Lee, H.; Archer, L.A. Functionalizing polymer surfaces by surface migration of copolymer additives: Role of additive molecular weight. *Polymer* **2002**, *43*, 2721–2728. [CrossRef]
3. Martinelli, E.; Galli, G.; Cwikel, D.; Murmur, A. Wettability and surface tension of amphiphilic polymer films: time-dependent measurements of the most stable contact angle. *Macromol. Chem. Phys.* **2012**, *213*, 1448–1456. [CrossRef]
4. Yasani, B.R.; Martinelli, E.; Galli, G.; Glisenti, A.; Mieszkin, S.; Callow, M.E.; Callow, J.A. A comparison between different fouling-release elastomer coatings containing surface-active polymers. *Biofouling* **2014**, *30*, 387–399. [CrossRef] [PubMed]
5. Lee, H.; Archer, L.A. Functionalizing polymer surfaces by field-induced migration of copolymer additives. 1. Role of surface energy gradients. *Macromolecules* **2001**, *34*, 4572–4579. [CrossRef]
6. Martinelli, E.; Fantoni, C.; Galli, G.; Gallot, B.; Glisenti, A. Low surface energy properties of smectic fluorinated block copolymer/SEBS blends. *Mol. Cryst. Liq. Cryst.* **2009**, *500*, 51–62. [CrossRef]

7. Inutsuka, M.; Yamada, N.L.; Ito, K.; Yokoyama, H. High density polymer brush spontaneously formed by the segregation of amphiphilic diblock copolymers to the polymer/water interface. *ACS Macro Lett.* **2013**, *2*, 265–268. [CrossRef]
8. Martinelli, E.; Hill, S.D.; Finlay, J.A.; Callow, M.E.; Callow, J.A.; Glisenti, A.; Galli, G. Amphiphilic modified-styrene copolymer films: antifouling/fouling release properties against the green alga *Ulva linza*. *Prog. Org. Coat.* **2016**, *90*, 235–242. [CrossRef]
9. Mielczarski, J.; Mielczarski, E.; Galli, G.; Morelli, A.; Martinelli, E.; Chiellini, E. The surface-segregated nanostructure of fluorinated copolymer-poly(dimethylsiloxane) blend films. *Langmuir* **2010**, *26*, 2871–2876. [CrossRef] [PubMed]
10. Sorgi, C.; Martinelli, E.; Galli, G.; Pucci, A. Julolidine-labelled fluorinated block copolymers for the development of two-layer films with highly sensitive vapochromic response. *Sci. China Chem.* **2018**, *61*, 947–956. [CrossRef]
11. Raffa, P.; Wever, D.A.Z.; Picchioni, F.; Broekhuis, A.A. Polymeric surfactants: Synthesis, properties, and links to applications. *Chem. Rev.* **2015**, *115*, 8504–8563. [CrossRef] [PubMed]
12. Galli, G.; Martinelli, E. Amphiphilic polymer platforms: Surface engineering of films for marine antibiofouling. *Macromol. Rapid Commun.* **2017**, *38*. [CrossRef] [PubMed]
13. Martin, C.; Aibani, N.; Callan, J.F.; Callan, B. Recent advances in amphiphilic polymers for simultaneous delivery of hydrophobic and hydrophilic drugs. *Ther. Deliv.* **2016**, *7*, 15–31. [CrossRef] [PubMed]
14. Nesvadba, P. Radical polymerization in industry. In *The Encyclopedia of Radicals in Chemistry, Biology and Materials*; John Wiley & Sons: Hoboken, NJ, USA, 2012.
15. Li, L.; Raghupathi, K.; Song, C.; Prasada, P.; Thayumanavan, S. Self-assembly of random copolymers. *Chem. Commun.* **2014**, *50*, 13417–13432. [CrossRef] [PubMed]
16. Matsumoto, K.; Terashima, T.; Sugita, T.; Takenaka, M.; Sawamoto, M. Amphiphilic random copolymers with hydrophobic/hydrogen-bonding urea pendants: Self-folding polymers in aqueous and organic media. *Macromolecules* **2016**, *49*, 7917–7927. [CrossRef]
17. Terashima, T.; Sugita, T.; Fukae, K.; Sawamoto, M. Synthesis and single-chain folding of amphiphilic random copolymers in water. *Macromolecules* **2014**, *47*, 589–600. [CrossRef]
18. Koda, Y.; Terashima, T.; Sawamoto, M. Multimode self-folding polymers via reversible and thermoresponsive self-assembly of amphiphilic/fluorous random copolymers. *Macromolecules* **2016**, *49*, 4534–4543. [CrossRef]
19. Guazzelli, E.; Masotti, E.; Biver, T.; Pucci, A.; Martinelli, E.; Galli, G. The self-assembly over nano- to submicro-length scales in water of a fluorescent julolidine-labeled amphiphilic random terpolymer. *J. Polym. Sci. Part A Polym. Chem.* **2018**, *56*, 797–804. [CrossRef]
20. Martinelli, E.; Guazzelli, E.; Galli, G.; Telling, M.T.F.; Dal Poggetto, G.; Immirzi, B.; Domenici, F.; Paradossi, G. Prolate and temperature-responsive self-assemblies of amphiphilic random copolymers with perfluoroalkyl and polyoxyethylene side chains in solution. *Macromol. Chem. Phys.* **2018**, *219*, 1800210. [CrossRef]
21. Galli, G.; Barsi, D.; Martinelli, E.; Glisenti, A.; Finlay, J.A.; Callow, M.E.; Callow, J.A. Copolymer films containing amphiphilic side chains of well-defined fluoroalkyl-segment length with biofouling-release potential. *RSC Adv.* **2016**, *6*, 67127–67135. [CrossRef]
22. Rufin, M.A.; Ngo, B.K.D.; Barry, M.E.; Page, V.M.; Hawkins, M.L.; Stafslien, S.J.; Grunlan, M.A. Antifouling silicones based on surface-modifying additive amphiphiles. *Green Mater.* **2017**, *5*, 4–13. [CrossRef]
23. Stafslien, S.J.; Christianson, D.; Daniels, J.; VanderWal, L.; Chernykha, A.; Chisholm, B.J. Combinatorial materials research applied to the development of new surface coatings XVI: Fouling-release properties of amphiphilic polysiloxane coatings. *Biofouling* **2015**, *31*, 135–149. [CrossRef] [PubMed]
24. Martinelli, E.; Del Moro, I.; Galli, G.; Barbaglia, M.; Bibbiani, C.; Mennillo, E.; Oliva, M.; Pretti, C.; Antonioli, D.; Laus, M. Photopolymerized network polysiloxane films with dangling hydrophilic/hydrophobic chains for the biofouling release of invasive marine serpulid *Ficopomatus enigmaticus*. *ACS Appl. Mater. Interfaces* **2015**, *7*, 8293–8301. [CrossRef] [PubMed]
25. Martinelli, E.; Pretti, C.; Oliva, M.; Glisenti, A.; Galli, G. Sol-gel polysiloxane films containing different surface-active trialkoxysilanes for the release of the marine foulant *Ficopomatus enigmaticus*. *Polymer* **2018**, *145*, 426–433. [CrossRef]
26. Noguer, A.C.; Olsen, S.M.; Hvilsted, S.; Kiil, S. Diffusion of surface-active amphiphiles in silicone-based fouling-release coatings. *Prog. Org. Coat.* **2017**, *106*, 77–86. [CrossRef]

27. Martinelli, E.; Gunes, D.; Wenning, B.M.; Ober, C.K.; Finlay, J.A.; Callow, M.E.; Callow, J.A.; Di Fino, A.; Clare, A.S.; Galli, G. Effects of surface-active block copolymers with oxyethylene and fluoroalkyl side chains on the antifouling performance of silicone-based film. *Biofouling* **2016**, *32*, 81–93. [CrossRef] [PubMed]
28. Patterson, A.L.; Wenning, B.; Rizis, G.; Calabrese, D.R.; Finlay, J.A.; Franco, S.C.; Zuckermann, R.N.; Clare, A.S.; Kramer, E.J.; Ober, C.K.; et al. Role of backbone chemistry and monomer sequence in amphiphilic oligopeptide- and oligopeptoid-functionalized PDMS- and PEO-based block copolymers for marine antifouling and fouling release coatings. *Macromolecules* **2017**, *50*, 2656–2667. [CrossRef]
29. Weinman, C.J.; Finlay, J.A.; Park, D.; Paik, M.Y.; Krishnan, S.; Sundaram, H.S.; Dimitriou, M.; Sohn, K.E.; Callow, M.E.; Callow, J.A.; et al. ABC triblock surface active block copolymer with grafted ethoxylated fluoroalkyl amphiphilic side chains for marine antifouling/fouling-release applications. *Langmuir* **2009**, *25*, 12266–12274. [CrossRef] [PubMed]
30. Gombotz, W.R.; Guanghui, W.; Horbett, T.A.; Hoffman, A.S. Protein adsorption to poly(ethylene oxide) surfaces. *J. Biomed. Mater. Res.* **1991**, *25*, 1547–1562. [CrossRef] [PubMed]
31. Krishnan, S.; Weinman, C.J.; Ober, C.K. Advances in polymers for anti-biofouling surfaces. *J. Mater. Chem.* **2008**, *18*, 3405–3413. [CrossRef]
32. Brady, R.F. Foul-release coatings for warships. *Def. Sci. J.* **2005**, *55*, 75–81. [CrossRef]
33. Lejars, M.; Marigaillan, A.; Bressy, C. Foul release coatings: A nontoxic alternative to biocidal antifouling coatings. *Chem. Rev.* **2012**, *112*, 4347–4390. [CrossRef] [PubMed]
34. Krishnan, S.; Wang, N.; Ober, C.K.; Finlay, J.A.; Callow, M.E.; Callow, J.A.; Hexemer, A.; Sohn, K.E.; Kramer, E.J.; Fischer, D.A. Comparison of the fouling release properties of hydrophobic fluorinated and hydrophilic PEGylated block copolymer surfaces: Attachment strength of the diatom *Navicula* and the green alga *Ulva*. *Biomacromolecules* **2006**, *7*, 1449–1462. [CrossRef] [PubMed]
35. Galli, G.; Martinelli, E.; Chiellini, E.; Ober, C.K.; Glisenti, A. Low surface energy characteristics of mesophase forming ABC, ACB triblock copolymers with fluorinated B blocks. *Mol. Cryst. Liq. Cryst.* **2005**, *441*, 211–226. [CrossRef]
36. Martinelli, E.; Galli, G.; Krishnan, S.; Paik, M.Y.; Ober, C.K.; Fischer, D.A. New poly(dimethylsiloxane)/poly(perfluorooctylethyl acrylate) block copolymers: Structure and order across multiple length scales in thin films. *J. Mater. Chem.* **2011**, *21*, 15357–15368. [CrossRef]
37. Wenning, B.M.; Martinelli, E.; Mieszkin, S.; Finlay, J.A.; Fischer, D.; Callow, J.A.; Callow, M.E.; Leonardi, A.K.; Ober, C.K.; Galli, G. Model amphiphilic block copolymers with tailored molecular weight and composition in PDMS-based films to limit soft biofouling. *ACS Appl. Mater. Interfaces* **2017**, *9*, 16505–16516. [CrossRef] [PubMed]
38. Martinelli, E.; Glisenti, A.; Gallot, B.; Galli, G. Surface properties of mesophase-forming fluorinated bicycloacrylate/polysiloxane methacrylate copolymers. *Macromol. Chem. Phys.* **2009**, *210*, 1746–1753. [CrossRef]
39. Sundaram, H.S.; Cho, Y.; Dimitriou, M.D.; Weinman, C.J.; Finlay, J.A.; Cone, G.; Callow, M.E.; Callow, J.A.; Kramer, E.J.; Ober, C.K. Fluorine-free mixed amphiphilic polymers based on PDMS and PEG side chains for fouling release applications. *Biofouling* **2011**, *27*, 589–602. [CrossRef] [PubMed]
40. Majumdar, P.; Stafslien, S.; Daniels, J.; Webster, D.C. High throughput combinatorial characterization of thermosetting siloxane-urethane coatings having spontaneously formed microtopographical surfaces. *J. Coat. Technol. Res.* **2007**, *4*, 131–138. [CrossRef]
41. Bodkhe, R.B.; Stafslien, S.J.; Cilz, N.; Daniels, J.; Thompson, S.E.M.; Callow, M.E.; Callow, J.A.; Webster, D.C. Polyurethanes with amphiphilic surfaces made using telechelic functional PDMS having orthogonal acid functional groups. *Prog. Org. Coat.* **2012**, *75*, 38–48. [CrossRef]
42. Martinelli, E.; Guazzelli, E.; Galli, G. Recent advances in designed non-toxic polysiloxane coatings to combat marine biofouling. In *Marine Coatings and Membranes*; Mittal, V., Ed.; Central West Publishing: Orange, NSW, Australia, 2019.
43. Shirley, D.A. High-resolution X-ray photoemission spectrum of the valence bands of gold. *Phys. Rev. B* **1972**, *5*, 4709. [CrossRef]
44. Moulder, J.F.; Stickle, W.F.; Sobol, P.E.; Bomben, K.D. *Handbook of X-ray Photoelectron Spectroscopy, Physical Electronics*; Perkin-Elmer Corp.: Eden Prairie, MN, USA, 1992.
45. McIntyre, N.S.; Chan, T.C. *Practical Surface Analysis 1*; Briggs, D., Seah, M.P., Eds.; Wiley: Chichester, UK, 1990; p. 485.

46. Marabotti, I.; Morelli, A.; Orsini, L.M.; Martinelli, E.; Galli, G.; Chiellini, E.; Lien, E.M.; Pettitt, M.E.; Callow, M.E.; Callow, J.A.; et al. Fluorinated/siloxane copolymer blends for fouling release: Chemical characterisation and biological evaluation with algae and barnacles. *Biofouling* **2009**, *25*, 481–493. [CrossRef] [PubMed]
47. Gudipati, C.S.; Greenleaf, C.M.; Johnson, J.A.; Pryoncpan, P.; Wooley, K.L. Hyperbranched fluoropolymer and linear poly(ethylene glycol) based amphiphilic crosslinked networks as efficient antifouling coatings: An insight into the surface compositions, topographies, and morphologies. *J. Polym. Sci. Part A Polym. Chem.* **2004**, *42*, 6193–6208. [CrossRef]
48. Krishnan, S.; Ayothi, R.; Hexemer, A.; Finlay, J.A.; Sohn, K.E.; Perry, R.; Ober, C.K.; Kramer, E.J.; Callow, M.E.; Callow, J.A.; et al. Anti-biofouling properties of comblike block copolymers with amphiphilic side chains. *Langmuir* **2006**, *22*, 5075–5086. [CrossRef] [PubMed]
49. Krishnan, S.; Paik, M.Y.; Ober, C.K.; Martinelli, E.; Galli, G.; Sohn, K.E.; Kramer, E.J.; Fischer, D.A. NEXAFS depth profiling of surface segregation in block copolymer thin films. *Macromolecules* **2010**, *43*, 4733–4743. [CrossRef]
50. Martinelli, E.; Pelusio, G.; Yasani, B.R.; Glisenti, A.; Galli, G. Surface chemistry of amphiphilic polysiloxane/triethyleneglycol-modified poly(pentafluorostyrene) block copolymer films before and after water immersion. *Macromol. Chem. Phys.* **2015**, *216*, 2086–2094. [CrossRef]

© 2019 by the authors. Licensee MDPI, Basel, Switzerland. This article is an open access article distributed under the terms and conditions of the Creative Commons Attribution (CC BY) license (http://creativecommons.org/licenses/by/4.0/).

Article

Argon Plasma Surface Modified Porcine Bone Substitute Improved Osteoblast-Like Cell Behavior

Cheuk Sing Choy [1,2], Eisner Salamanca [3], Pei Ying Lin [3], Haw-Ming Huang [1,3,4], Nai-Chia Teng [1,5], Yu-Hwa Pan [3,6,7,8,*] and Wei-Jen Chang [3,9,*]

1. Department of Community Medicine, En Chu Kong Hospital, New Taipei City 237, Taiwan; prof.choy@gmail.com (C.S.C.); hhm@tmu.edu.tw (H.-M.H.); dianaten@tmu.edu.tw (N.-C.T.)
2. Yuanpei University of Medical technology, Hsin Chu, Taipei 300, Taiwan
3. School of Dentistry, College of Oral Medicine, Taipei Medical University, Taipei 110, Taiwan; eisnergab@hotmail.com (E.S.); payinglin53@gmail.com (P.Y.L.)
4. Graduate Institute of Biomedical Materials & Tissue Engineering, College of Oral Medicine, Taipei Medical University, Taipei 110, Taiwan
5. Dental Department, Taipei Medical University Hospital, Taipei 110, Taiwan
6. Department of General Dentistry, Chang Gung Memorial Hospital, Taipei 105, Taiwan
7. Graduate Institute of Dental & Craniofacial Science, Chang Gung University, Taoyuan 333, Taiwan
8. School of Dentistry, College of Medicine, China Medical University, Taichung 404, Taiwan
9. Dental Department, Taipei Medical University, Shuang-Ho hospital, Taipei 235, Taiwan
* Correspondence: shalom.dc@msa.hinet.net (Y.-H.P.); cweijen1@tmu.edu.tw (W.-J.C.); Tel.: +886-2-2736-1661 (ext. 5148); Fax: +886-2-2736-2295 (Y.-H.P. & W.-J.C.)

Received: 14 December 2018; Accepted: 16 February 2019; Published: 19 February 2019

Abstract: Low-temperature plasma-treated porcine grafts (PGPT) may be an effective means for treating demanding osseous defects and enhance our understanding of plasma-tissue engineering. We chemically characterized porcine grafts under low-temperature Argon plasma treatment (CAP) and evaluated their biocompatibility in-vitro. Our results showed that PGPT did not differ in roughness, dominant crystalline phases, absorption peaks corresponding to phosphate band peaks, or micro-meso pore size, compared to non-treated porcine grafts. The PGPT Ca/P ratio was 2.16; whereas the porcine control ratio was 2.04 ($p < 0.05$). PGPT's [C 1s], [P 2p] and [Ca 2p] values were 24.3%, 5.6% and 11.0%, respectively, indicating that PGPT was an apatite without another crystalline phase. Cell viability and alkaline phosphatase assays revealed enhanced proliferation and osteoblastic differentiation for the cells cultivated in the PGPT media after 5 days ($p < 0.05$). The cells cultured in PGPT medium had higher bone sialoprotein and osteocalcin relative mRNA expression compared to cells cultured in non-treated porcine grafts ($p < 0.05$). CAP treatment of porcine particles did not modify the biomaterial's surface and improved the proliferation and differentiation of osteoblast-like cells.

Keywords: argon plasma treatment; porcine bone graft; biological apatite; chemical properties; in-vitro behavior

1. Introduction

Reconstruction of osseous defects, as performed in the fields of Periodontics, maxillofacial surgery, and orthopedics, can often be achieved by using autogenous bone. Autotransplantation of human bone has been documented since the early 19th century, and autologous bone grafts are considered the gold standard for bone regeneration due to their osteogenic, osteoinductive, osteoconductive, and osseointegrative characteristics [1,2]. However, it is challenging to use only autogenous bone when these defects are large, complex, or require adequate bone volume at the desired location. Availability (which in some cases is shallow) and autograft acquisitions carry a considerable patient burden, including additional surgical incisions, increased postoperative morbidity, weakened donor bone sites, and potential complications [3]. These adverse effects necessitate the use of alternative materials that mimic the physicochemical and biological performance of natural bone-derived apatite [4–7]. Examples of such alternatives include allografts, alloplastic materials, or xenografts [8]. Unfortunately, fresh bone allografts carry a risk of rejection by the host immune system. Alloplastic materials are either more complex or costly than biological apatite that is directly prepared from animal hard tissues (e.g., bovine, porcine, and cuttlefish bone) [9].

The first reported transfer of fresh bone during cross-species transplantation was described by van Meekeren in 1668 [10]. Nowadays, porcine bone is considered to most closely resemble human bone in terms of their macrostructure and microstructure [11], chemical composition, and remodeling rate [12]. Porcine grafts are also abundant, making it an excellent candidate bone graft material for reconstructing osseous defects [13]. To improve the performance of the porcine bone graft material, various attempts have been made in order to modify the graft material. For example, the addition of fluoride [9] may enable both the fluorapatite formation as well as the antibacterial function of the materials [14].

Plasma discharge applications are based on different geometries and use various electrode materials [15]. In plasma discharge, a source gas is dissociated and ionized, and various particles (e.g., electrons, ions, atoms, radicals, and ultraviolet (UV)) are inactivated following contact with a biological system. Common gases such as Ar, N_2, and O_2 are used for gas plasma sterilization, thus assuring that no toxic chemicals remain on the objects after treatment [16]. A much-debated topic in the field of tissue engineering is how to define the plasma treatment "dose". Plasma appears to operate over multiple pathways, and the type of device and target tissue will affect the treatment outcomes [17].

Tissue engineering in reconstructive surgery demands the 3-dimensional (3D) growth of osteoblasts and chondroblasts onto suitable carriers [2,18]. This necessitates the fabrication of multifunctional scaffolds that meet structural, mechanical, and nutritional requirements. These scaffolds are used to direct 3D tissue ingrowth for repairing large, complex, and multi-tissue defects [19,20]. A 2017 study found a lack of common understanding regarding the interaction between plasma and living cells, tissues, and organisms. This knowledge gap represents a significant obstacle to developing large-scale clinical trials of plasma-related medical devices and procedures [17]. The use of plasma-treated porcine grafts (PGPT) could enable better 3D reconstruction of challenging osseous defects. PGPT may enhance our understanding of plasma-tissue engineering further. Consequently, we sought to physically and chemically characterize porcine grafts under low-temperature Argon plasma treatment, described previously as cold atmospheric plasma (CAP) [21], and evaluate their biocompatibility in-vitro.

2. Materials and Methods

2.1. Sample Preparation

Porcine grafts were described previously in another study performed by the same laboratory. Briefly, the porcine graft particles ranged from 500–1000 μm in size and consisted of cortical porcine bone, originating from special pathogen-free (SPF) porcine long bones (Agriculture Technology Research Institute, Hsinchu, Taiwan). First, the bones were immersed in 0.5 N hydrochloric acid (HCl) for 60 min to remove the organic matrix. Subsequently, the porcine bones were heated with a ramp rate of 5 °C/30 s, and were settled at 800 °C for 2 h. Afterward, for another 2 h, with a different ramp rate of 5 °C/min, the grafts were settled at 1000 °C to remove residual soft tissues and proteins. Next, the pieces were cooled to room temperature, milled to 500–1000 μm particle size, and sterilized using γ-rays [11]. Forty samples, measuring 200 mg each, were used for the control porcine grafts (porcine control) and PGPT.

2.2. Low-Temperature Argon Plasma Treatment

The Plasma Jet (PJ; AST Products, Inc., North Billerica, MA, USA) was used to subject porcine particles to low-temperature Argon plasma treatment. The APT was carried out at a power level of 80 W, with a frequency of 13.56 MHz, under a 100 m Torr of pressure over 15 min with Argon plasma placed 10 mm from the porcine graft particles.

2.3. Surface Topography Evaluations

Scanning electron microscopy (SEM) images (SU3500, Hitachi, Ltd., Kyoto, Japan) were used to evaluate and compare the surface morphologies of PGPT and the porcine control. Sample particles were prepared for imaging using 25 nm-thick layers of Au/Pd sputter surface coating with a sputtering apparatus (IB-2; Hitachi, Ltd., Tokyo, Japan). High-resolution images were taken at a low pressure and an accelerating voltage of 15.0 kV. The electron beam was focused to a fine point (primary electron) and magnified to 500× and 1500× by secondary electrons.

2.4. Energy Dispersive Spectrometry

The samples' surface topography probe measurements were taken with an electron beam covering a 70 μm spectrum, operating at 15 kV. An Energy Dispersive X-Ray Micro Analyzer machine was used for the evaluation (EX-250, HORIBA, Kyoto, Japan). Energy dispersive spectrometry (EDS) was useful for analyzing the percentage weight for each element of the PGPT and porcine control particle grafts ($n = 6$).

2.5. X-Ray Photoelectron Spectroscopy

Elemental and chemical surface analyses were performed on PGPT and porcine controls using X-ray Photoelectron Spectroscopy (XPS) measurements, following the protocol by Silversmit et al. The measurements were recorded with a Perkin–Elmer Phi ESCA 5500 system equipped with a monochromated 450 W Al Kα source (Quantera II, ULVAC-PHI Inc, Kanagawa, Japan). The base pressure of the ESCA system was less than 1×10^{-7} Pa. All experiments were recorded with a 220 W source power and an angular acceptance of $\pm 7°$. The analyzer axis was oriented at an angle of 45° with the specimen surface. Wide-scan spectra were measured over a binding energy range of 0–1400 eV and a pass energy of 187.85 eV. We recorded the C 1s, O 1s, Ca 2p, and P 2p core levels [22].

2.6. X-Ray Diffraction Analysis

Powder X-ray diffraction (XRD) was used to analyze the crystalline structures and chemical compositions of PGPT and porcine controls. Diffraction patterns were collected on a PANalytical X'Pert3 Pro system (X'Pert3 Powder, PANalytical Co. Ltd., Almelo, The Netherlands) operated at 60 kV and a 45 mA current with Mo Kα (0.71073 Å) source. Samples were scanned over a range of $10° \leq 2\theta \leq 70°$.

2.7. Fourier Transform Infrared Spectroscopy (FTIR) Characterization

Perkin–Elmer Spectrum One Fourier Transform Infrared spectroscopy (Perkin–Elmer Corp., Waltham, MA, USA) was used to collect all spectra. Dry PGPT and porcine control particles were equilibrated at 50% relative humidity, at room temperature, and clamped directly onto the crystal for analysis [23]. The nominal resolution of 4.00 and a number of sample scans equal to 1000 was collected in a range of 450–4000 cm^{-1}. A computer running Perkin–Elmer 3.02 software was used to record data.

2.8. Cell Culture and Seeding

MG-63 osteoblast-like cells were purchased from the Bioresource Collection and Research Center (BCRC, Hsinchu, Taiwan). The cells were expanded in Dulbecco's modified Eagle's medium (DMEM; HyClone, Logan, UT, USA) supplemented with L-glutamine (4 mmol/L), 10% fetal bovine serum, and 1% penicillin-streptomycin at 37 °C in a humidified atmosphere containing 95% air and 5% CO_2. The confluent cells were sub-cultured to the next passage using 0.05% trypsin—EDTA, up to passage 4. Once 90% confluence cell density was reached, the concentration was adjusted to 1×10^4 cells/mL and the samples were aliquoted into 24-well Petri dishes (Nunclon; Nunc, Roskilde, Denmark). The same day, DMEM medium was mixed with CPG, porcine graft, or HA/β-TCP at a concentration of 1 g/10 mL. Twenty-found hours later, each test well's medium was removed and substituted for the test media, consisting of the previously described DMEM + CPG, porcine graft, or HA/β-TCP. The same DMEM media first described was used for the control wells. The medium in all wells was changed every 3 days.

2.9. Cell Cytotoxicity

Cell cytotoxicity was assessed on days 1, 3, and 5 after adding medium and particle bone substitute to the test wells. Tests were performed according to the Cell Proliferation Reagent manufacturer's instructions (WST-1 Kit, Roche Applied Science, Mannheim, Germany). In brief, the cell medium, prepared as described previously, was replaced with 500 µl of fresh medium. Later, the cells were moved to a 96 well microtiter plate (5×10^4 cells/well) for a final volume of 100 µL of the culture medium, from which any remaining particle grafts were absent. Afterward, the cells were incubated for 24 h and 10 µL WST-1 reagent was added to each well, after which the cells were again incubated for 4 h in the same standard culture conditions. Next, each plate was settled for 1 min on a shaker to mix its contents. Subsequently, we used a microplate reader at OD = 420–480 nm, a reference wavelength of 650 nm to measure absorbance samples. The percent cytotoxicity was calculated from the following equation: % cytotoxicity = $(100 \times (control-sample))/control$ [24]. Each test was repeated 3 times ($n = 6$).

2.10. Alkaline Phosphatase Assay

After cell culture and seeding, alkaline phosphatase activity was performed on days 1, 3, and 5. The cells were washed twice with phosphate-buffered saline (PBS). PBS was removed using suction, and 300 µL of Triton X-100 (BioShop, Canada Inc. Burlington, ON, Canada) was added at a concentration of 0.05%. To induce rupture, the cells were subjected to 3 cycles of 5 min at 37 °C and 5 min at −4°C, after which the samples were placed into 96-well plates. Alkaline phosphatase (ALP) activities were determined by following the Thermo Scientific 1-Step p-nitrophenyl phosphate disodium salt (PNPP) manufacturer's instructions. PNPP was supplied pre-mixed with a substrate buffer and ready-to-use at room temperature. The 1-Step PNPP was gently mixed. Next, 100 µL of the mixture was added to each 96-well and mixed thoroughly by gently agitating the plate. The 96-well plates were incubated at room temperature for 30 min. To stop the reaction, 50 µL of 3M NaOH was added and mixed thoroughly by gently agitating the plate. The absorbance of each well was measured at 405 nm ($n = 6$) using a Multiskan™ GO Microplate Spectrophotometer (Thermo Fisher Scientific, Waltham, MA, USA). Enzymatic activity was normalized to total protein concentration using bovine serum albumin (BSA; Roche, Basel, Switzerland). The standard Bradford (Sigma) method was used to do protein measurements. The ALP activity was compared by plotting OD intensity [25].

2.11. Real-Time Polymerase Chain Reaction (PCR)

The assay for 5 days was done after the cell culture and seeding protocol. A NanoDrop ND-1000 spectrophotometer (CapitalBio Nano Q, Beijing, China) was used to quantify the total RNA. Later, RNA was processed with the Novel Total RNA Mini Kit, Cat. No. NR-200 (NovelGene, Molecular Biotech, Taipei, Taiwan) according to the manufacturer's instructions. The cells were trypsinized, harvested, and resuspended in 100 µL PBS and subjected to cell lysis by adding 400 µL NR Buffer and 4 µL S-mercaptoethanol to the sample. RNA binding was performed with 400 µL 70% ethanol and centrifuged at 13,000 rpm. Afterward, the sample was washed and eluted with 50 µL RNase-free water.

Real-time polymerase chain reaction (PCR) was used to quantify expression. Gene expression levels were normalized to the expression of the housekeeping gene GAPDH and expressed as fold changes, relative to the expression of the cell culture in DMEM only. The delta-delta calculation method was used to perform quantification. A Primer-BLAST from the United States National Library of Medicine was used to design forward and reverse primers and probes for bone sialoprotein (Bsp) and osteocalcin (OC) genes [26].

2.12. Statistical Analyses

All measurements are presented as mean ± standard deviation, and normality of the results was analyzed using The Jarque–Bera test. Differences between the PGPT and the porcine control groups were identified using the Student t-test and considered significant when $p < 0.05$. Microsoft Excel Professional Plus 2016 (Microsoft Software, Redmond, WA, USA) was used to perform all data analyses.

3. Results and Discussion

The aim of the present study was to physically and chemically characterize porcine grafts under low-temperature Argon plasma treatment and evaluate their biocompatibility in-vitro. Porcine grafts were used previously in animal studies (e.g., Bone Regeneration for Sheep's Iliac Crestal Defects), where a corticocancellous porcine bone with a 250–1000 microns particulate mix was used as a scaffold to induce bone regeneration. Here, porcine bone was found to be highly biocompatible and capable of inducing faster and greater bone formation. After four months of healing, bone formation was found around the graft particles within the defects [27].

Porcine grafts have also been used successfully in human studies. Successful in-socket preservation, with the combination of a porcine xenograft and collagen membrane, was utilized to maintain successfully the vertical and horizontal dimensions of four-wall bone defects. The authors reported sufficient bone volume for implant placement in all sites prior to implant placement [28]. In another multicenter single-blind randomized control trial, pre-hydrated collagenated cortico-cancellous porcine bone was compared with cortical porcine bone. During the 4-month analysis, both test groups showed reduced bone loss compared to naturally healing sockets. However, the two grafting materials were not able to preserve the alveolar crest, and final results were approximately 30% less than the estimates after healing [29]. Another similar multicenter randomized controlled clinical trial showed that porcine bone (used with guided bone regeneration) was more effective than the control group without grafting after tooth extraction [30].

3.1. SEM Surface Morphological Observations

No significant differences were revealed on the surface morphology of the PGPT and non-treated porcine control surfaces. In addition, PGPT showed no difference in roughness, characterized by isotropic and irregular surface patterns, nor a micro-meso pores size, compared to the porcine control. The PGPT particle surface (Figure 1A,B) had the attributes of the non-treated porcine particle (Figure 1C,D), with rough surfaces and microporous structures.

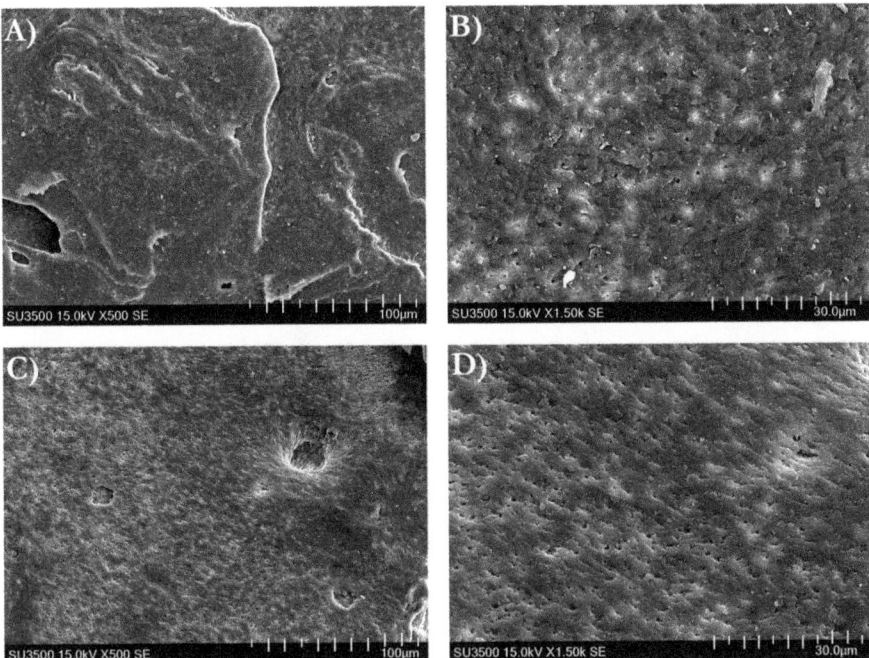

Figure 1. The scanning electron microscope images of the representative surface morphologies. (**A**) The rough surface of the plasma-treated porcine graphs (PGPT) is seen at 500× magnification. (**B**) PGPT micro- and mesopores are seen at 1500× magnification. (**C**) Porcine control similar to rough surface PGPT is seen at 500× magnification. (**D**) Porcine control with the same size micro- and mesopores as the PGPT are seen at 1500× magnification.

3.2. Element's Weight Percentage on Particles Surfaces

EDS analyses showed that PGPT particle element weight (wt %) contained 37.19 ± 4.93 wt % calcium, 34.37 ± 8.38 wt % oxygen, 17.19 ± 1.79 wt % phosphorus, 4.75 ± 4.36 wt % gold, 4.54 ± 2.62 wt % carbon, and 0.60 ± 0.10 wt % magnesium. By comparison, elemental concentrations in porcine control particles had less calcium with only 30.69 ± 4.11 wt % ($p < 0.05$), 15.03 ± 0.38 wt % phosphorus ($p < 0.05$), 31.13 ± 6.98 wt % oxygen, and 1.13 ± 0.17 wt % sodium compared to the plasma-treated porcine particles. The porcine control had a higher level of carbon (6.02 ± 6.49 wt %), magnesium (1.28 ± 0.95 wt %), and gold (14.72 ± 2.76 wt %). The Ca/P ratio in the PGPT particles was 2.16, higher than the hydroxyapatite (HA) value. In contrast, the Ca/P ratio of the porcine control was 2.04, with a statistically significant difference between both Ca/P ratios ($p < 0.05$) (Table 1, Figure 2).

Table 1. The results of the elemental analysis by energy dispersive spectrometry.

Element Weight %	PGPT	Porcine Control
Ca	37.19 ± 4.93	30.69 ± 4.11
P	17.19 ± 1.79	15.03 ± 0.38
O	34.37 ± 8.38	31.13 ± 6.98
C	4.54 ± 2.62	6.02 ± 6.49
Mg	0.60 ± 0.10	1.28 ± 0.95
Au	4.75 ± 4.36	14.72 ± 2.76
Na	1.35 ± 0.53	1.13 ± 0.17

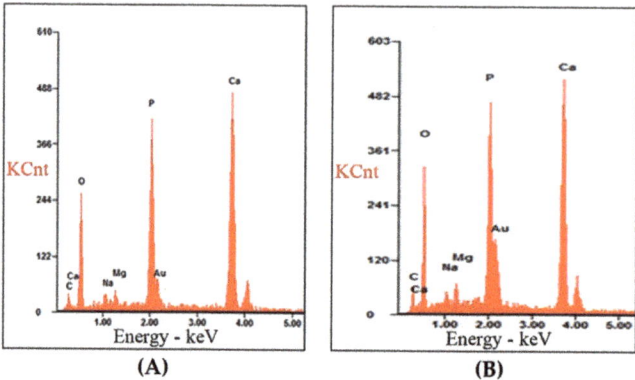

Figure 2. The energy dispersive spectra. (**A**) PGPT. (**B**) Porcine control.

3.3. XPS Peak Fitting Results

Surface chemistry and atomic concentrations (as percentages), using XPS analysis, showed that the C 1s values were 24.3% for the PGPT and 28.5% for the porcine control, indicating a decrease in the number of carbonate groups built into the HAp structure. Moreover, the O 1s values were 51.1% (PGPT) and 47.9% (control). The P 2p values were 5.6% (PGPT) and 7.1% (control). The mean Ca 2p values were 11.0% (PGPT) and 11.11% (control). The Mg 1s levels were 6.1% (PGPT) and 2.3% (control). Na 1s was lower (1.4%) for PGPT compared with the porcine control (3.2%) (Figure 3).

XPS analysis supported the EDS results and indicated that PGPT values were in concordance with Seo et al. who fabricated HA bioceramics from recycled pig bones by heating to 1000 °C. The HA was composed of calcium, phosphate, carbon and magnesium ions in similar percentage concentrations as the present study's results [31]. In the present study, the Ca/P ratio for the PGPT particles was 2.16, higher than the HA value. In contrast, the Ca/P ratio of the porcine control was 2.04, with a statistically significant difference between both Ca/P ratios ($p < 0.05$). (Table 1, Figure 2) Both Ca/P ratios correspond to high crystalline apatite particles [32,33]. The PGPT higher Ca/P ratio over porcine control ($p < 0.05$) can be directly related to the low-temperature Argon plasma treatment. The Ca/P ratio of both porcine grafts in the present study was higher than the one found by other authors. This can be attributed to the lower temperatures used during the porcine graft preparation in the other studies [31,34].

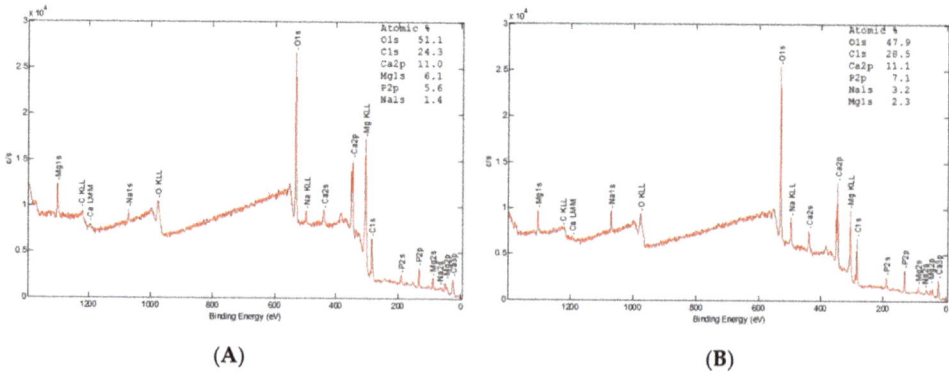

Figure 3. The X-ray photoelectron spectroscopy (XPS) spectra used to determine atomic composition (%). (**A**) PGPT. (**B**) Porcine control. Both particle grafts present Ca, P, O, and C.

3.4. X-Ray Diffraction Patterns

XRD measurements were compared to diffraction patterns of pure hydroxyapatite, attributing the peaks only to this calcium apatite [31,34]. The porcine control graft remained highly crystalline, as exhibited by the same sharp peak patterns obtained before and after the CAP treatment. The higher-intensity peaks in the patterns correspond to that found in apatites with high crystallinity (Figure 4).

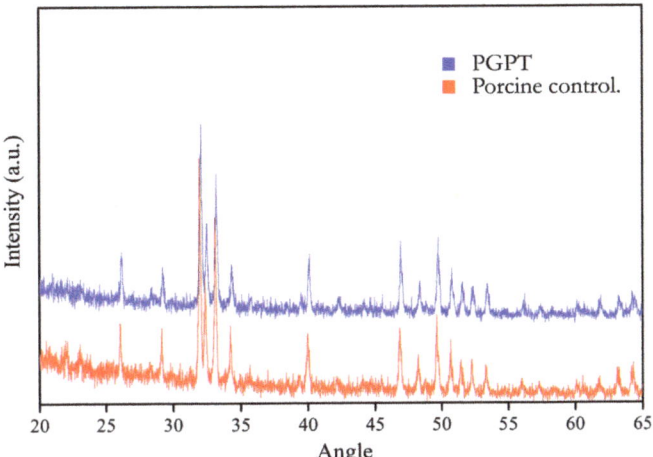

Figure 4. The X-ray diffraction patterns. Both porcine particle grafts have the same dominant crystalline phases generating the same intense, sharp peaks.

3.5. Fourier Transform Infrared Spectroscopy Profiles

FTIR spectra of sintered PGPT and porcine control particles revealed similar results to those found in the XRD analysis. All samples were characterized as apatite without any other types of crystalline phases. Both samples had similar patterns, with pronounced peaks between 473 and 700 cm^{-1} [35], and absorption peaks ascribed to phosphatase band peaks at 962, 1051, and 1089 cm^{-1} (Figure 5).

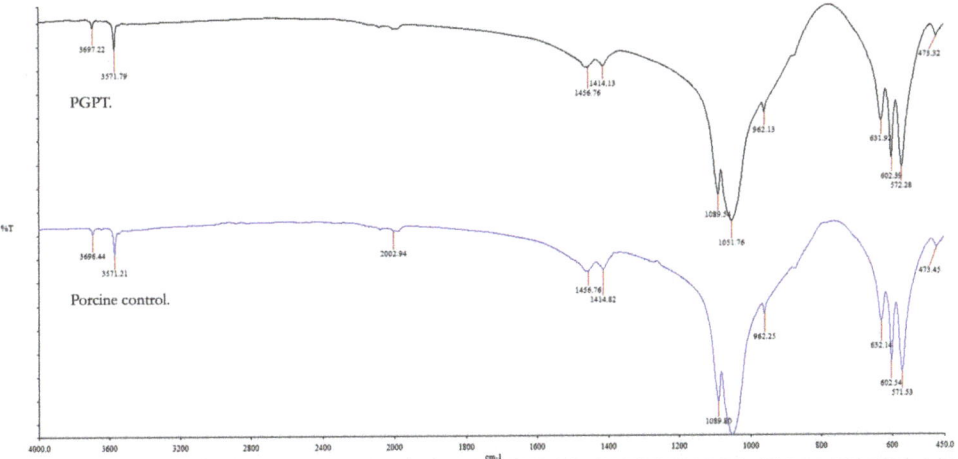

Figure 5. The Fourier Transform Infrared Spectroscopy Characterization. Fourier Transform Infrared Spectroscopy (FTIR) spectra of both PGPT and the porcine control with the same absorption peaks corresponding to the phosphate band peaks.

The results of the FTIR and XRD characterization analysis of sintered PGPT and porcine control particles agreed with a study done by Figueiredo et al. which compared the physicochemical properties of bone substitutes used in dentistry vs. calcined human bone. In their study, the diffractograms in the XRD of the porcine graft and natural human bone spectrum presented the more intense characteristic peaks of HA, with coincident peak positions and relative intensities. In their FTIR results, porcine and human graft results showed very similar spectra to the typical bands originated by the HA mineral. More intense phosphate stretching bands were observed around 1010 and 560 cm^{-1} [36]. These were very similar to the FTIR results in the present study, where the same phosphate bands were observed at around 1051 and 572 cm^{-1}.

3.6. Cell Proliferation Assessment

Cell proliferation OD values on the porcine CAP treated particles were 0.9 ± 0.05 (day 1), 1.57 ± 0.02 (day 3), and 1.61 ± 0.08 (day 5). Conversely, cell proliferation OD values on the non-treated porcine particles were 1.04 ± 0.04 (day 1), 1.38 ± 0.03 (day 3), and 1.52 ± 0.01 (day 5). Cells cultured in the control media had OD values of 1.13 ± 0.04 (day 1), 1.46 ± 0.03 (day 3), and 1.41 ± 0.02 (day 5). MG-63 cells presented with spreading attachments on the PGPT surfaces, improving cell proliferation compared to the non-treated porcine particle surfaces on days 3 and 5 ($p < 0.05$) (Figure 6).

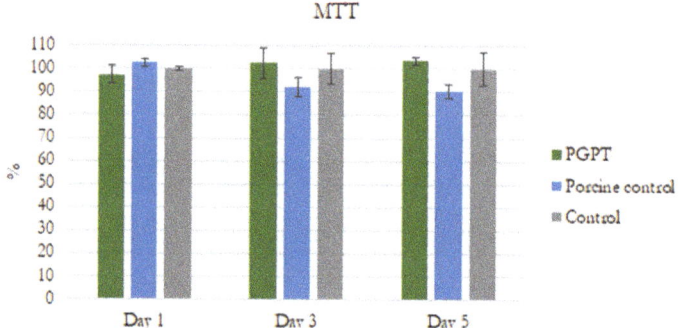

Figure 6. The cell proliferation at days 1, 3, and 5 ($p < 0.05$).

3.7. Osteoblast Differentiation.

During the time-dependent analysis, the PGPT exhibited higher ALP activity. Significantly greater differences were observed with PGPT than with the non-treated porcine particles and controls on days 3 and 5 ($p < 0.05$) (Figure 7).

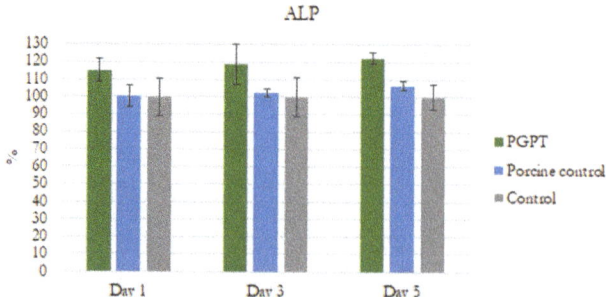

Figure 7. The Alkaline Phosphatase (ALP) analysis. Cold Atmospheric Plasma (CAP) porcine treated particles stimulated more alkaline phosphatase production than did porcine control particles in osteoblast-like cells ($p < 0.05$).

After 5 days, the results of real-time PCR revealed that cells cultured with PGPT and porcine control, compared with the control cells, had elevated Bsp and OC genes. In addition, cells cultured with PGPT had a higher relative mRNA expression for both genes than did the cells cultured with porcine control ($p < 0.05$) (Figure 8). To the best of our knowledge, this is the first time that bone sialoprotein and osteocalcin genes have been measured in cells cultured with PGPT. For a better understanding of the porcine bone graft bio-physical surface treatment results within osteoblast-like cells, further analyses of these genes, together with other genes, are necessary.

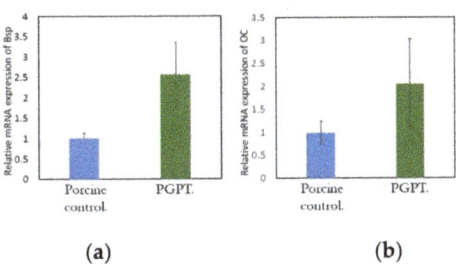

Figure 8. The real-time PCR. (**A**) Bone sialoprotein mRNA expression of cells cultured at 5 days. (**B**) Osteocalcin of cells cultured at 5 days ($p < 0.05$).

Plasma has gained great scientific and industrial importance. A great advantage of using low-temperature plasmas under atmospheric pressure is our ability to chemically "design" the plasma. For example, plasma composition can be varied depending on the desired effect. In addition, plasma acts rapidly, effectively, and penetrates the smallest openings and hollow spaces [37]. Surface modification of polymers with low pressure is often used to improve coating adhesion, wettability, printability, biocompatibility, and other surface-related properties of polymers or other materials [38,39]. Plasma treatment has also been used for titanium surface modification [40,41]. Sarma et al. examined the biomimetic growth of HA nanocrystals on Ti and sputtered TiO_2 substrates [42]. Yi et al. prepared akermanite ($Ca_2MgSi_2O_7$) bioactive coatings using a plasma-spraying technique, with bioactive ceramic coatings on titanium (Ti) alloys [43]. The key question is how to best design and operate the plasma source for optimal biological applications, and this question remains a challenge in the field. This challenge has two parts: first, the physics and chemistry of different plasma devices are far from being fully understood. Second, the mechanisms by which plasma alters biological cells, tissues, and organisms are not well-established [17]. In the present study, we wanted to focus on the effects of setting Argon plasma treatment with a power level of 80 W, a frequency of 13.56 MHz, under a 100 mTorr pressure. During 15 min within low-temperature Argon plasma at a distance of 10 mm from the porcine graft particles, we found that this treatment did not modify the biomaterial's surface (Figures 1–5); however, it improved osteoblast-like cell proliferation and differentiation (Figures 6–8). This could be explained by the removal of toxic elements (e.g., carbon) from the particles' surfaces, as EDS and XPS showed reduced carbon in PGPT. Another explanation could be the removal of micro-nano particles left during porcine particle production. Argon plasma treatment cleansed the porcine particles during their production process and led to stronger and faster interactions with cells.

In previous studies, porcine particle grafts exhibited osteoconductive properties of a bioactive bone graft material: able to provide an appropriate scaffold, allowing vascularization, promoting calcified tissue deposition, cellular infiltration, and attachment [44,45]. In the present study, PGPT produced superior results compared to the porcine control. Therefore, depending on clinical needs, PGPT can potentially be put to use in daily clinical practice [11,13]. Further studies are needed to achieve a better understanding of low-temperature plasma treatment over porcine particles for bone regeneration.

4. Conclusions

The present study found that the low-temperature plasma treatment of porcine graft particles did not modify the biomaterial's surface, as demonstrated by the SEM, EDS, XPS, XRD, and FTIR results. Furthermore, PGPT demonstrated higher biocompatibility by enhancing osteoblast-like cell proliferation and differentiation, according to WST-1, ALP, and real-time PCR. Further studies are needed to understand the use of PGPT for bone regeneration better.

Author Contributions: Conceptualization, C.S.C. and W.-J.C.; Methodology, E.S.; Software, P.Y.L.; Validation, H.-M.H., N.-C.T. and Y.-H.P.; Formal Analysis, C.S.C.; Investigation, P.Y.L.; Resources, W.-J.C.; Data Curation, E.S.; Writing—Original Draft Preparation, C.S.C.; Writing—Review and Editing, W.-J.C. and E.S.; Visualization, H.-M.H.; Supervision, N.-C.T.; Project Administration, Y.-H.P.; Funding Acquisition, Y.-H.P. and W.-J.C.

Funding: This research received no external funding.

Acknowledgments: The authors would like to thank Enago (www.enago.tw) for the English language review.

Conflicts of Interest: The authors declare no conflicts of interest.

References

1. Gamblin, A.-L.; Brennan, M.A.; Renaud, A.; Yagita, H.; Lézot, F.; Heymann, D.; Trichet, V.; Layrolle, P. Bone tissue formation with human mesenchymal stem cells and biphasic calcium phosphate ceramics: The local implication of osteoclasts and macrophages. *Biomaterials* **2014**, *35*, 9660–9667. [CrossRef]
2. HjØrting-Hansen, E. Bone grafting to the jaws with special reference to reconstructive preprosthetic surgery. *Mund-, Kiefer-und Gesichtschirurgie* **2002**, *6*, 6–14.
3. Burchardt, H. The biology of bone graft repair. *Clin. Orthop. Relat. Res.* **1983**, *174*, 28–42. [CrossRef]
4. Nguyen, N.K.; Leoni, M.; Maniglio, D.; Migliaresi, C. Hydroxyapatite nanorods: Soft-template synthesis, characterization and preliminary in vitro tests. *J. Biomater. Appl.* **2013**, *28*, 49–61. [CrossRef]
5. Wei, M.; Evans, J.; Bostrom, T.; Grøndahl, L. Synthesis and characterization of hydroxyapatite, fluoride-substituted hydroxyapatite and fluorapatite. *J. Mater. Sci.* **2003**, *14*, 311–320.
6. Spence, G.; Patel, N.; Brooks, R.; Rushton, N. Carbonate substituted hydroxyapatite: Resorption by osteoclasts modifies the osteoblastic response. *J. Biomed. Mater.* **2009**, *90*, 217–224. [CrossRef]
7. Okada, S.; Ito, H.; Nagai, A.; Komotori, J.; Imai, H. Adhesion of osteoblast-like cells on nanostructured hydroxyapatite. *Acta Biomater.* **2010**, *6*, 591–597. [CrossRef]
8. Sheikh, Z.; Najeeb, S.; Khurshid, Z.; Verma, V.; Rashid, H.; Glogauer, M. Biodegradable materials for bone repair and tissue engineering applications. *Materials* **2015**, *8*, 5744–5794. [CrossRef]
9. Qiao, W.; Liu, Q.; Li, Z.; Zhang, H.; Chen, Z. Changes in physicochemical and biological properties of porcine bone derived hydroxyapatite induced by the incorporation of fluoride. *Sci. Technol. Adv. Mater.* **2017**, *18*, 110–121. [CrossRef]
10. Van Meek'ren, J. *Heel en Geneeskonstige Aanmerkingen*; Casparus Commelijn: Amsterdam, The Netherlands, 1668.
11. Salamanca, E.; Lee, W.-F.; Lin, C.-Y.; Huang, H.-M.; Lin, C.-T.; Feng, S.-W.; Chang, W.-J. A novel porcine graft for regeneration of bone defects. *Materials* **2015**, *8*, 2523–2536. [CrossRef]
12. Pearce, A.; Richards, R.; Milz, S.; Schneider, E.; Pearce, S. Animal models for implant biomaterial research in bone: A review. *Eur. Cell Mater.* **2007**, *13*, 1–10. [CrossRef]
13. Salamanca, E.; Hsu, C.-C.; Huang, H.-M.; Teng, N.-C.; Lin, C.-T.; Pan, Y.-H.; Chang, W.-J. Bone regeneration using a porcine bone substitute collagen composite in vitro and in vivo. *Sci. Rep.* **2018**, *8*, 984. [CrossRef]
14. Łukomska-Szymańska, M.; Zarzycka, B.; Grzegorczyk, J.; Sokołowski, K.; Półtorak, K.; Sokołowski, J.; Łapińska, B. Antibacterial properties of calcium fluoride-based composite materials: In vitro study. *BioMed Res. Int.* **2016**, *2016*, 1048320.
15. Nur, M.; Kinandana, A.; Winarto, P.; Muhlisin, Z. Study of an atmospheric pressure plasma jet of Argon generated by column dielectric barrier discharge. *J. Phys.* **2016**, *776*, 012102. [CrossRef]
16. Von Keudell, A.; Awakowicz, P.; Benedikt, J.; Raballand, V.; Yanguas-Gil, A.; Opretzka, J.; Flötgen, C.; Reuter, R.; Byelykh, L.; Halfmann, H. Inactivation of bacteria and biomolecules by low-pressure plasma discharges. *Plasma Process. Polym.* **2010**, *7*, 327–352. [CrossRef]

17. Adamovich, I.; Baalrud, S.D.; Bogaerts, A.; Bruggeman, P.; Cappelli, M.; Colombo, V.; Czarnetzki, U.; Ebert, U.; Eden, J.; Favia, P. The 2017 plasma roadmap: Low temperature plasma science and technology. *J. Phys. D Appl. Phys.* **2017**, *50*, 323001. [CrossRef]
18. Syed, M.R.; Khan, M.; Sefat, F.; Khurshid, Z.; Zafar, M.S.; Khan, A.S. Bioactive glass and glass fiber composite: biomedical/dental applications. In *Biomedical, Therapeutic and Clinical Applications of Bioactive Glasses*; Elsevier: Amsterdan, The Netherlands, 2019; pp. 467–495.
19. Lee, J.-Y.; Choi, B.; Wu, B.; Lee, M. Customized biomimetic scaffolds created by indirect three-dimensional printing for tissue engineering. *Biofabrication* **2013**, *5*, 045003. [CrossRef]
20. Khurshid, Z.; Husain, S.; Alotaibi, H.; Rehman, R.; Zafar, M.S.; Farooq, I.; Khan, A.S. Novel techniques of scaffold fabrication for bioactive glasses. In *Biomedical, Therapeutic and Clinical Applications of Bioactive Glasses*; Elsevier: Amsterdam, The Netherlands, 2019; pp. 497–519.
21. Graves, D.B. Low temperature plasma biomedicine: A tutorial review. *Phys. Plasmas* **2014**, *21*, 080901. [CrossRef]
22. Silversmit, G.; Depla, D.; Poelman, H.; Marin, G.B.; De Gryse, R. Determination of the V2p XPS binding energies for different vanadium oxidation states (V^{5+} to V^{0+}). *J. Electron Spectrosc. Relat. Phenom.* **2004**, *135*, 167–175. [CrossRef]
23. Warren, F.J.; Gidley, M.J.; Flanagan, B.M. Infrared spectroscopy as a tool to characterise starch ordered structure—A joint FTIR–ATR, NMR, XRD and DSC study. *Carbohydr. Polym.* **2016**, *139*, 35–42. [CrossRef]
24. Ngamwongsatit, P.; Banada, P.P.; Panbangred, W.; Bhunia, A.K. WST-1-based cell cytotoxicity assay as a substitute for MTT-based assay for rapid detection of toxigenic Bacillus species using CHO cell line. *J. Microbiol. Methods* **2008**, *73*, 211–215. [CrossRef]
25. Sila-Asna, M.; Bunyaratvej, A.; Maeda, S.; Kitaguchi, H.; Bunyaratavej, N. Osteoblast differentiation and bone formation gene expression in strontium-inducing bone marrow mesenchymal stem cell. *Kobe J. Med. Sci.* **2007**, *53*, 25–35.
26. Sollazzo, V.; Palmieri, A.; Scapoli, L.; Martinelli, M.; Girardi, A.; Alviano, F.; Pellati, A.; Perrotti, V.; Carinci, F. Bio-Oss® acts on stem cells derived from peripheral blood. *Oman Med. J.* **2010**, *25*, 26. [CrossRef]
27. Scarano, A.; Lorusso, F.; Ravera, L.; Mortellaro, C.; Piattelli, A. Bone regeneration in iliac crestal defects: An experimental study on sheep. *BioMed Res. Int.* **2016**, *2016*, 4086870. [CrossRef]
28. Kivovics, M.; Szabó, B.T.; Németh, O.; Tari, N.; Dőri, F.; Nagy, P.; Dobó-Nagy, C.; Szabó, G. Microarchitectural study of the augmented bone following ridge preservation with a porcine xenograft and a collagen membrane: Preliminary report of a prospective clinical, histological, and micro-computed tomography analysis. *Int. J. Oral Maxillofac. Surg.* **2017**, *46*, 250–260. [CrossRef]
29. Barone, A.; Toti, P.; Menchini-Fabris, G.B.; Derchi, G.; Marconcini, S.; Covani, U. Extra oral digital scanning and imaging superimposition for volume analysis of bone remodeling after tooth extraction with and without 2 types of particulate porcine mineral insertion: A randomized controlled trial. *Clin. Implant Dent. Relat. Res.* **2017**, *19*, 750–759. [CrossRef]
30. Barone, A.; Toti, P.; Quaranta, A.; Alfonsi, F.; Cucchi, A.; Negri, B.; Di Felice, R.; Marchionni, S.; Calvo-Guirado, J.L.; Covani, U. Clinical and histological changes after ridge preservation with two xenografts: Preliminary results from a multicentre randomized controlled clinical trial. *J. Clin. Periodontol.* **2017**, *44*, 204–214. [CrossRef]
31. Seo, D.S.; Hwang, K.H.; Yoon, S.Y.; Lee, J.K. Fabrication of hydroxyapatite bioceramics from the recycling of pig bone. *J. Ceram. Process. Res.* **2012**, *13*, 586–589.
32. Ramesh, S.; Tan, C.; Hamdi, M.; Sopyan, I.; Teng, W. The influence of Ca/P ratio on the properties of hydroxyapatite bioceramics. In Proceedings of the International Conference on Smart Materials and Nanotechnology in Engineering, Harbin, China, 1–4 July 2007; p. 64233A.
33. Lin, J.C.; Kuo, K.; Ding, S.; Ju, C.-P. Surface reaction of stoichiometric and calcium-deficient hydroxyapatite in simulated body fluid. *J. Mater. Sci. Mater. Med.* **2001**, *12*, 731–741. [CrossRef]
34. Figueiredo, A.; Coimbra, P.; Cabrita, A.; Guerra, F.; Figueiredo, M. Comparison of a xenogeneic and an alloplastic material used in dental implants in terms of physico-chemical characteristics and in vivo inflammatory response. *Mater. Sci. Eng. C* **2013**, *33*, 3506–3513. [CrossRef]
35. Fujisawa, K.; Akita, K.; Fukuda, N.; Kamada, K.; Kudoh, T.; Ohe, G.; Mano, T.; Tsuru, K.; Ishikawa, K.; Miyamoto, Y. Compositional and histological comparison of carbonate apatite fabricated by dissolution–precipitation reaction and Bio-Oss®. *J. Mater. Sci. Mater. Med.* **2018**, *29*, 121. [CrossRef]

36. Figueiredo, M.; Henriques, J.; Martins, G.; Guerra, F.; Judas, F.; Figueiredo, H. Physicochemical characterization of biomaterials commonly used in dentistry as bone substitutes—comparison with human bone. *J. Biomed. Mater. Res. Part B* **2010**, *92*, 409–419. [CrossRef]
37. Heinlin, J.; Morfill, G.; Landthaler, M.; Stolz, W.; Isbary, G.; Zimmermann, J.L.; Shimizu, T.; Karrer, S. Plasma medicine: Possible applications in dermatology. *Journal der Deutschen Dermatologischen Gesellschaft* **2010**, *8*, 968–976. [CrossRef]
38. Slepička, P.; Vasina, A.; Kolská, Z.; Luxbacher, T.; Malinský, P.; Macková, A.; Švorčík, V. Argon plasma irradiation of polypropylene. *Nucl. Instrum. Methods Phys. Res. Sect. B* **2010**, *268*, 2111–2114. [CrossRef]
39. Salamanca, E.; Pan, Y.-H.; Tsai, A.I.; Lin, P.-Y.; Lin, C.-K.; Huang, H.-M.; Teng, N.-C.; Wang, P.D.; Chang, W.-J. Enhancement of osteoblastic-like cell activity by glow discharge plasma surface modified hydroxyapatite/β-tricalcium phosphate bone substitute. *Materials* **2017**, *10*, 1347. [CrossRef]
40. Chang, Y.C.; Feng, S.W.; Huang, H.M.; Teng, N.C.; Lin, C.T.; Lin, H.K.; Wang, P.D.; Chang, W.J. Surface analysis of titanium biological modification with glow discharge. *Clin. Implant Dent. Relat. Res.* **2015**, *17*, 469–475. [CrossRef]
41. Chang, Y.-C.; Lee, W.-F.; Feng, S.-W.; Huang, H.-M.; Lin, C.-T.; Teng, N.-C.; Chang, W.J. In vitro analysis of fibronectin-modified titanium surfaces. *PLoS ONE* **2016**, *11*, e0146219. [CrossRef]
42. Sarma, B.K.; Das, A.; Barman, P.; Pal, A.R. Biomimetic growth and substrate dependent mechanical properties of bone like apatite nucleated on Ti and magnetron sputtered TiO_2 nanostructure. *J. Phys. D Appl. Phys.* **2016**, *49*, 145304. [CrossRef]
43. Yi, D.; Wu, C.; Ma, X.; Ji, H.; Zheng, X.; Chang, J. Preparation and in vitro evaluation of plasma-sprayed bioactive akermanite coatings. *Biomed. Mater.* **2012**, *7*, 065004. [CrossRef]
44. LeGeros, R.; Lin, S.; Rohanizadeh, R.; Mijares, D.; LeGeros, J. Biphasic calcium phosphate bioceramics: Preparation, properties and applications. *J. Mater. Sci.* **2003**, *14*, 201–209.
45. Pagliani, L.; Andersson, P.; Lanza, M.; Nappo, A.; Verrocchi, D.; Volpe, S.; Sennerby, L. A collagenated porcine bone substitute for augmentation at Neoss implant sites: A prospective 1-year multicenter case series study with histology. *Clin. Implant Dent. Relat. Res.* **2012**, *14*, 746–758. [CrossRef]

© 2019 by the authors. Licensee MDPI, Basel, Switzerland. This article is an open access article distributed under the terms and conditions of the Creative Commons Attribution (CC BY) license (http://creativecommons.org/licenses/by/4.0/).

Article

Characteristics of AISI 420 Stainless Steel Modified by Low-Temperature Plasma Carburizing with Gaseous Acetone

Ruiliang Liu [1],* and Mufu Yan [2]

[1] Key Laboratory of Superlight Material and Surface Technology of Ministry of Education, College of Material Science and Chemical Engineering, Harbin Engineering University, Harbin 150001, China
[2] National Key Laboratory for Precision Hot Processing of Metals, School of Materials Science and Engineering, Harbin Institute of Technology, Harbin 150001, China; yanmufu@hit.edu.cn
* Correspondence: liuruiliang@hrbeu.edu.cn; Tel.: +86-451-8251-8731

Received: 27 December 2018; Accepted: 23 January 2019; Published: 26 January 2019

Abstract: In this research work, low-temperature carburizing of AISI 420 martensitic stainless steel was conducted at 460 °C for different amounts of time using an acetone source. The microstructure and phase structure of the carburized layers were characterized by optical microscope and X-ray diffraction. The properties of the carburized layers were tested with a microhardness tester and an electrochemical workstation. The results indicate uniform layers are formed on martensitic stainless steel surfaces, and the carburized layers are mainly composed of carbon "expanded" α ($α_C$) and Fe_3C phases. The property tests indicated that after plasma–carburizing, the hardness of the stainless steel surface can reach up to 850 $HV_{0.1}$. However, the corrosion resistance of stainless steel decreased slightly, and the corrosion characteristic of stainless steel was altered from pitting to general corrosion. The semiconductor characteristic of the passivation film on stainless steel was transformed from the *p*-type for untreated specimens to the n-type for carburized specimens.

Keywords: martensitic stainless steel; low temperature plasma carburizing; carbon source; microstructure; corrosion behavior

1. Introduction

Stainless steel is widely used for its inherent corrosion resistance. However, unmodified stainless steel can be problematic when it is used in moving parts in some harsh environments, such as aerospace, electric power stations, ships, and ocean engineering [1]. At present, surface modification has become an important solution to improve surface properties of stainless steel to meet demands in the harsh environments mentioned above [2–4]. There are various surface modification methods that have been utilized to improve the properties of stainless steel, of which thermo-chemical diffusion treatment strategies have drawn significant consideration for their simple process, low cost, and other advantages [5]. Among these, low-temperature nitriding, carburizing, and nitrocarburizing treatments are the commonly used processes that can improve the hardness and wear–resistance of stainless steels without reducing their corrosion resistance [6,7]. Thus, these treatments have received extensive research attention [7–9].

Regarding the low-temperature carburization of stainless steel, most research is focused on studying the microstructure and mechanical properties of the modified layers [10–14]. A literature survey indicated that the corrosion property studies of the carburized layers were mainly conducted on austenitic stainless steels [15–17], and only a few were related to the carburized layers on martensitic stainless steels [18,19]. Notably, most low-temperature carburization of stainless steel was conducted using CH_4 or CO gas carbon sources [10–14,20].

Interestingly, when bearing steel (M50NiL) was plasma-carburized with an acetone source, diamond-like carbon/Fe_3C-containing carburized layers were formed, and the layers possessed self-lubricating and anti-corrosion properties [21–23]. Recently, authors have also proved that the low-temperature carburization of AISI 431 stainless steel could be successfully conducted with gaseous acetone. The carburized layer thickness could reach up to 45 μm with a hardness higher than 1051 $HV_{0.1}$, while corrosion resistance decreased only slightly [24]. Since the alloying elements in stainless steel have an important influence on the microstructure and properties of the low-temperature plasma surface alloyed layer [25], there is great need to study the plasma carburization of other kinds of stainless steels with an acetone source.

Therefore, based on previous research [15–20], AISI 420 martensitic stainless steel was plasma-carburized at 460 °C for 4–12 h using an acetone source. Then, a preliminary study of the microstructure and properties of the carburized layer was conducted, and the effects of the alloy elements were especially considered. Since the corrosion resistance of the carburized layer is one of the important concerns for stainless steel, the present investigation primarily focused on the corrosion behaviors of the carburized layer.

2. Materials and Experimental Methods

Commercially available AISI 420 martensitic stainless steel with the chemical composition (wt %) of 0.16%–0.25% C, ≤0.60% Si, ≤0.80% Mn, ≤0.035% P, ≤0.030% S, ≤0.75% Ni, 12.0%–14.0% Cr, and Fe in balance was used in the present investigation. The as-received AISI 420 stainless steel rod was first machined to the size of Φ 25 mm × 5 mm, and then the steel was austenitized at 1050 °C for 1 h and oil-cooled.

Plasma-carburizing treatments were conducted in a 30 kW home-made pulse plasma multi-diffusing unit [9]. The process parameters were: temperature, 460 °C; time, 4–12 h; pressure, 200–300 Pa; voltage, 650 V; and carburizing atmosphere, H_2:CH_3COCH_3 = 4:1. Acetone (CH_3COCH_3) was heated into vapor and then inputted into the furnace along with hydrogen (H_2). The ageing treatment of stainless steel could be conducted simultaneously during plasma carburization.

The carburized specimens were cross-sectioned and set in bakelite, and then the cross section was grinded with 240–2000# sandpapers and polished with Al_2O_3 polishing powder. After that, the specimens were light-etched using Marble's reagent [9], and then were observed by a metallographic microscope (OM, CMM-33E, Shanghai Changfang Optical Instrument CO., LTD., Shanghai, China). The thicknesses of the carburized layers were roughly estimated based on OM observations because of an obvious boundary between layer and matrix, and then were proved by hardness profiles on the cross sections.

The identification of the phases present in the carburized layers was carried out by X'Pert Pro X-ray diffraction (PANalytical, Almelo, The Netherlands). Test conditions were: Cu $K\alpha$ radiation (λ = 1.5406 nm); voltage, 40 kV; current, 40 mA; speed, 0.7°/s; and scanning range, 20–100°.

Surface hardness and cross-sectional hardness profile of the carburized specimens were obtained using a microhardness tester (HV-1000, Fangyuan Instrument CO., LTD., Jinan, China). The applied load was 100 gf, and the hold time was 15 s.

The corrosion behaviors of the carburizing layers were studied by the potentiodynamic polarization curves and electrochemical impedance spectroscopy (EIS) using electrochemical testing equipment (Chi660e, Shanghai Chenhua Instruments Co., Ltd., Shanghai, China) in 3.5 wt % NaCl solution. Based on the standard of GB/T 24196-2009 [26], the potentiodynamic polarization curves of the specimens were tested. A three-electrode system was also used in the present investigation, including the working electrode (W) of specimen, the auxiliary electrode (C) of a platinum wire, and the reference electrode (R) of a saturated calomel electrode (SCE). The size of the test surface was about 1 cm^2 in area. For polarization curve tests, scanning started from the open circuit potential of −300 mV with a scanning speed of 5 mV/s. Scanning stopped when the current density reached 10 mA/cm^2. In the EIS tests, the specimens were submerged in solution for 120 s under a voltage of 250 mV to gain a stable passivation membrane. The excitation signal had a sine wave with an amplitude of 5 mV, and the test frequency was 10^{-2}–10^5 Hz. Mott–Schottky (M–S) curve tests were used to characterize the semiconductor characteristics of the film formed on the carburized layer, where the test potential range was 1000–600 mV, and test speed was 50 mV/step.

3. Results and Discussion

3.1. Phase Structure of the Carburized Layer

The phase compositions of the carburized layers on AISI 420 stainless steel are shown in Figure 1. It is well-known that the only phase on an untreated steel surface is the α'-Fe phase. After plasma carburization, carbon "expanded" α'-Fe (α_C) phases and Fe_3C phases formed on the stainless steel surface. The α_C phase formation was evidenced by the broadening of the original martensite peaks and their displacement to lower angles [12,13,18]. Moreover, with the increase of carburizing time, the peak intensities of α_C phases showed almost no change. The phase composition in the carburized layer played an important role in the properties of the carburized layer, as shown in the following parts.

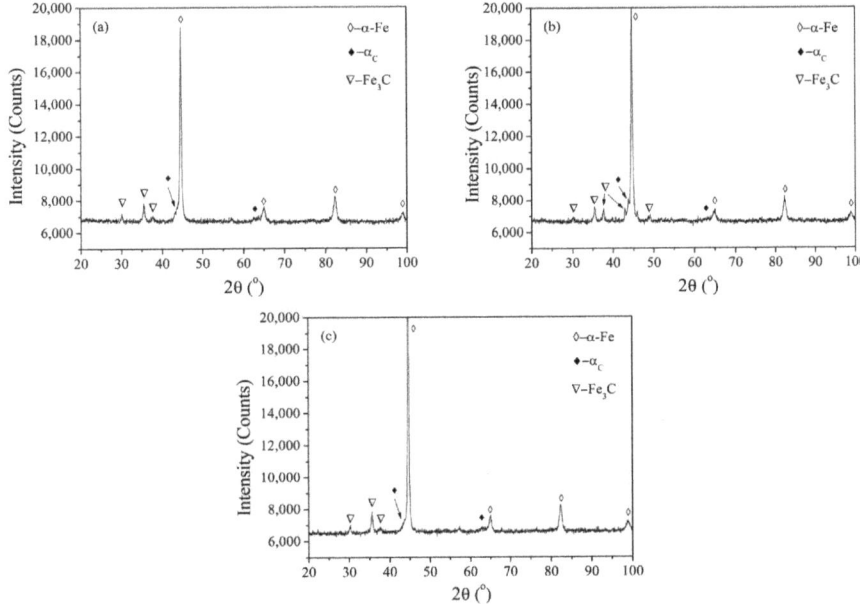

Figure 1. XRD patterns of AISI 420 stainless steel plasma carburized at 460 °C: (**a**) 4 h; (**b**) 8 h; and (**c**) 12 h.

3.2. Microstructure and Hardness Profile of the Carburized Layer

The cross-sectional optical micrographs and microhardness profiles of low-temperature-carburized layers on AISI 420 stainless steels are presented in Figure 2. One can see in Figure 2a–c that prominent plasma-carburized layers were formed on the stainless steel surfaces after light erosion, and the layer thickness increased with time. Based on the OM observation, the carburized layer thicknesses were determined to be about 26, 42, and 62 μm for the specimens plasma-carburized at 460 °C for 4, 8, and 12 h, respectively. The layer thicknesses obtained on AISI 420 stainless steel were higher than those on AISI 431 stainless steel in the same plasma-carburizing condition [24]. The present investigation also proved that the alloying elements had obvious effects on the microstructures of the carburized layers, as shown in [25]. Moreover, the etchant (Marble's reagent) did not attack the carburized layers, which is commonly an empirical indication of an improvement in corrosion resistance in the etching medium. However, when the carburizing time increased up to 12 h (Figure 2c), the microstructure showed a porous state with some black spots, which can be caused by the precipitation and grain growth of the carbides [18,24].

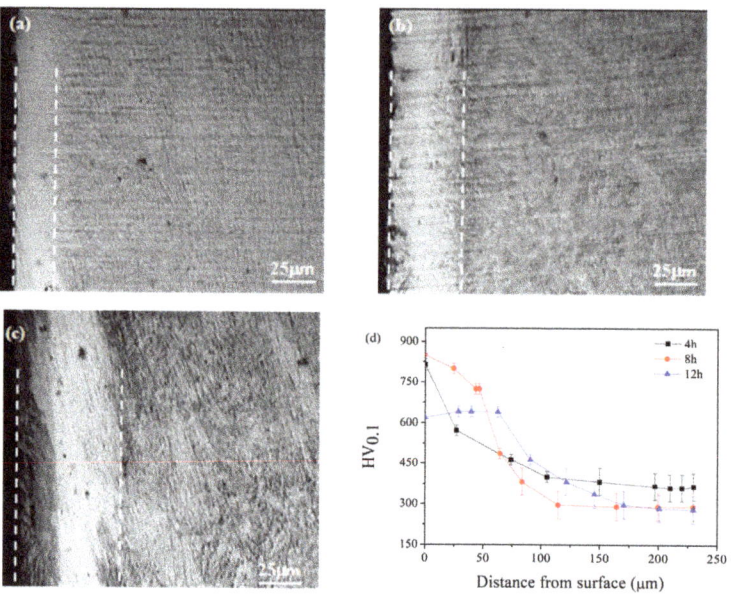

Figure 2. (**a**–**c**) Cross section microstructures. (**d**) Microhardness profiles of AISI 420 stainless steel plasma carburized at 460 °C for 4, 8, and 12 h.

As shown in References [21–23], a diamond-like carbon/Fe_3C-containing carburized layer could be formed on M50NiL steel after plasma–carburizing with acetone. The fine microstructure characterization of the carburized layer on AISI 420 stainless steel will be characterized in the near future.

In addition, it can be seen from Figure 2d that the surface hardness of stainless steel was improved after plasma carburizing. The surface hardnesses of the untreated, 4, 8, and 12 h treated specimens were about 300, 810, 850, and 620 $HV_{0.1}$, respectively. That is, the surface hardness of the stainless steel was improved up to 2.8 times more than that of the untreated steel, which should be attributed to the higher lattice distortion strengthening in "expanded" α'-Fe lattices [11] and high hardness of the Fe_3C phase. Moreover, the surface hardness increased first and then decreased with duration. The reasons for this could be mainly due to the grain growth of carbides and microstructure deterioration for the 12 h carburized layer, as shown in Figure 2c.

It can also be seen that the bulk hardnesses changed with time, which is almost in agreement with the ageing effect observed for each treatment temperature [9,12]. Thus, the ageing of the martensitic stainless steel can be conducted simultaneously with the plasma carburization treatment. Compared with the surface hardnesses of the carburized layers obtained on AISI 431 stainless steel under the same carburization conditions, the hardnesses did not show much variation between 4 and 12 h carburized layers, but the highest hardness decreased from 1050 to 850 $HV_{0.1}$ [24]. Moreover, compared with AISI 420 stainless steel plasma carburized at 450 °C with a CH_4 source, the carburized layer thicknesses and hardnesses were almost the same and showed the same variation trends with the present investigation [12,13].

3.3. Corrosion Behavior of the Carburized Layer

Figure 3 shows the typical polarization curves of stainless steel carburized at 460 °C for 4–12 h, and the corresponding corrosion parameters are given in Table 1. It can be seen from Table 1 that for the untreated specimen and the specimens carburized for 4, 8, and 12 h, the corresponding corrosion potentials (E_{corr}) were −0.38 V (SCE), −0.56 V (SCE), −0.56 V (SCE), and −0.58 V (SCE), and the corrosion current densities were 3.19×10^{-6} A/cm^2, 2.54×10^{-5} A/cm^2, 1.99×10^{-5} A/cm^2, and 2.99×10^{-5} A/cm^2, respectively. Moreover, there was no obvious pitting phenomenon (i.e., no evident passivation region) for the carburized specimens. The corrosion resistance of the martensitic stainless steel was slightly worse than that of the untreated one, and the specimen carburized for 8 h had the better corrosion resistance at the present test condition, which is consistent with plasma carburization of AISI 431 stainless steel [24]. However, this disagrees with the corrosion test results obtained from low-temperature carburization of AISI 316 austenitic stainless steel [15,16]. The main reasons could be the multi-phase microstructure characteristics shown in XRD results, because the microstructure had a significant influence on the corrosion behavior of the steel [6,18]. In fact, some other researchers also found this kind of corrosion phenomenon in the low-temperature ion nitriding of martensitic stainless steel—for example, Corengia et al. [19] found that ion nitriding at 673–773 K (400–500 °C) reduced the corrosion resistance of AISI 410 martensitic stainless steel.

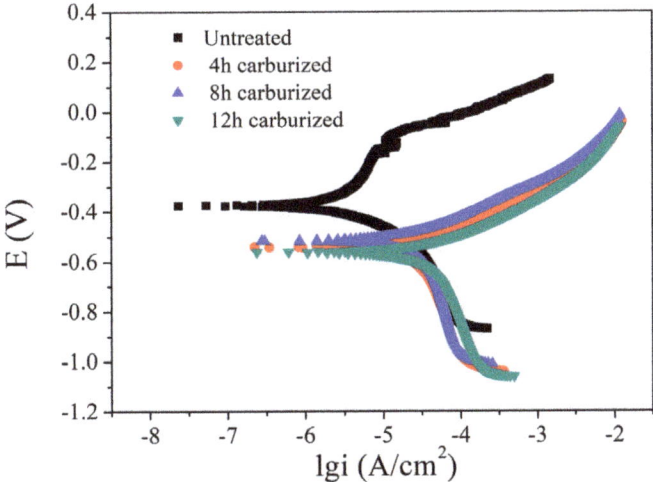

Figure 3. Polarization curves of AISI 420 stainless steel plasma carburized at 460 °C for 4–12 h.

Table 1. Parameters obtained from polarization curves of AISI 420 stainless steel plasma carburized at 460 °C for 4–12 h.

Specimen	E_{corr} (V)	I_{corr} (A/cm^2)
Untreated	−0.38	3.19×10^{-6}
4 h	−0.56	2.54×10^{-5}
8 h	−0.56	1.99×10^{-5}
12 h	−0.58	2.99×10^{-5}

To further study the stabilities of the passive films formed during corrosion of the carburized layer, EIS measurements were carried out. Figure 4 depicts the Nyquist and Bode curves of the untreated specimen and specimens plasma-carburized at 460 °C for 4–12 h. Electrical equivalent circuits were obtained based on ZView software (Version 3.0a), and the values of each part could be directly measured. The corresponding data are shown in Table 2. In the table, CPE is an abbreviation for constant phase angle element [27], the impedance value of which is the function of angle frequency (ω), and its amplitude angle is independent of frequency. There are two parameters: CPE-T and CPE-P. CPE-T is called double-layer capacitance, CPE-P is a dispersion index and is normally used to characterize dispersion effects. The simulation results of the percentage errors were within 5%, that is, the data-fitting results were good and the chosen model could be used to characterize the specimens' corrosion equivalent circuit. It can be seen from the figure that the impedance spectra were all a single capacitive arc with only one time constant. The bigger the capacitance arc radius, the higher the stability of the film [14]. The biggest capacitance arc radius, from the specimen carburized for 8 h, indicated that the passive film that formed on this specimen possessed the highest stability in a corrosive environment. These results are consistent with the plasma carburization of AISI 431 stainless steel [24].

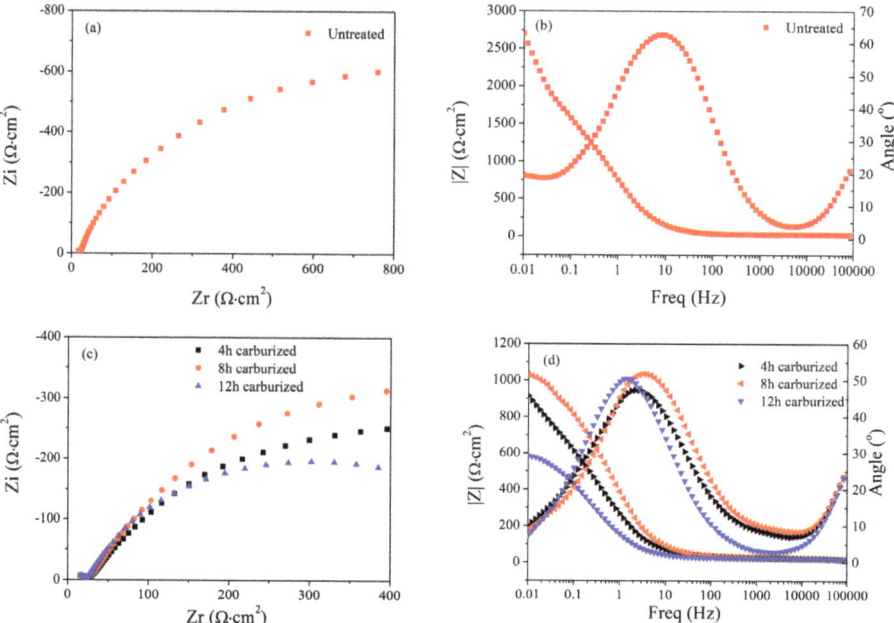

Figure 4. Nyquist and Bode plots of AISI 420 stainless steel. (**a**,**b**) Untreated specimen, (**c**,**d**) Carburized specimens.

Table 2. Fitting parameter values for the untreated and the carburized AISI 420 stainless steel. CPE-P: dispersion index; CPE-T: double-layer capacitance.

Specimen	R_s (Ω)		CPE-T (F)		CPE-P (F)		R_p (Ω)	
	Value	Error%	Value	Error%	Value	Error%	Value	Error%
Untreated	21.95	0.19	0.00023	0.5	0.84	0.13	1506	0.93
4 h	21.21	0.66	0.0015	1.38	0.62	0.59	1227	4.5
8 h	22.66	0.45	0.00078	0.99	0.68	0.35	1006	1.89
12 h	23.01	0.52	0.0016	1.24	0.72	0.52	786.5	3.41

In addition, it can be seen in Table 2 that for the untreated specimen, the membrane resistance (R_p) and the capacitance (CPE-T) were 1506 Ω and 0.00023 F. For the specimens plasma-carburized for 4, 8, and 12 h, the membrane resistances (R_p) were 1227, 1006, and 786.5 Ω, while the capacitances were 0.0015, 0.00078, and 0.0016 F, respectively. The larger the value of R_p, the smaller the corrosion rate of the film. All the above values indicated that, after plasma carburization at 460 °C, the stabilities of passive films formed on the specimens were all lower than those of untreated specimen, and decreased slightly with time. The passive film was the least stable for the stainless steel carburized for 12 h, which was likely due to the microstructure deterioration as shown before.

Finally, it should be pointed out that the corrosion resistance of the carburized layer showed a slight decrease under the present investigation condition. As such, optimization of the process parameters will be important work in the next steps, even though some reports also found that the low-temperature treatments were not necessary to keep or improve the corrosion resistance of stainless steels [19]. Considering that plasma processing parameters have an important influence on the microstructure of the carburized layers, deeper experiments are being conducted aiming to evaluate the effects of the temperature and acetone gas content to obtain better corrosion resistance.

Based on the Mott–Schottky (M–S) theory, it is well-known that the space charge layer capacitance (C) of the passivation film is a function of the electrode potential [15,16]. For an n-type semiconductor, the value of C can be represented by Equation (1):

$$\frac{1}{C^2} = \frac{2}{\varepsilon\varepsilon_0 q N_d}(E - E_{fb} - \frac{kT}{q}) \tag{1}$$

For the p-type semiconductor, the value of C can be represented by Equation (2):

$$\frac{1}{C^2} = \frac{-2}{\varepsilon\varepsilon_0 q N_a}(E - E_{fb} - \frac{kT}{q}) \tag{2}$$

where N_d is the carrier charge concentration of the donor and N_a is the carrier charge concentration of the acceptor.

According to the slope of the linear part in the M–S curve, the electronic structure type of the passive film can be determined. When the value is positive, it possesses the n-type semiconductor characteristic, otherwise, it possesses the p-type semiconductor characteristic [15,16]. The charge concentration of donor or acceptor in the surface space is directly proportional to the reciprocal of the slope of the M–S curve, and can be estimated by Equation (3):

$$N_d(N_a) = \frac{2}{e\varepsilon\varepsilon_0 S} \tag{3}$$

Figure 5 plots the M–S curves of the untreated specimen and specimens plasma–carburized at 460 °C for 4–12 h. For the untreated specimen, the slope of the linear part (0.15–0.50 V) in the curve was negative, the film formed on the untreated specimen possessed the p-type semiconductor property, and the donor charge concentration was calculated to be about 2.59×10^{22} cm^{-3}. On the other hand, for the specimens after plasma carburization at 460 °C, the slope of the linear parts (0.15–0.50 V) in the curves were all positive, which all corresponded to n-type semiconductor properties [24], and the donor charge concentrations for specimens carburized for 4, 8, and 12 h were calculated to be about 3.02×10^{21}, 2.56×10^{21}, and 6.66×10^{21} cm^{-3}, respectively.

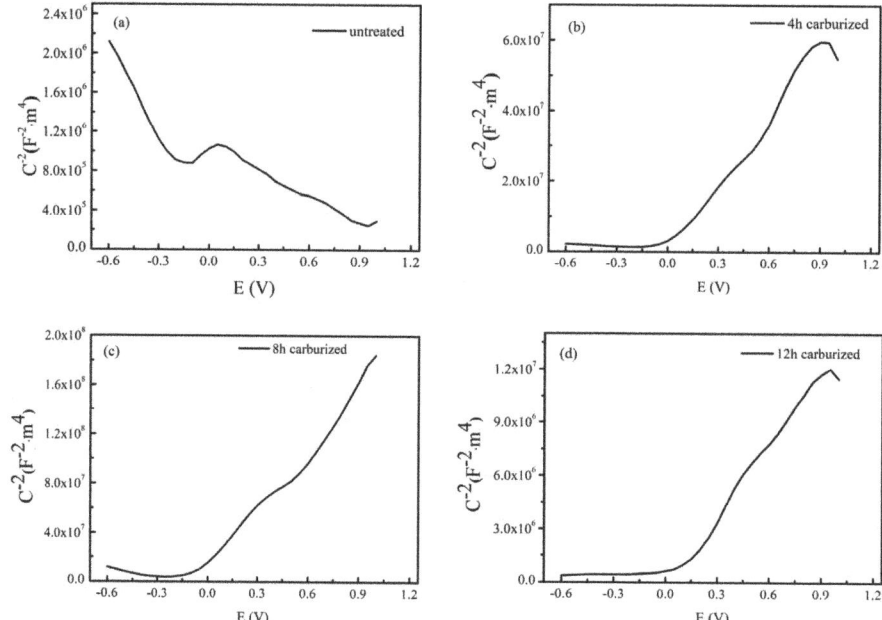

Figure 5. Mott–Schottky (M–S) plots of the untreated and the carburized AISI 420 stainless steel: (**a**) untreated; (**b**) 4 h carburized; (**c**) 8 h carburized; (**d**) 12 h carburized.

4. Conclusions

- After plasma carburization at 460 °C with gaseous acetone, uniform layers were formed on the AISI 420 martensitic stainless steel surface, and the carburized layer was mainly composed of carbon "expanded" α (αc) and some Fe_3C phases.
- The hardness of the carburized layer on stainless steel could be improved up to 850 $HV_{0.1}$, which was about 2.8 times higher than that of the untreated one.
- The corrosion resistance of stainless steel after plasma carburization showed a slight decrease under the present test conditions, and the corrosion characteristic of the stainless steel was altered from pitting to general corrosion. The semiconductor characteristic of the passivation film on stainless steel was transformed from the *p*-type for the untreated specimen to the n-type for carburized specimens.

Author Contributions: Conceptualization, R.L.; Methodology R.L.; Investigation, R.L.; Resource, R.L.; Data Curation, R.L. and M.F.; Writing-Original Draft Preparation; Writing and Editing, R.L. and M.F.; Supervision, R.L. and M.F.; Project Administration, R.L.; Funding Acquisition, R.L.

Funding: This research was funded by the National Natural Science Foundation of China (Nos. 51401062, 51871071).

Acknowledgments: The authors would like to thank A. Xu for his experimental assistance.

Conflicts of Interest: The authors declare no conflict of interest.

References

1. Bahrami, A.; Mousavi Anijdan, S.H.; Taheri, P.; Yazdan Mehr, M. Failure of AISI 304H stainless steel elbows in a heat exchanger. *Eng. Fail. Anal.* **2018**, *90*, 397–403. [CrossRef]
2. Sun, J.; Yao, Q. Fabrication of microalloy nitrided layer on low carbon steel by nitriding combined with surface nano-alloying pretreatment. *Coatings* **2016**, *6*, 63. [CrossRef]

3. Yamaguchi, T.; Hagino, H. Formation of a titanium-carbide-dispersed hard coating on austenitic stainless steel by laser alloying with a light-transmitting resin. *Vacuum* **2018**, *155*, 23–28. [CrossRef]
4. Adachi, S.; Ueda, N. Wear and corrosion properties of cold-sprayed AISI 316L coatings treated by combined plasma carburizing and nitriding at low temperature. *Coatings* **2018**, *8*, 456. [CrossRef]
5. Nikolov, K.; Bunk, K.; Jung, A.; Kaestner, P.; Bräuer, G.; Klages, C.-P. High-efficient surface modification of thin austenitic stainless steel sheets applying short-time plasma nitriding by means of strip hollow cathode method for plasma thermochemical treatment. *Vacuum* **2014**, *110*, 106–113. [CrossRef]
6. Li, Y.; He, Y.; Xiu, J.; Wang, W.; Zhu, Y.; Hu, B. Wear and corrosion properties of AISI 420 martensitic stainless steel treated by active screen plasma nitriding. *Surf. Coat. Technol.* **2017**, *329*, 184–192. [CrossRef]
7. Alphonsa, J.; Raja, V.S.; Mukherjee, S. Development of highly hard and corrosion resistant A286 stainless steel through plasma nitrocarburizing process. *Surf. Coat. Technol.* **2015**, *280*, 268–276. [CrossRef]
8. Kim, S.K.; Yoo, J.S.; Priest, J.M.; Fewell, M.P. Characteristics of martensitic stainless steel nitrided in a low-pressure RF plasma. *Surf. Coat. Technol.* **2003**, *163*, 380–385. [CrossRef]
9. Liu, R.; Qiao, Y.; Yan, M.; Fu, Y. Layer growth kinetics and wear resistance of martensitic precipitation hardening stainless steel plasma nitrocarburized at 460 °C with rare earth addition. *Met. Mater. Int.* **2013**, *19*, 1151–1157. [CrossRef]
10. Ernst, F.; Cao, Y.; Michal, G.M.; Heuer, A.H. Carbide precipitation in austenitic stainless steel carburized at low temperature. *Acta Mater.* **2007**, *55*, 1895–1906. [CrossRef]
11. Hummelshøj, T.S.; Christiansen, T.L.; Somers, M.A.J. Lattice expansion of carbon-stabilized expanded austenite. *Scripta Materialia* **2010**, *63*, 761–763. [CrossRef]
12. Scheuer, C.J.; Cardoso, R.P.; Zanetti, F.I.; Amaral, T.; Brunatto, S.F. Low-temperature plasma carburizing of AISI 420 martensitic stainless steel: Influence of gas mixture and gas flow rate. *Surf. Coat. Technol.* **2012**, *206*, 5085–5090. [CrossRef]
13. Scheuer, C.J.; Cardoso, R.P.; Mafra, M.; Brunatto, S.F. AISI 420 martensitic stainless steel low-temperature plasma assisted carburizing kinetics. *Surf. Coat. Technol.* **2013**, *214*, 30–37. [CrossRef]
14. Rovani, A.C.; Breganon, R.; De Souza, G.S.; Brunatto, S.F.; Pintaúde, G. Scratch resistance of low-temperature plasma nitrided and carburized martensitic stainless steel. *Wear* **2017**, *376–377*, 70–76. [CrossRef]
15. Sun, Y. Corrosion behaviour of low temperature plasma carburised 316L stainless steel in chloride containing solutions. *Corros. Sci.* **2010**, *52*, 2661–2670. [CrossRef]
16. Sun, Y. Depth-profiling electrochemical measurements of low temperature plasma carburised 316L stainless steel in 1 M H_2SO_4 solution. *Surf. Coat. Technol.* **2010**, *204*, 2789–2796. [CrossRef]
17. Sun, Y. Tribocorrosion behavior of low temperature plasma carburized stainless steel. *Surf. Coat. Technol.* **2013**, *228*, S342–S348. [CrossRef]
18. Li, C.X.; Bell, T. Corrosion properties of plasma nitrided AISI 410 martensitic stainless steel in 3.5% NaCl and 1% HCl aqueous solutions. *Corros. Sci.* **2006**, *48*, 2036–2049. [CrossRef]
19. Corengia, P.; Ybarra, G.; Moina, C.; Cabo, A.; Broitman, E. Microstructure and corrosion behaviour of DC-pulsed plasma nitrided AISI 410 martensitic stainless steel. *Surf. Coat. Technol.* **2004**, *187*, 63–69. [CrossRef]
20. Rong, D.S.; Gong, J.M.; Jiang, Y.; Peng, Y. Effect of CO concentration on paraequilibrium gas carburization of 316L stainless steel. *Trans. Mater. Heat Treat.* **2015**, *36*, 204–209. (In Chinese)
21. Yang, Y.; Yan, M.; Zhang, Y.; Zhang, C.; Wang, X. Self-lubricating and anti-corrosion amorphous carbon/Fe_3C composite coating on M50NiL steel by low temperature plasma carburizing. *Surf. Coat. Technol.* **2016**, *304*, 142–149. [CrossRef]
22. Yang, Y.; Yan, M.; Zhang, Y.; Li, D.; Zhang, C.; Zhu, Y.; Wang, Y. Catalytic growth of diamond-like carbon on Fe_3C-containing carburized layer through a single-step plasma-assisted carburizing process. *Carbon* **2017**, *122*, 1–8. [CrossRef]
23. Yang, Y.; Yan, M.; Zhang, S.; Guo, J.; Jiang, D.; Li, D. Diffusion behavior of carbon and its hardening effect on plasma carburized M50NiL steel: Influences of treatment temperature and duration. *Surf. Coat. Technol.* **2018**, *333*, 96–103. [CrossRef]
24. Liu, R.; Wei, C.; Xu, A.; Yan, M.; Qiao, Y. Preparation and properties of "expanded" α phase layer on AISI 431 stainless steel. *Trans. Mater. Heat Treat.* **2017**, *38*, 165–172. (In Chinese)

25. Buhagiar, J.; Li, X.; Dong, H. Formation and microstructural characterisation of S-phase layers in Ni-free austenitic stainless steels by low-temperature plasma surface alloying. *Surf. Coat. Technol.* **2009**, *204*, 330–335. [CrossRef]
26. *GB/T 24196-2009 Corrosion of Metals and Alloys—Electrochemical Test Methods—Guidelines for Conducting Potentiostatic and Potentiodynamic Polarization Measurements*; Standardization Administration of the People's Republic of China: Beijing, China, 2009.
27. Orazem, M.E.; Tribollet, B. *Electrochemical Impedance Spectroscopy*; John Wiley & Sons: Hoboken, NJ, USA, 2008.

© 2019 by the authors. Licensee MDPI, Basel, Switzerland. This article is an open access article distributed under the terms and conditions of the Creative Commons Attribution (CC BY) license (http://creativecommons.org/licenses/by/4.0/).

Article

Preparation and Corrosion Resistance of ETEO Modified Graphene Oxide/Epoxy Resin Coating

Chunling Zhang, Xueyan Dai, Yingnan Wang, Guoen Sun, Peihong Li, Lijie Qu, Yanlong Sui and Yanli Dou *

Key Laboratory of Automobile Materials, Ministry of Education, College of Materials Science and Engineering, Jilin University, Changchun 130022, China; clzhang@jlu.edu.cn (C.Z.); xydai18@mails.jlu.edu.cn (X.D.); wangyn15@mails.jlu.edu.cn (Y.W.); sge@jlu.edu.cn (G.S.); liph18@mails.jlu.edu.cn (P.L.); qulj17@mails.jlu.edu.cn (L.Q.); suiyl17@mails.jlu.edu.cn (Y.S.)
* Correspondence: douyl@jlu.edu.cn; Tel.: +86-431-8509-5170

Received: 10 December 2018; Accepted: 11 January 2019; Published: 15 January 2019

Abstract: Improving the corrosion resistance of epoxy resin coatings has become the focus of current research. This study focuses on synthesizing a functionalized silane coupling agent (2-(3,4-epoxycyclohexyl)ethyl triethoxysilane) to modify the surface of graphene oxide to address nanomaterial agglomeration and enhance the coating resistance of the epoxy resin coating to corrosion by filling the coating with functionalized graphene oxide. Functionalized graphene oxide and coatings filled with functionalized graphene oxide were characterized by Fourier transform infrared spectroscopy, X-ray diffraction, X-ray photoelectron spectroscopy, scanning electron microscopy, and transmission electron microscopy. The corrosion performance of each coating was studied by electrochemical impedance spectroscopy and a salt spray test. Results showed that the incorporation of functionalized graphene oxide enhances the corrosion protection performance of the epoxy composite coating, and the composite coating exhibited the best anticorrosion performance when the amount of functionalized graphene oxide was 0.7 wt %.

Keywords: epoxy resin; silane coupling agent; graphene oxide; surface modification; anticorrosion performance

1. Introduction

Corrosion protection has gained considerable attention because corrosion has brought great harm to the industry [1]. At present, metal protection methods are mainly categorized into alloy protection, electroplating protection, electrochemical protection, and organic coating protection [2]. Among them, organic coating protection is widely used in several industrial fields owing to its low price, easy construction and excellent performance [3]. An organic coating separates a corrosive medium from a metal substrate and protects the metal substrate with a filler added to the coating and to the film-forming material [4,5]. As a high performance resin, epoxy resin is widely used in aerospace, industrial manufacturing, construction, and chemical fields [6,7]. Epoxy-based coatings have high chemical resistance, good barrier properties, good adhesion, low film shrinkage, high cross-linking density, easy curing, and low toxicity and are thus used to replace highly toxic and carcinogenic zinc/chromate composite coating [8–10].

However, with the deteriorating environment and the diversity of production, pure epoxy resin anticorrosion coatings have become increasingly difficult to meet the requirements, which also limits the application of epoxy resin anticorrosion coating [11]. Therefore, methods for modifying epoxy resin coatings, including structural modification [12,13], rubber modification [14], resin modification [15], and filler modification [16–23] are necessary. During the curing, the volatilization of a solvent can lead to the appearance of micropores, microcracks and diffusion channels in the coatings. These defects cause the easy absorption of moisture and reduction of the barrier and the adhesion properties of the coatings, thereby accelerating corrosion [24,25]. For this problem, various fillers have been used as reinforcements to improve coating performance. The modification of epoxy resin coatings with fillers can reduce the costs of coating production and the shrinkage and fluidity of the coatings and improve the chemical stability, thermal conductivity and mechanical properties of the coatings. The commonly used fillers are roughly classified into three types, namely, flaky particles, fibrous fillers and nanomaterials. Flaky particles mainly include aluminum powder and mica flakes [26]. They increase the corrosion resistance and improve the wear resistance of the coating. Fiber-based fillers, like carbon fiber and glass fiber, have excellent thermal stability, corrosion resistance, wear resistance, low density, and good interfacial compatibility [27,28]. However, the length of fibers is extremely difficult to limit; hence, its application is still limited. Nanomaterials generally refer to materials that have at least one dimension in a three-dimensional space with a nanometer size (1–100 nm). The addition of nanomaterials to the coating greatly enhances the adhesion, mechanical properties and corrosion resistance of the latter [29].

As an excellent nanomaterial, graphene oxide (GO) has gradually demonstrated broad prospects and important functions in the field of coating fabrication. GOs have a highly specific surface area and unique properties [30,31] and can enhance the barrier and corrosion resistance of coatings. The distribution and compatibility of GO plays a decisive role in the performance of composite coatings. GOs have a strong van der Waals force and contain hydrophilic oxygen-containing groups on their surfaces and are thus incompatible with coating matrixes [32]. Moreover, these features result in agglomeration, which in turn increases the amount of micropores and microcracks. The large amount of oxygen-containing groups, that is, hydroxyl, carboxyl and epoxide groups [32], on GO surfaces can provide suitable reaction sites for modification. Therefore, surface modification of GO is a promising route to improve compatibility. Li et al. [33] studied the properties of silane-coupling-agent-modified GO composite epoxy resin coatings. Their results showed that modified GO remarkably enhanced the mechanical properties of epoxy resin.

Cyclohexene oxide and its derivatives are active chemical raw materials. They are easily soluble in alcohols, ketones and various organic solvents. The epoxide groups of the molecular structure in cyclohexene oxide are sufficiently active to participate in a variety of reactions. For example, it can react with curing agents containing amino groups in coatings. Cyclohexene oxide is also a good reactive diluent of epoxy resin, which can improve the compatibility with epoxy resin after modifying the fillers in the coating system. At present, there is little research on the modification of fillers by cyclohexene oxide and its derivatives. In this work, we synthesized a cyclohexene oxide-based silane coupling agent, namely 2-(3,4-epoxycyclohexyl)ethyl triethoxysilane (ETEO), by hydrosilylation reaction [34], and used it to modify the surface of GO. Then, GO and the modified GO were added into epoxy resin coatings, and the effects of the modified GO on mechanical properties and corrosion resistance of the epoxy resin composite coatings were investigated. Finally, we determined the optimum ratio of the epoxy resin coatings with different filler contents to obtain the composite coating with the best anticorrosion performance.

2. Materials and Methods

2.1. Materials

Karstedt catalyst solution, 4-vinyl-1-cyclohexene-1,2-epoxide (VCHO), triethoxysilane (TES), and graphite (99.95%, metal basis) were Aladdin products (Shanghai, China). Potassium permanganate, concentrated sulfuric acid, hydrogen peroxide, sodium nitrate, hydrochloric acid, and toluene were purchased from Beijing Chemical Works (Beijing, China). A commercially available epoxy resin (E-44) and polyamide curing agent (650) were supplied by Wuxi Dic Epoxy Co., Ltd., Wuxi, China. The steel panels (Q235, 80 mm × 40 mm × 0.2 mm) were purchased from Biuged Laboratory Instruments (Guangzhou) Co., Ltd., Guangzhou, China.

2.2. Preparation of 2-(3,4-Epoxycyclohexyl)Ethyl Triethoxysilane (ETEO)

ETEO was prepared as described in our previous work [34]. In short, ETEO was synthesized through a hydrosilylation reaction with VCHO and TES. The reaction equation is shown in Figure 1.

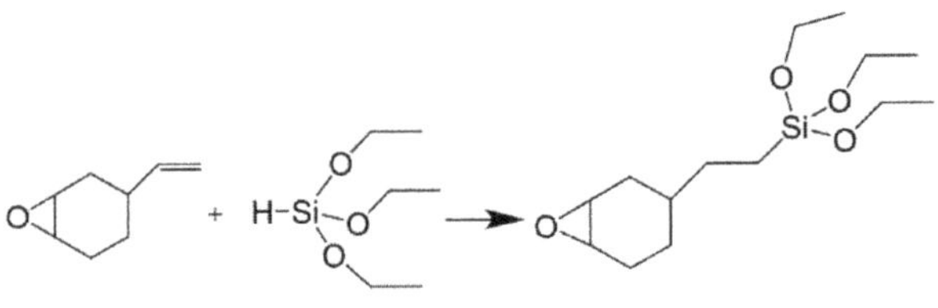

Figure 1. Schematic reaction of the 4-vinyl-1-cyclohexene-1,2-epoxide (VCHO) and triethoxysilane (TES).

2.3. Preparation of ETEO-Modified GO (ETEO-GO)

In this experiment, the graphene oxide was prepared by the Hummers method [35]. GO (0.2 g) was ultrasonically mixed with deionized water (40 mL) for 1 h. Then, the suspension was placed in a water bath at 60 °C and continuously stirred. The obtained ETEO (1.2 g) was gradually added, and the system was reacted under reflux for 12 h. The suspension was filtered after the reaction, unreacted ETEO was washed away with deionized water, and the product was dried at 60 °C for 24 h. Figure 2 shows the diagram of the synthesis of ETEO-modified GO.

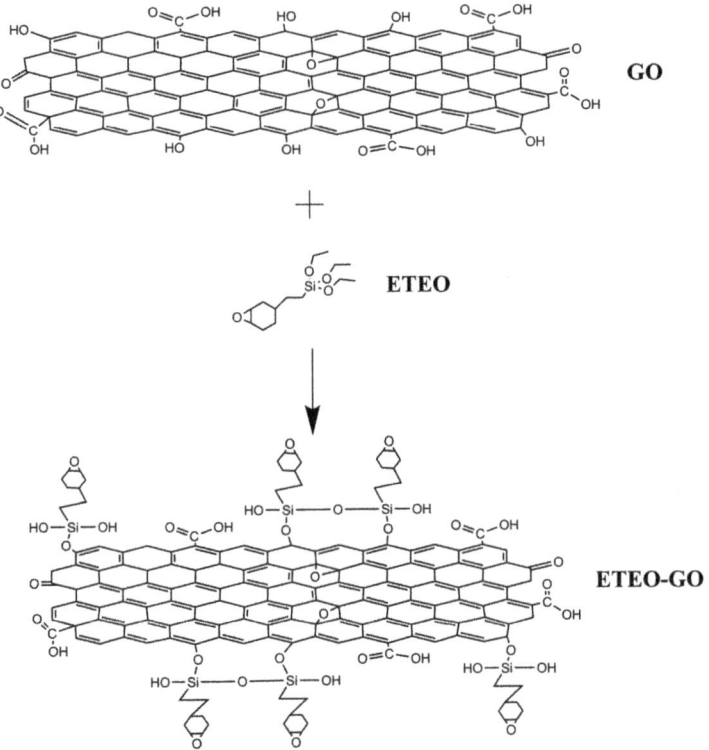

Figure 2. Synthesis reaction of GO and 2-(3,4-epoxycyclohexyl)ethyl triethoxysilane (ETEO).

2.4. Preparation of Coatings

The metal substrate was wiped clean with ethanol and acetone and dried before spraying. The coatings were made by taking the desired quantity of filler (0.7 wt % GO and 0.1 wt %, 0.4 wt %,

0.7 wt %, and 1.0 wt % ETEO-GO) with epoxy resin. In addition, the pure epoxy coating (EP) was used as a blank control group. Then, a curing agent was added into the compound and stirred at a high speed for 30 min (Table 1). After spraying, the samples were kept at room temperature for 72 h and then cured in an oven at 60 °C for 4 h. The dry thickness of the coatings were 120 ± 10 µm.

Table 1. Formation of coatings.

Sample	Epoxy Resin (g)	Curing Agent (g)	Filler (wt %)
Pure epoxy (EP)	20	16	–
GO	20	16	GO (0.7 wt %)
EGO1	20	16	ETEO-GO (0.1 wt %)
EGO2	20	16	ETEO-GO (0.4 wt %)
EGO3	20	16	ETEO-GO (0.7 wt %)
EGO4	20	16	ETEO-GO (1.0 wt %)

2.5. Characterization

Fourier transform infrared spectroscopy (FTIR) was recorded using KBr disks on Fourier transform infrared spectrometer (Nicolet Nexus 670, Nicolet, Thermo Fisher Scientific, Waltham, MA, USA). In all cases, the scans were carried out in the spectral range from 4000 cm^{-1} to 400 cm^{-1} with a resolution of 4 cm^{-1}. The phase crystalline structures of the GO and ETEO-GO were characterized by XRD analysis using Japanese D/max 2500PC X-ray diffractometer (Rigaku, Tokoy, Japan) with a scan range of 8°–90° and a scan speed of 5°/min. The chemical bonding of ETEO on the GO surface was studied by X-ray photoelectron spectroscopy (XPS, ESCALAB 250, American Thermoelectricity, Waltham, MA, USA) using Al-Kα as the source. The surface morphology of the GO and ETEO-GO and the cross-section morphology of the coatings were investigated by SEM (JSM-6700F, JEOL, Tokyo, Japan). The surface morphology was also studied by TEM (JEM-2100F, JEOL).

2.6. Mechanical Performance Test

The mechanical properties of the coating were investigated by hardness and impact resistance. The hardness of the coating was tested using a pencil hardness tester according to the GB/T 6739-2006 standard [36], and a pencil with a certain hardness was applied to the dried film. The scratches, indicated by the hardest pencil mark, do not cause damage to the coating film; the coating impact resistance was measured with a DuPont impactor according to ASTM-D2794 standard [37] and the test piece was placed face up under the instrument head. The test results were expressed in terms of the maximum drop height of the 1 kg weight that caused the coating to break.

2.7. Contact Angle Measurements

The contact angle of the coating was characterized using an OCA20 optical contact angle measuring instrument (Data physics, Filderstadt, Germany). The test environment was at room temperature, and the volume of a single drop of water was 2 µL. The value of each sample in the test was the average of three test points per plane, and the measurement accuracy was ±0.1°. The test procedure for all samples was performed after the water droplets were stable for 30 s.

2.8. Coating Electrochemical Impedance Test

The electrochemical impedance of the coating was tested using a PARSTAT 2273 electrochemical workstation (Ametek, Princeton, NJ, USA). The test frequency range was 10^5–10^{-2} Hz, the test voltage was 20 mV, the exposed area of the steel plate was 21 cm^2 with the NaCl solution, and the electrolytic cell was a three-electrode electrolytic cell (the counter electrode was platinum electrode, the reference electrode was saturated calomel electrode).

2.9. Salt Spray Test

According to the "Neutral Salt Spray Test Standard" (GB/T 10125-1997 [38]), the coating was continuously sprayed using a salt spray corrosion test chamber (ZhongkeMeiqi Technology Co., Ltd., Beijing, China). The coated panel was placed in a test chamber at an inclination angle of 15°, and the etching solution was 5% aqueous NaCl solution, which was scribed at the surface of the coating to accelerate the observation of the degree of corrosion of the coating.

3. Results and Discussion

3.1. FTIR Analysis

To confirm the functionalization of GO sheets with ETEO, we examined the FTIR spectra of the GO and ETEO-GO samples, as shown in Figure 3. The FTIR spectrum of GO revealed the C–OH stretching as a broad peak at 3427 cm^{-1}, the C=O stretching vibrations of carboxyl at 1730 and 1384 cm^{-1}, and C=C skeletal vibrations at 1629 cm^{-1}. After modification of GO with ETEO, the new bands appeared at 2923 and 1021 cm^{-1}, which were assigned to the methylene of ETEO and the Si–O–C bands between GO and ETEO. It can still be seen that the stretching vibration of the epoxide group appeared at 1070 and 830 cm^{-1}. The appearance of the above characteristic peaks proved that ETEO successfully modified GO.

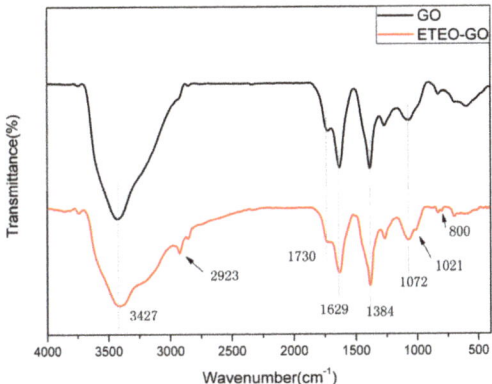

Figure 3. Fourier transform infrared spectroscopy (FTIR) spectroscopy of GO and ETEO-GO.

3.2. XRD Analysis

The XRD spectra of GO and ETEO-GO samples are shown in Figure 4. The sharp diffraction peak appeared at 10.7°, which belonged to the (001) crystal face for the GO sample, indicating that the structure of GO was highly ordered [39]. From Bragg's law, the interlayer distance of GO was 0.826 nm. The diffraction peak of the (001) crystal face disappeared, representing an increase of the interlayer spacing between the graphene oxide sheets. After functionalization of GO with ETEO, the XRD pattern of the ETEO-GO sample showed no significant diffraction peaks, and the diffraction peak of GO completely disappeared. This phenomenon may indicate that the ETEOs on the GO surfaces have been able to exfoliate away the lamellae and produce a significantly less compact structure [40]. That was why the XRD shows no diffraction alongside the measurable angel span. The ETEO destroyed the periodic structure of the GO and prevented the aggregation of the graphene sheets.

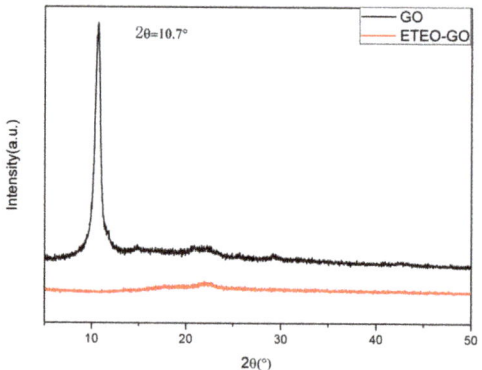

Figure 4. XRD pattern spectra of GO and ETEO-GO.

3.3. XPS Analysis

The surface elements and valence structures of GO and ETEO-GO were analyzed by XPS. From the results shown in Figure 5, the main element orbitals of GO were C 1s and O 1s. XPS spectra of GO and ETEO-GO revealed that the C:O:Si ratio of GO was 60:40:0 and that of ETEO-GO was 59:33:8, also indicating that GO has been functionalized with ETEO. The C 1s high-resolution spectra of GO sample (Figure 5a) provided a more detailed description, where the binding energies at 287.9, 286.7, 285.4, and 284.5 eV correspond to the groups of O–C=O, C=O, C–O, and C–C bond, respectively [32]. Figure 5c shows the XPS spectrum of ETEO-GO. The main elemental orbitals of ETEO-GO included C 1s, O 1s, and Si 2p. Compared with the GO sample, the C 1s signal of ETEO-GO in Figure 5d exhibited new peak at 285.6 (C–O/C–O–Si). The new bond indicated that the surface modification of GO by ETEO had successfully caused the change of the valence structure of the element.

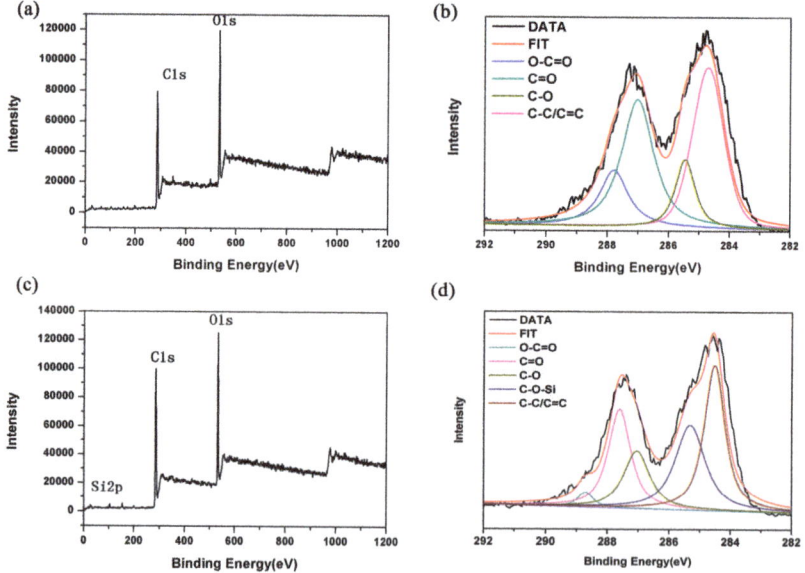

Figure 5. XPS survey spectra: (**a**) GO, (**c**) ETEO-GO; and high resolution spectra: (**b**) C 1s of GO, (**d**) C 1s of ETEO-GO.

3.4. Morphology Analysis

Scanning electron microscopy and transmission electron microscopy were used to observe the morphologies of GO and ETEO-GO in Figure 6 and Figure S1. GO exhibited a lamellar structure in Figure 6a, and a stacking phenomenon occurred between the lamellas. The ETEO-GO sample (Figure 6b) exhibited a very rough surface, as well as a fluffy, homogeneous deposited, and folded morphology. The reason might be that the GO had high surface energy, resulting in high mutual attraction between lamellas. After modification by ETEO, the surface energy of ETEO-GO was reduced due to the presence of siloxane groups, resulting in a decrease of mutual attraction between the ETEO-GO lamellas and preventing the lamellas from being separated due to excessive interaction force. As can be seen from Figure 6c, GO has a typical lamella-like structure, and the surface is very wrinkled. In Figure 6d, the surface of ETEO-GO nanomaterials was very rough and did not appear in a simple lamella form. Combined with the conclusions drawn from FTIR, XRD and XPS, the ETEO-GO nanomaterial was successfully prepared.

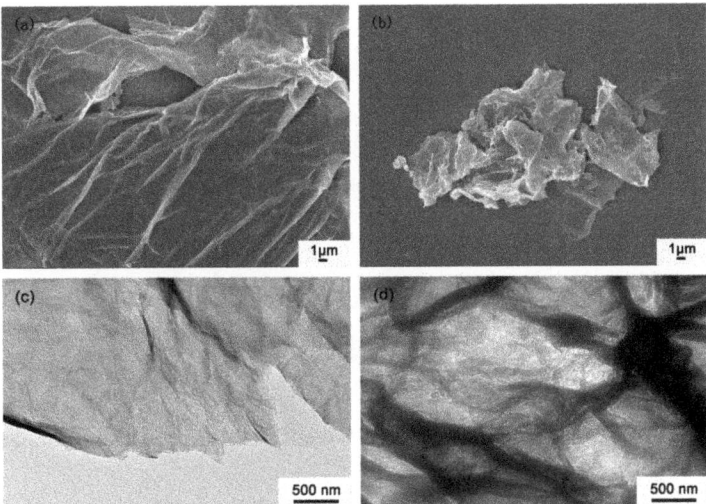

Figure 6. Morphologies of GO and ETEO-GO: (**a**) SEM image of GO; (**b**) SEM image of ETEO-GO; (**c**) TEM images of GO; (**d**) TEM images of ETEO-GO.

3.5. Morphology Analysis of Coating Cross-Section

The cross-section SEM image of the coating is shown in Figures 7 and S2. Figure 7a shows the cross-sectional morphology of the EP coating. The cross-section of EP was very uniform and smooth with no other impurities. Figure 7b shows the cross-sectional SEM image of the GO coating. The GO had agglomeration in the coating, leading to a decrease in the compactness of the coating. Therefore, many cracks and holes appeared in the cross-section. The distribution of ETEO-GO in the two coatings (Figure 7c,d) did not agglomerate, and the fracture mode of the coating section was still dominated by brittle fracture. When the content of ETEO-GO was increased to 0.7 wt % (Figure 7e), the cross-sectional morphology of the coating changed from smooth to rough, and the presence of ETEO-GO was fluffy, helping the ETEO-GO fill in the epoxy coating and increasing the compactness of the coating. In Figure 7f, the amount of ETEO-GO added was increased to 1 wt %, and cracks and holes appeared in the cross section of the coating. The ETEO-GO was agglomerated, showing that the high content of ETEO-GO would decrease the compactness of the coating.

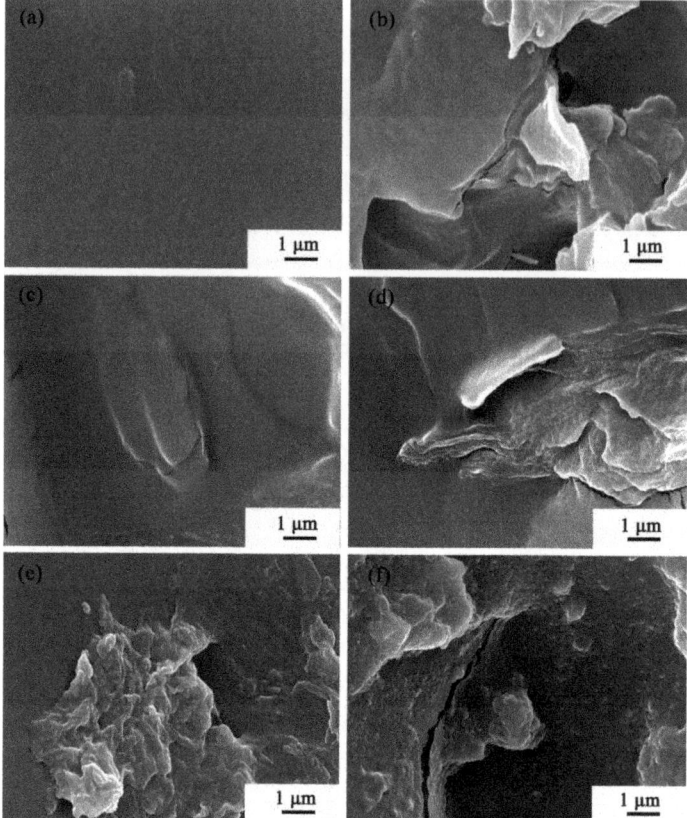

Figure 7. SEM images of cross-section microstructures: (**a**) EP; (**b**) GO; (**c**) EGO1; (**d**) EGO2; (**e**) EGO3; and (**f**) EGO4.

3.6. Mechanical Properties

Table 2 shows the results of the coating hardness tests. After the addition of GO, the hardness of the GO coating increased from 2H to 3H, while the hardness of the EGO3 coating with the same mass fraction increased by two grades. This result was mainly because the GO agglomerated in the coating and decreased the hardness due to the interaction force between the lamellas. With the increase of the amount of ETEO-GO added, the hardness of the coatings was continuously enhanced. When the addition amount was increased to 0.7 wt %, the hardness of the coating reached a maximum of 4H and did not continue to increase as the ETEO-GO continued to increase. This result may be attributed to the fact that ETEO-GO, as a rigid nanomaterial, can be evenly distributed in the epoxy resin to enhance the hardness of the coating. When ETEO-GO was added in excess to cause agglomeration, it still acted as a filler to enhance the strength of the coating and ensure that the coating retains its original strength.

Table 2. Hardness of EP, GO, and EGO coatings.

Sample	Hardness (H)
EP	2
GO	3
EGO1	2
EGO2	3
EGO3	4
EGO4	4

Table 3 lists the results of the coating impact resistance tests. As shown in Table 3, when GO and ETEO-GO were added as fillers to the coatings, the impact resistance of the coatings was higher than that of the EP coating. It showed that the addition of fillers can improve the coating impact resistance. When the addition amount was 0.7 wt %, the impact resistance of the ETEO-GO coating was greatly improved, more so than that of GO coating. The reason might be that ETEO-GO had good dispersibility in epoxy resin. In addition, ETEO-GO can be arranged in parallel in the matrix resin, and the oxygen-containing groups on the surface can be crosslinked with epoxy resin. These can increase the uniformity of the coating. Also, the good dispersion of ETEO-GO in the epoxy resin matrix effectively filled the micropores and microcracks presented in the epoxy resin itself, thereby reducing the defects of the coatings. When the coatings were subjected to external impact, ETEO-GO would reduce the local stress concentration of the coatings due to the fewer amounts of defects and enhance the impact resistance of the coatings. When the addition amount of ETEO-GO was changed, the EGO3 coating showed the best impact resistance. As mentioned, ETEO-GO produced agglomeration in the EGO4 coating, causing cracks and holes in the coating, which resulted in the lower impact resistance of the EGO4 coating than EGO3.

Table 3. Impact resistance of EP, GO, and EGO coatings; the values are the mean of five replicates and (±) correspond to the standard.

Sample	Height (cm)
EP	10.2 ± 0.5
GO	38.7 ± 1.6
EGO1	19.4 ± 0.4
EGO2	22.4 ± 0.5
EGO3	≥50
EGO4	40.1 ± 0.9

3.7. Wetting Performance

The contact angle can reflect the wettability of the coatings and affect the corrosion resistance [41]. Figure 8 shows a photograph of the contact angles of the composite coatings. The contact angle of the GO coating was reduced to 56.2° relative to the EP coating. This was mainly because the surface of the GO had a large number of oxygen-containing functional groups [42], and when the surface of the coating was in contact with water, the oxygen-containing functional groups were easily combined with water molecules by making hydrogen bonding [39], resulting in a phenomenon in which the value of the contact angle decreased. The ETEO molecule had a cyclohexene oxide group containing several methylene groups. Besides, Si–O–Si structure was formed after ETEO was grafted to the GO surface, which caused an increase in the non-polarity of the GO surface. Therefore, the surface energy of the ETEO-GO coating was lower than the GO coating, and the contact angle was higher as shown in Figure 8. In addition, due to the poor dispersion of GO, the surface energy of the coating may be uneven. The region where GO agglomerated was more likely to form hydrogen bonds. Thus, the contact angle was decreased and the surface energy was increased.

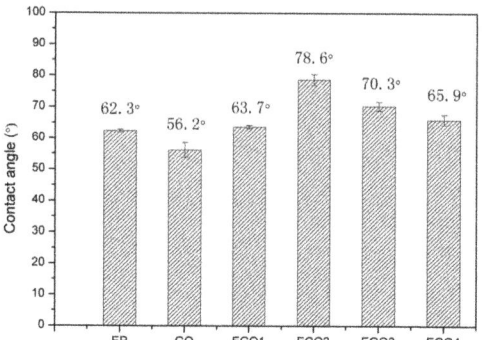

Figure 8. Water contact angle test results of different coatings: EP, GO, EGO1, EGO2, EGO3, and EGO4.

3.8. Electrochemical Impedance Spectroscopy

The coatings were immersed in a 3.5 wt % NaCl solution for 1 day, 10 days and 50 days. The electrochemical protection properties of the coating on the steel were observed, and the results are shown in Figures 9 and 10. In Figure 9, the phase angle values of all coatings were close to 90° in the 10–10^5 Hz frequency range, and the low-frequency impedance modulus exceeded 10^9 $\Omega\cdot cm^2$, where the EGO3 coating was the highest in all samples, exceeding 10^{11} $\Omega\cdot cm^2$. After 10 days, the low-frequency impedance modulus of the EP coating and the GO coating were reduced to 2×10^7 and 4×10^8 $\Omega\cdot cm^2$, respectively, while the EGO3 coating remained above 10^9 $\Omega\cdot cm^2$. The phase angle of the EGO3 coating was close to 90° in the frequency range of 10–10^5 Hz. On the contrary, the phase angle value of the EP coating was obviously reduced, and the phase angle value of the GO coating in the intermediate frequency phase decreased rapidly. This result indicated that the defects caused by corrosion occurred in the two coatings at this time. The low-frequency impedance modulus of the EGO3 coating was still the highest among the all coatings after 50 days. Figure 10 shows the Nyquist diagrams of the coatings. The corrosion resistance of the coating can be measured by the radius of capacitance arc. After the immersion time reached 50 days, the EP coating was shown as two capacitance arcs with two relaxation times. The coating was contacted with the corrosive medium, and the corrosion protection effect failed. The curve radius of the EGO3 coating was always the largest in the Nyquist diagram of different immersed times, and the protection effect was the best. The data in the Bode and Nyquist diagrams shows that the corrosive medium passed through defects and holes of coating into coating/substrate interface, resulting in delamination of coating. The addition of GO filled in some coating defects. Given that the GO was not uniformly dispersed in the coating, the corrosion resistance was not greatly improved. The EGO3 coating still had good corrosion protection to the steel substrate after immersion for 50 days, which might be attributed to the organic functional groups on the surface of ETEO-GO that promoted its uniform dispersion in the coating to fill the holes and corrosion channels. The corrosion resistance of composite coatings for different levels of ETEO-GO can also be illustrated by the low-frequency impedance modulus. With the increase of ETEO-GO content, the corrosion resistance of EGO composite coatings first increased and then decreased, which may be caused by the agglomeration of ETEO-GO in the coating.

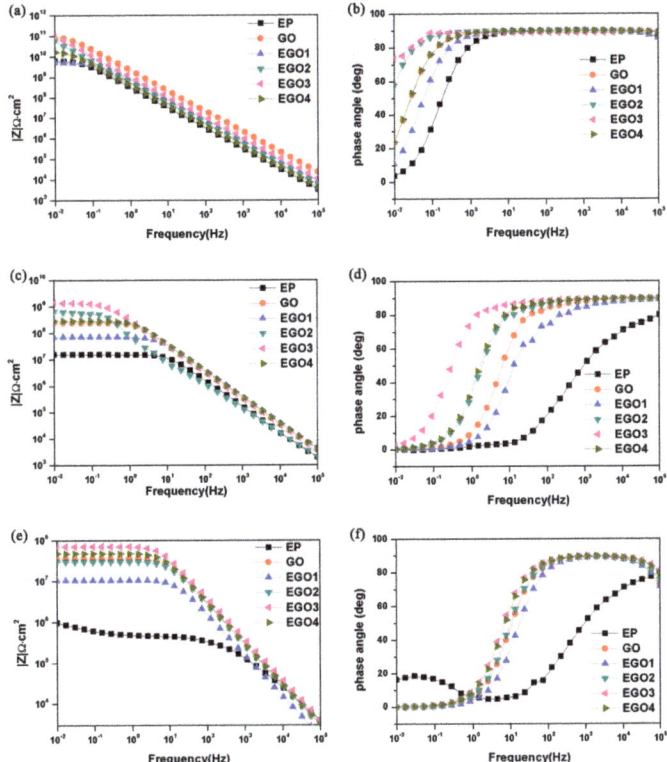

Figure 9. Bode diagrams of coatings immersed in 3.5 wt % NaCl solution for (**a,b**) 1 day; (**c,d**) 10 days and (**e,f**) 50 days.

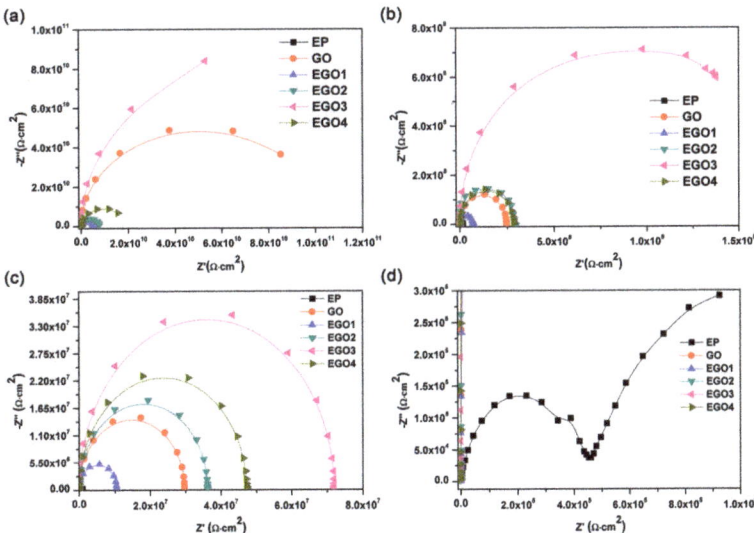

Figure 10. Nyquist diagrams of coatings immersed in 3.5 wt % NaCl solution for (**a**) 1 day, (**b**) 10 days and (**c**) 50 days. (**d**) Enlarged view of the red area in (**c**).

3.9. Salt Spray Test

Figure 11 is a digital photograph of coatings after a 400-h salt spray test. Rusts appeared on the surface of the EP coating, and a large area of blisters was generated on the surface, indicating that the corrosion medium had penetrated the EP coating completely to the steel substrate. However, a significant agglomeration of GO was observed in the GO coating, mainly because the uneven dispersion of GO in the coating resulted in different densities on the surface of the GO coating. The agglomeration of GO increased the number of micropores and microcracks in the coating. Micropores and microcracks can promote the generation of corrosion channels. Thus, a large amount of corrosive medium diffused through the corrosion channels to the substrate, then the substrate began to corrode, and corrosion products appeared. The corrosion products would destroy the original structure of the coating and enlarge the corrosion channels, which would eventually lead to more serious corrosion. Lowered corrosion products were created in the scratch section of the coatings reinforced by ETEO-GO. These results confirmed that surface modification of GO with ETEO improved the barrier and protection performance of the coatings. Besides, corrosion decreased with increasing ETEO-GO contents. The surface of the EGO3 and EGO4 coatings showed only a small amount of corrosion products and no blisters, showing better corrosion protection effects.

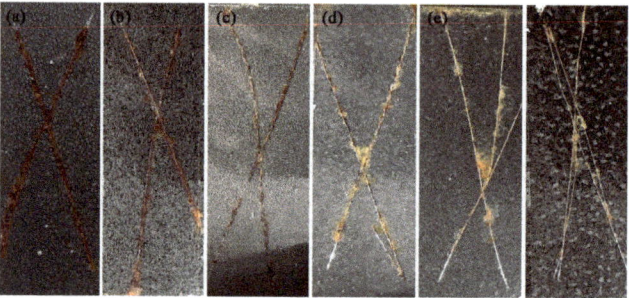

Figure 11. Photograph of samples after salt spray test for 400 h: (**a**) EP; (**b**) GO; (**c**) EGO1; (**d**) EGO2; (**e**) EGO3; and (**f**) EGO4.

3.10. Analysis of Corrosion Protection Mechanism

The results of EIS experiments and salt spray tests showed that the corrosion resistance of the composite coatings was obviously improved after the addition of GO and ETEO-GO. The reason might be the reduction of cracks and micropores in the coating after the addition of GO and ETEO-GO. It has been shown that the addition of ETEO modified the GO surface and can significantly increase the layer spacing of GO or completely exfoliate away the lamellae. Therefore, the ETEO-GO was better dispersed in the coating system. During the curing process, ETEO-GO molecules with epoxide groups formed covalent bonds with the amino groups in the low molecular polyamide curing agents, which can enhance the crosslink density of the composite coating and increase the compatibility of the fillers with the coating matrix. Thereby, the barrier of the fillers can be increased against the penetration of the corrosive medium. Moreover, the cyclohexene oxide groups which existed in the ETEO molecules and Si–O–Si structures which formed after the modification can increase in the non-polarity of the GO surface and lower the surface energy of the ETEO-GO coatings.

4. Conclusions

ETEO silane coupling agent was synthesized by VCHO and TES through the hydrosilylation method, and the surface modification of GO was carried out by using ETEO. The periodic lamella structure of GO was grafted with ETEO and then destroyed and transformed into a fluffy state with a rough surface, which prevented the agglomeration of GO. The addition of ETEO-GO improved the mechanical properties of the coatings and imparted a certain hydrophobicity to the coatings.

The results of EIS and salt spray tests showed that the coatings containing ETEO-GO had better corrosion resistance than the pure epoxy coating and the coatings with GO. The coating with 0.7 wt % ETEO-GO showed the most excellent corrosion resistance.

Supplementary Materials: The following are available online at http://www.mdpi.com/2079-6412/9/1/46/s1, Figure S1: TEM images of (a) GO and (b) ETEO-GO, Figure S2: SEM images of cross-section microstructures for smaller magnifications: (a) EP; (b) GO; (c) EGO1; (d) EGO2; (e) EGO3; and (f) EGO4.

Author Contributions: Conceptualization, C.Z., Y.D. and X.D.; Formal Analysis, C.Z., X.D. and G.S.; Data Curation, X.D., Y.W. and L.Q.; Writing—Original Draft Preparation, X.D. and Y.W.; Writing—Review and Editing, X.D., Y.S. and P.L.

Funding: This research was funded by the Jilin Province Natural Science Foundation of China (No. 20180101197jc).

Conflicts of Interest: The authors declare no conflict of interest.

References

1. Zheludkevich, M.L.; Tedim, J.; Ferreira, M.G.S. "Smart" coatings for active corrosion protection based on multi-functional micro and nanocontainers. *Electrochim. Acta* **2012**, *82*, 314–323. [CrossRef]
2. Tallman, D.E.; Spinks, G.; Dominis, A.; Wallace, G.G. Electroactive conducting polymers for corrosion control. *J. Solid State Electrochem.* **2002**, *6*, 73–84. [CrossRef]
3. Nguyen, T.N.; Hubbard, J.B.; McFadden, G.B. Mathematical model for the cathodic blistering of organic coatings on steel immersed in electrolytes. *J. Coat. Technol.* **1991**, *63*, 43–52.
4. Grundmeier, G.; Schmidt, W.; Stratmann, M. Corrosion protection by organic coatings: Electrochemical mechanism and novel methods of investigation. *Electrochim. Acta* **2000**, *45*, 2515–2533. [CrossRef]
5. Babić, R.; Metikoš-Huković, M.; Radovčić, H. The study of coal tar epoxy protective coatings by impedance spectroscopy. *Prog. Org. Coat.* **1994**, *23*, 275–286. [CrossRef]
6. Pascault, J.P.; Williams, R.J.J. Relationships between glass-transition temperature and conversion-analyses of limiting cases. *Polym. Bull.* **1990**, *24*, 115–121. [CrossRef]
7. Chmielewska, D.; Sterzyński, T.; Dudziec, B. Epoxy compositions cured with aluminosilsesquioxanes: Thermomechanical properties. *J. Appl. Polym. Sci.* **2014**, *131*, 8444–8452. [CrossRef]
8. Arman, S.Y.; Ramezanzadeh, B.; Farghadani, S.; Mehdipour, M.; Rajabi, A. Application of the electrochemical noise to investigate the corrosion resistance of an epoxy zinc-rich coating loaded with lamellar aluminum and micaceous iron oxide particles. *Corros. Sci.* **2013**, *77*, 118–127. [CrossRef]
9. Gharagozlou, M.; Ramezanzadeh, B.; Baradaran, Z. Synthesize and characterization of a novel anticorrosive cobalt ferrite nanoparticles dispersed in silica matrix ($CoFe_2O_4$-SiO_2) to improve the corrosion protection performance of epoxy coating. *Appl. Surf. Sci.* **2016**, *377*, 86–98. [CrossRef]
10. Shanker, A.K.; Cervantes, C.; Loza-Tavera, H.; Avudainayagam, S. Chromium toxicity in plants. *Environ. Int.* **2005**, *31*, 739–753. [CrossRef]
11. Hodgkin, J.H.; Simon, G.P.; Varley, R.J. Thermoplastic toughening of epoxy resins: A critical review. *Polym. Adv. Technol.* **1998**, *9*, 3–10. [CrossRef]
12. Na, T.; Liu, X.; Jiang, H.; Zhao, L.; Zhao, C. Enhanced thermal conductivity of fluorinated epoxy resins by incorporating inorganic filler. *React. Funct. Polym.* **2018**, *128*, 84–90. [CrossRef]
13. Tao, Z.; Yang, S.; Ge, Z.; Chen, J.; Fan, L. Synthesis and properties of novel fluorinated epoxy resins based on 1,1-bis(4-glycidylesterphenyl)-1-(3′-trifluoromethylphenyl)-2,2,2-trifluoroethane. *Eur. Polym. J.* **2007**, *43*, 550–560. [CrossRef]
14. Chikhi, N.; Fellahi, S.; Bakar, M. Modification of epoxy resin using reactive liquid (ATBN) rubber. *Eur. Polym. J.* **2002**, *38*, 251–264. [CrossRef]
15. Chaudhary, S.; Surekha, P.; Kumar, D.; Rajagopal, C.; Roy, P.K. Amine-functionalized poly(styrene) microspheres as thermoplastic toughener for epoxy resin. *Polym. Compos.* **2015**, *36*, 174–183. [CrossRef]
16. Conde, A.; Durán, A.; de Damborenea, J.J. Polymeric sol–gel coatings as protective layers of aluminium alloys. *Prog. Org. Coat.* **2003**, *46*, 288–296. [CrossRef]
17. Dhoke, S.K.; Khanna, A.S. Effect of nano-Fe_2O_3 particles on the corrosion behavior of alkyd based waterborne coatings. *Corros. Sci.* **2009**, *51*, 6–20. [CrossRef]

18. Dolatzadeh, F.; Moradian, S.; Jalili, M.M. Influence of various surface treated silica nanoparticles on the electrochemical properties of SiO_2/polyurethane nanocoatings. *Corros. Sci.* **2011**, *53*, 4248–4257. [CrossRef]
19. Kozhukharov, S.; Tsaneva, G.; Kozhukharov, V.; Gerwann, J.; Schem, M.; Schmidt, T.; Veith, M. Corrosion protection properties of composite hybrid coatings with involved nanoparticles of zirconia and ceria. *J. Univ. Chem. Technol. Metall.* **2008**, *43*, 73–80.
20. Radhakrishnan, S.; Siju, C.R.; Mahanta, D.; Patil, S.; Madras, G. Conducting polyaniline–nano-TiO_2 composites for smart corrosion resistant coatings. *Electrochim. Acta* **2009**, *54*, 1249–1254. [CrossRef]
21. Ramezanzadeh, B.; Attar, M.M. Studying the corrosion resistance and hydrolytic degradation of an epoxy coating containing zno nanoparticles. *Mater. Chem. Phys.* **2011**, *130*, 1208–1219. [CrossRef]
22. Schem, M.; Schmidt, T.; Gerwann, J.; Wittmar, M.; Veith, M.; Thompson, G.E.; Molchan, I.S.; Hashimoto, T.; Skeldon, P.; Phani, A.R.; et al. CeO_2-filled sol–gel coatings for corrosion protection of AA2024-T3 aluminium alloy. *Corros. Sci.* **2009**, *51*, 2304–2315. [CrossRef]
23. Yu, H.J.; Wang, L.; Shi, Q.; Jiang, G.H.; Zhao, Z.R.; Dong, X.C. Study on nano-$CaCO_3$ modified epoxy powder coatings. *Prog. Org. Coat.* **2006**, *55*, 296–300. [CrossRef]
24. Wan, Y.-J.; Tang, L.-C.; Gong, L.-X.; Yan, D.; Li, Y.-B.; Wu, L.-B.; Jiang, J.-X.; Lai, G.-Q. Grafting of epoxy chains onto graphene oxide for epoxy composites with improved mechanical and thermal properties. *Carbon* **2014**, *69*, 467–480. [CrossRef]
25. Yi, H.; Chen, C.; Zhong, F.; Xu, Z. Preparation of aluminum oxide-coated carbon nanotubes and the properties of composite epoxy coatings research. *High Perform. Polym.* **2014**, *26*, 255–264. [CrossRef]
26. Kalenda, P.; Kalendová, A.; Štengl, V.; Antoš, P.; Šubrt, J.; Kváča, Z.; Bakardjieva, S. Properties of surface-treated mica in anticorrosive coatings. *Prog. Org. Coat.* **2004**, *49*, 137–145. [CrossRef]
27. Dong, Y.; Li, S.; Zhou, Q. Self-healing capability of inhibitor-encapsulating polyvinyl alcohol/polyvinylidene fluoride coaxial nanofibers loaded in epoxy resin coatings. *Prog. Org. Coat.* **2018**, *120*, 49–57. [CrossRef]
28. Spainhour, L.K.; Wootton, I.A. Corrosion process and abatement in reinforced concrete wrapped by fiber reinforced polymer. *Cem. Concr. Compos.* **2008**, *30*, 535–543. [CrossRef]
29. Shi, X.; Nguyen, T.A.; Suo, Z.; Liu, Y.; Avci, R. Effect of nanoparticles on the anticorrosion and mechanical properties of epoxy coating. *Surf. Coat. Technol.* **2009**, *204*, 237–245. [CrossRef]
30. Stankovich, S.; Dikin, D.A.; Dommett, G.H.; Kohlhaas, K.M.; Zimney, E.J.; Stach, E.A.; Piner, R.D.; Nguyen, S.T.; Ruoff, R.S. Graphene-based composite materials. *Nature* **2006**, *442*, 282–286. [CrossRef]
31. Liang, J.; H, Y.; Zhang, L.; Wang, Y.; Ma, Y.; Guo, T.; Chen, Y. Molecular-level dispersion of graphene into poly(vinyl alcohol) and effective reinforcement of their nanocomposites. *Adv. Funct. Mater.* **2009**, *19*, 2297–2302. [CrossRef]
32. Dreyer, D.R.; Park, S.; Bielawski, C.W.; Ruoff, R.S. The chemistry of graphene oxide. *Chem. Soc. Rev.* **2010**, *39*, 228–240. [CrossRef] [PubMed]
33. Li, Z.; Wang, R.; Young, R.J.; Deng, L.; Yang, F.; Hao, L.; Jiao, W.; Liu, W. Control of the functionality of graphene oxide for its application in epoxy nanocomposites. *Polymer* **2013**, *54*, 6437–6446. [CrossRef]
34. Wang, Y.N.; Dai, X.Y.; Xu, T.L.; Qu, L.J.; Zhang, C.L. Preparation and anticorrosion properties of silane grafted nano-silica/epoxy composite coating. *Chem. J. Chin. Univ.* **2018**, *39*, 1564–1572.
35. Hummers, W.S.; Offeman, R.E. Preparation of graphitic oxide. *J. Am. Chem. Soc.* **1958**, *80*, 1339. [CrossRef]
36. *GB/T 6739-2006 Paints and Varnishes–Determination of Film Hardness by Pencil Test*; Standardization Administration of China: Beijing, China, 2006. (In Chinese)
37. *ASTM D2794-93 Standard Test Method for Resistance of Organic Coatings to the Effects of Rapid Deformation (Impact)*; ASTM International: West Conshohocken, PA, USA, 2010.
38. *GB/T 10125-1997 Corrosion Tests in Artificial Atmospheres–Salt Spray Tests*; Standardization Administration of China: Beijing, China, 2006. (In Chinese)
39. Ramezanzadeh, B.; Ahmadi, A.; Mahdavian, M. Enhancement of the corrosion protection performance and cathodic delamination resistance of epoxy coating through treatment of steel substrate by a novel nanometric sol-gel based silane composite film filled with functionalized graphene oxide nanosheets. *Corros. Sci.* **2016**, *109*, 182–205. [CrossRef]
40. Ganjaee Sari, M.; Shamshiri, M.; Ramezanzadeh, B. Fabricating an epoxy composite coating with enhanced corrosion resistance through impregnation of functionalized graphene oxide-co-montmorillonite nanoplatelet. *Corros. Sci.* **2017**, *129*, 38–53. [CrossRef]

41. Zhang, D.; Wang, L.; Qian, H.; Li, X. Superhydrophobic surfaces for corrosion protection: A review of recent progresses and future directions. *J. Coat. Technol. Res.* **2016**, *13*, 11–29. [CrossRef]
42. Kang, W.-S.; Rhee, K.Y.; Park, S.-J. Influence of surface energetics of graphene oxide on fracture toughness of epoxy nanocomposites. *Compos. Part B Eng.* **2017**, *114*, 175–183. [CrossRef]

© 2019 by the authors. Licensee MDPI, Basel, Switzerland. This article is an open access article distributed under the terms and conditions of the Creative Commons Attribution (CC BY) license (http://creativecommons.org/licenses/by/4.0/).

Article

Cost-Effective Surface Modification of Carbon Cloth Electrodes for Microbial Fuel Cells by Candle Soot Coating

Bor-Yann Chen, Yuan-Ting Tsao and Shih-Hang Chang *

Department of Chemical and Materials Engineering, National I-Lan University, I-Lan 260, Taiwan; bychen@niu.edu.tw (B.-Y.C.); r0523006@ms.niu.edu.tw (Y.-T.T.)
* Correspondence: shchang@niu.edu.tw; Tel.: +886-3-935-7400

Received: 7 November 2018; Accepted: 16 December 2018; Published: 17 December 2018

Abstract: This study explored an economically-feasible and environmentally friendly attempt to provide more electrochemically promising carbon cloth anodes for microbial fuel cells (MFCs) by modifying them with candle soot coating. The sponge-like structure of the deposited candle soot apparently increased the surface areas of the carbon cloths for bacterial adhesion. The super-hydrophilicity of the deposited candle soot was more beneficial to bacterial propagation. The maximum power densities of MFCs configured with 20-s (13.6 ± 0.9 mW·m^{-2}), 60-s (19.8 ± 0.2 mW·m^{-2}), and 120-s (17.6 ± 0.8 mW·m^{-2}) candle-soot-modified carbon cloth electrodes were apparently higher than that of an MFC configured with an unmodified electrode (10.2 ± 0.2 mW·m^{-2}). The MFCs configured with the 20- and 120-s candle-soot-modified carbon cloth electrodes exhibited lower power densities than that of the MFC with the 60-s candle-soot-modified carbon cloth electrode. This suggested that the insufficient residence time of candle soot led to an incomplete formation of the hydrophilic surface, whereas protracted candle sooting would lead to a thick deposited soot film with a smaller conductivity. The application of candle soot for anode modification provided a simple, rapid, cost-effective, and environment-friendly approach to enhancing the electron-transfer capabilities of carbon cloth electrodes. However, a postponement in the MFC construction may lead to a deteriorated hydrophilicity of the candle-soot-modified carbon cloth.

Keywords: microbial fuel cells; candle soot; carbon cloth electrode; surface modification

1. Introduction

Microbial fuel cells (MFCs) are an environmentally friendly option for alternative-energy applications, as they can convert chemically bound energy into biomass-based electricity by electroactive bacteria during a wastewater treatment [1–4]. MFCs can also be applied in the removal of toxic pollutants, in environmental sensors, in harvesting the energy stored in marine sediments, in bioremediation, and in desalination [3]. Recently, MFCs are utilized as a simultaneous power source of self-powered electrochemical biosensors, because no potentiostat, power for the potentiostat, and/or power for the signaling device are needed [5]. However, there are still some challenges that need to be resolved in the practical applications of MFCs, including low power generation, the cost of anode materials for large-scale applications, system development, and energy recovery [6,7]. Furthermore, the low extracellular electron transfer efficiency between the microorganism and the electrode is the main bottleneck limiting the practical applications of MFCs [8]. Therefore, it is important to improve electrode properties by a surface treatment to enhance the extracellular electron transfer efficiency at the anode. Electrochemically active bacteria generate bioelectricity through three mechanisms: A direct electron transport through membrane-bound proteins, conductive nanowires, and indirect shuttles through redox mediators [9]. As bacteria play a crucial role in the generation of bioelectricity, the characteristics of anode electrodes are crucial for bacterial attachment to have a power generation capability in the cost-effective operations of MFCs [10,11]. Compared to other materials, carbonaceous electrodes are typically suitable as anodes of MFCs, owing to their high conductivities, good biocompatibilities, excellent chemical stabilities, and relatively low costs [12,13]. However, the undesirable hydrophobicity of the carbonaceous electrodes normally leads to a high resistance of electron transfer and low bioelectricity-generating efficiency.

To solve such technical problems, numerous studies demonstrated that appropriate modifications upon the carbonaceous electrode surface seem to effectively improve electron-transfer characteristics and power-generating performances of MFCs [10–14]. For example, Cheng and Logan [14] used an ammonia treatment to increase the positive charges on the surfaces of carbon cloth electrodes and obtained a maximum power density of 1970 mW·m^{-2}. Feng et al. [15] reported that the power generation of MFCs could be improved by an acid soaking of carbon fibers and approached a maximum power density of 1370 mW·m^{-2}. Lowy et al. [16] and Tang et al. [17] demonstrated that the performances of MFCs could be improved by an electrochemical oxidation treatment of graphite electrodes. Lowy et al. [16] reported that the quinone-modification of previously oxidized graphite electrodes yielded an increase of the kinetic activity by a factor of 218. Tang et al. [17] showed that MFCs with electrochemically oxidized graphite felt anodes produced a current of 1.13 mA, 39.5% higher compared with that of MFCs containing untreated anodes. Chang et al. [18] reported that carbon cloth electrodes exhibited superior surface and electrochemical properties after a modification by atmospheric-pressure plasma jets. According to their study, the maximum power density of the MFCs could be increased from 2.38 to 7.56 mW·m^{-2} after modification. In addition, the surface properties of anode electrodes can be improved by coating them with carbon nanotubes [19], ferric oxides [20], Au nanoparticles [21], goethite nanowhiskers [22], NiO nanoflake arrays [23], reduced graphene oxides [24], and tungsten carbide [25]. However, most of these surface modification methods are time consuming, less economically feasible, or not environmentally appropriate, due to the use of chemicals that are potentially harmful to the environment.

Candle soot particles are tiny, unburned carbon that originated from the incomplete combustion of easily available candles. Recently, candle soot has been widely implemented in solar and fuel cell applications owing to its low cost, rapid and simple preparation, non-toxicity, high specific surface area, and good conductivity [26–29]. For example, Wei et al. [26] have developed cost-efficient, environmentally stable clamping solar cells by using candle soot for the hole extraction from ambipolar perovskites. Kakunuri and Sharma [27] reported a simple and inexpensive approach to synthesizing a fractal-like interconnected network of carbon nanoparticles from candle soot, used as an anode material for a high-rate lithium-ion battery. Khalakhan et al. [28] reported that the elementary preparation, high specific surface area, good conductivity, and hydrophobicity make candle soot a promising material for the support of the proton-exchange-membrane fuel-cell catalyst. Liang et al. [29] reported that ultrafine soot particles formed in the flame tip region of a candle are composed of elemental carbon and ash, have a large specific surface area, and are hydrophilic. Evidently, the hydrophilicity and large specific surface area of the flame-tip soot particles suggested that they were likely promising for the surface modification of carbonadoes electrodes in MFCs. Nevertheless, no extensive studies on the applications of candle soot have been reported for MFCs. Singh et al. [30] reported for the first time the use of candle soot to modify the electrodes of MFCs. They successfully fabricated an ultrafine stainless steel wire disk deposited by carbon nanoparticles derived from candle soot as the electrode of an MFC and demonstrated that such modified MFC could provide a high capability for bioenergy extraction. Although stainless steels own excellent corrosion resistances, long-term interactions between stainless steel and living organisms might still cause corrosion of the steel's chromium oxide layer, leading to the release of metal ions and the inhibition of microbial growth. Here, carbon nanoparticles were coated with candle soot, but they were directly deposited on the surfaces of carbon cloths. The surface properties of the candle-soot-modified carbon cloths with various duration of times were studied for the maximal power generation of MFCs.

2. Materials and Methods

2.1. Construction of MFCs

Membrane-free air-breathing cathode single-chamber MFCs were adopted as described elsewhere [31]. The MFCs were constructed in cylindrical tubes made of poly(methyl methacrylate) with an operating volume of 220 mL. The anodes of the MFCs were carbon cloths (without waterproofing or catalyst) with projected areas of approximately 22.9 cm^2. The sizing of the air cathodes had dimensions approximately equal to those of the anodes and comprised a polytetrafluoroethylene diffusion layer on the air-facing side. Both carbon cloth and polytetrafluoroethylene diffusion layer were purchased from CeTech, Taichung, Taiwan. Figure 1 shows the photography of the membrane-free air-breathing cathode single-chamber MFC used in this study. Some of the carbon cloth anodes were directly placed into the candle flame tip region for 20, 60, and 120 s prior to the construction of MFCs. To consider the economic feasibility for sustainable development, the candles used were without further treatment and purchased from a common grocery store in I-Lan, Taiwan.

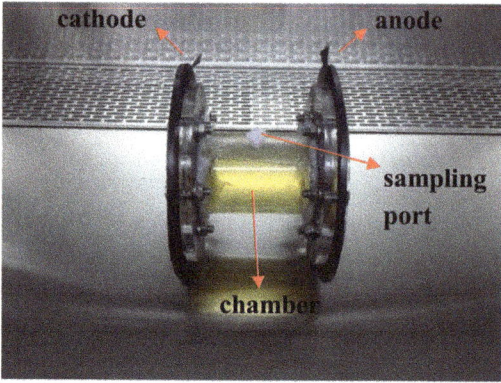

Figure 1. Photography of the membrane-free air-breathing cathode single-chamber microbial fuel cells (MFC).

2.2. Experimental Operations

Acclimation step: the microbe used in MFCs was *Aeromonas hydrophila* NIU01. The Luria–Bertani (LB) broth medium (tryptone: 10 g·L^{-1}, yeast extract: 5 g·L^{-1}, and sodium chloride: 10 g·L^{-1}) was used as the culture medium in the MFCs. Approximately 5 mL of a concentrated O/N cultured biomass was mixed with 0.2 × LB in MFCs for acclimatization. For a serial acclimation, approximately 5 mL of the cell broth were replaced by an impulse injection with a fresh sterile 8.8 × LB medium every 48 h. The output voltage of the MFC was continuously monitored to determine whether stable bioelectricity-generating profiles were achieved to guarantee success of electrochemical acclimatization. Then, the steady-state output power generation of MFCs was approximately achieved after approx. 30 days acclimation.

Experimental step: The batch-fed mode of MFC operation with impulse injection of energy substrate was carried out at 25 °C every 7 days. That is, 5.0 mL of 8.8 × LB broth laden was supplemented to MFCs to maintain culture medium in 0.2 × LB for inspection. Approx. 1 h after impulse injection of energy substrate, the supplemented medium was considered to be well-distributed in MFC and then electrochemical analysis of MFCs was conducted.

2.3. Characterizations

The surface and cross-sectional morphologies of the candle-soot-modified carbon cloths were measured with a scanning electron microscope (SEM) (5136MM, Tescan, Kohoutovice, Czech Republic). The surface wettabilities of the candle-soot-modified carbon cloths were measured using the sessile drop method by a contact-angle instrument (FTA125, First Ten Ångstroms, Portsmouth, OH, USA). Digital images of deionized-water droplets with volumes of approximately 10 µL were captured after the droplet on the film reached a steady state (approximately 5 s) to determine the equilibrium sessile contact angle. The average contact angle was determined by averaging 7 random measurements at different locations on the film, excluding outlier-data values. The surface chemical compositions of the candle-soot-modified carbon cloths were analyzed using an X-ray photoelectron spectrometer (XPS) (K-Alpha, Thermo Scientific, Waltham, MA, USA) with a monochromatic Al Kα radiation source (1468.6 eV). Survey spectra of the specimens were acquired in the range of 0–1000 eV with steps of 1 eV. C1s and O1s spectra of the specimens were measured in steps of 0.05 eV. The power-generating capabilities of the MFC were evaluated using an electrochemical workstation (ZIVE SP1, WonAtech, Seoul, Korea). The MFC voltage was automatically measured with an external resistance of R_{out} = 1000 Ω for comparison purpose. The power and current densities of each MFC were determined by linear sweep voltammetry (LSV) measurements at a scan rate of 0.1 mV$\cdot$s^{-1}; the corresponding voltages were recorded using a multimeter (ZIVE SP1, WonAtech, Seoul, Korea). All of the MFC experimental tests were carried out at ambient temperature. The average power density was calculated from 7 replicated-measurements. The internal resistance of the MFC was measured by electrochemical impedance spectroscopy (EIS) at open-circuit voltage conditions in a frequency range of 0.005–100,000 Hz at an amplitude of 10 mV.

3. Results

3.1. Surface Morphologies

Figure 2a–d exhibits top-view SEM images of the unmodified carbon cloth and those modified by 20, 60, and 120 s of candle sooting, respectively. Figure 2a shows that the unmodified carbon cloth comprises smooth carbon fibers with diameters of ca. 10 µm. Figure 2b shows the diameters of the carbon fibers increased to approximately 20 µm after 20 s of candle sooting, indicating that some candle soot particles were deposited on the surfaces of the carbon fibers. Figure 2c reveals that the carbon fibers of the carbon cloth modified by 60 s of candle sooting were apparently thicker than those modified for 20 s. It clearly suggested that more abundant soot particles were deposited on the surface of the carbon cloth. Figure 2d shows that the carbon cloth was densely covered by soot particles after the 120 s of candle sooting. In addition, the morphologies of the carbon fibers became characterless. According to Figure 2a–d, it was concluded that the number of deposited soot particles significantly increased with the residence time for candle soot. Figure 2e,f shows magnified SEM images of the carbon cloths modified by 60 s (Figure 2c) and 120 s (Figure 2d) of candle sooting, respectively. Figure 2e,f demonstrates that both deposited soot particles exhibited sponge-like structures. This suggests that the surface areas of the carbon cloths could be effectively increased for a microbial attachment after the modification by candle soot, if the biotoxicity potency of modified cloths were not significantly augmented.

Figure 3a–d presents cross-sectional SEM images of the unmodified carbon cloth and those modified by 20, 60, and 120 s of candle sooting, respectively. Figure 3a shows that the thickness of the unmodified carbon cloth was approximately 200 µm. Figure 3b reveals that some soot particles were deposited on the surface of the carbon cloth; however, they seemed to not be very uniformly distributed. Figure 3c,d shows that more abundant soot particles were deposited on the surfaces of the carbon cloths, forming dense candle soot films after the candle sootings for 60 and 120 s, respectively. Figure 3 also elucidates that carbon fibers in the carbon cloths were not fractured or attenuated during candle soot modification.

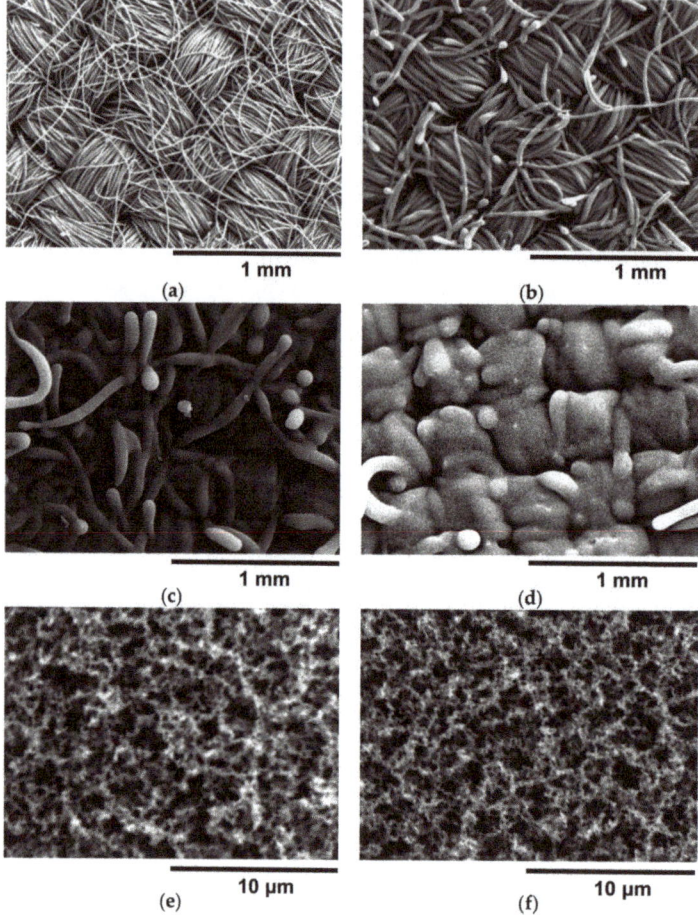

Figure 2. Top-view SEM images of the (**a**) unmodified carbon cloth and those modified by candle sooting for (**b**) 20 s, (**c**) 60 s, and (**d**) 120 s. (**e**,**f**) Magnified views of (**c**) and (**d**), respectively.

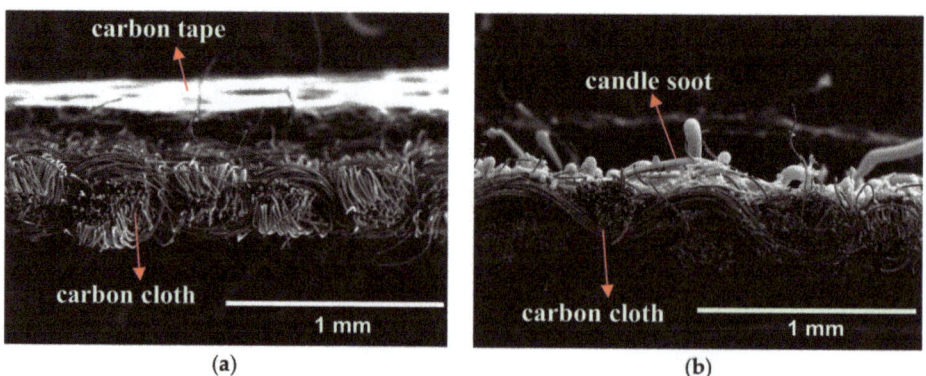

Figure 3. *Cont.*

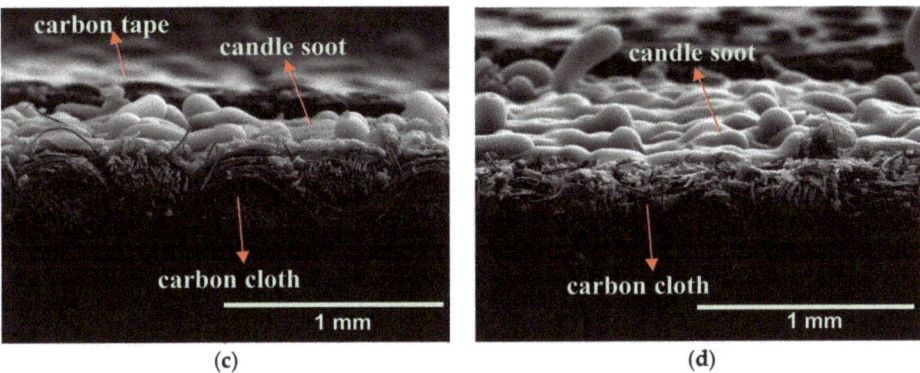

Figure 3. Cross-sectional SEM images of the (**a**) unmodified carbon cloth and those modified by candle sooting for (**b**) 20 s, (**c**) 60 s, and (**d**) 120 s.

3.2. Wettability

Figure 4a shows the water contact angle measurement results of the unmodified and candle-soot-modified carbon cloths. As shown in Figure 4a, unmodified carbon cloth exhibited a high-water contact angle of approximately 131.5° ± 1.9°, indicating that the surface of the as-received carbon cloth was highly hydrophobic. After 20 s of candle sooting, the water contact angle of the carbon cloth decreased to approximately 89.9° ± 26.3°. The high standard deviation (±26.3°) could be explained by the fact that the deposited soot particles were not uniformly distributed on the surface of the carbon cloth (Figure 3b). Figure 4a also shows that the water contact angles of the carbon cloths modified by 60 and 120 s of candle sooting approached the value of zero, suggesting that the carbon cloths tended to be highly hydrophilic with a sufficient time of candle sooting.

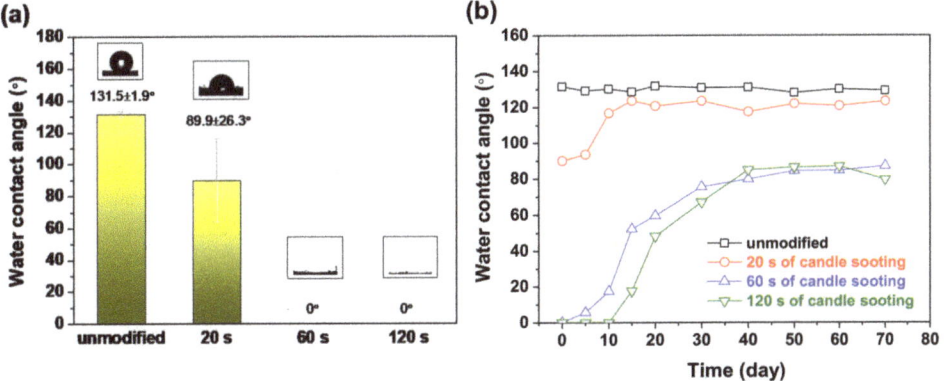

Figure 4. (**a**) Water contact angles of the unmodified carbon cloth and those modified by candle sooting for 20, 60, and 120 s. (**b**) Water contact angles of the unmodified and candle-soot-modified carbon cloths as a function of the standing time.

To evaluate the feasible duration to stably maintain such a hydrophilicity of the modified carbon cloths, the altered carbon cloths were exposed in ambient environment after the candle sooting and the water contact angles of these cloths were determined every five days. Apparently, as Figure 4b revealed, the water contact angles of the unmodified and candle-soot-modified carbon cloths could be found as a function of exposure time. As shown in Figure 4b, the water contact angle of the untreated carbon cloth was maintained at approximately 130° throughout 70 days. On the other hand, the water contact angle of the 20-s candle-soot-modified carbon cloth gradually increased from approximately 90° to 120° during exposure in the presence of ambient air for 10 days. Although the water contact angles of the 60- and 120-s candle-soot-modified carbon cloths were nearly 0° immediately after surface modification, their high hydrophilicities began to deteriorate after exposure to ambient air for five days. Henceforth, their water contact angles gradually increased to approximately 80° after 40 days. Nonetheless, the 60- and 120-s candle-soot-modified carbon cloths were still more hydrophilic than the unmodified and 20-s candle-soot-modified carbon cloths after 70 days. This result seemed to suggest that the coating of candle soot onto carbon cloths was more likely a physical and less likely a chemical attachment.

3.3. XPS Measurements

Figure 5 presents survey XPS of the unmodified and candle-soot-modified carbon cloths. Each specimen exhibited signals only at approximately 284 and 532 eV corresponding to C1s and O1s, respectively. Figure 6a–d shows C1s high-resolution XPS of the unmodified carbon cloth and those modified by 20, 60, and 120 s of candle sooting, respectively. As shown in Figure 6a, the C1s characteristic peak of the unmodified carbon cloth can be deconvoluted into a major sp^3 C–C peak at approximately 284.8 eV and minor C–O peak at approximately 286.1 eV. Figure 6b reveals that the C1s characteristic peak of the 20-s candle-soot-modified carbon cloth comprises C–C and C–O peaks, as those of the unmodified carbon cloth. Additional sp^2 C–C and C=O peaks appeared at approximately 284.4 and 288.7 eV, respectively. Figure 6c,d shows that the C1s characteristic peaks of the 60- and 120-s candle-soot-modified carbon cloths, respectively, were similar to that of the 20-s candle-soot-modified carbon cloth.

Figure 7a–d shows O 1s high-resolution XPS of the unmodified carbon cloth and those modified by 20, 60, and 120 s of candle sooting, respectively. As shown in Figure 7a, the signal peak for the O1s characteristic of the unmodified carbon cloth, was insignificant. The origin of the oxygen signal was likely attributed to the residual contamination on the carbon cloth surface. On the contrary, as shown in Figure 7b–d, the 20, 60, and 120-s candle-soot-modified carbon cloths exhibited significant O1s characteristic peaks, which could be distributed into C–O, C=O, and C–OH peaks at approximately 531, 532, and 533 eV, respectively. According to Figures 6 and 7, it was concluded that the chemical bonding on the surface of the carbon cloth was effectively modified from major sp^3 C–C and minor C–O to abundant sp^2 C–C, C–O, C=O, and C–OH after the candle soot modification. Nevertheless, the candle soot residence time seemed not to significantly influence on the constitution of the chemical bonding.

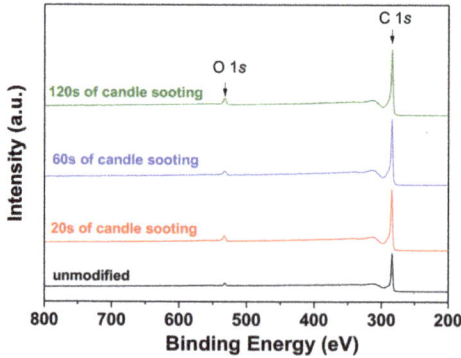

Figure 5. Survey XPS of the unmodified and candle-soot-modified carbon cloths.

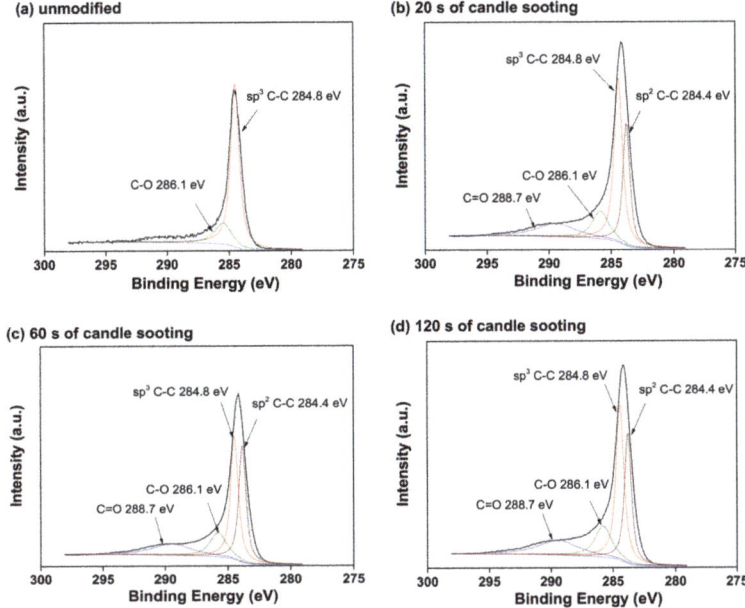

Figure 6. C 1s XPS of the surfaces of the (**a**) unmodified carbon cloth and those modified by candle sooting for (**b**) 20 s, (**c**) 60 s, and (**d**) 120 s.

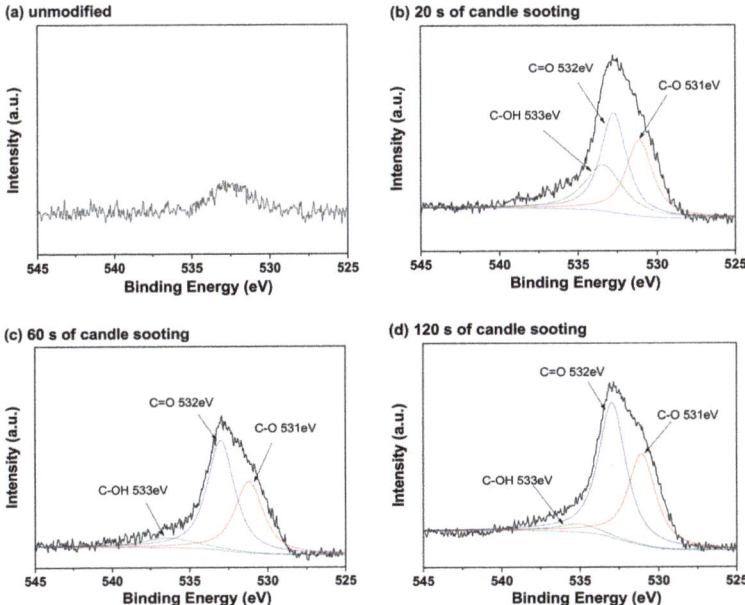

Figure 7. O1s XPS of the surfaces of the (**a**) unmodified carbon cloth and those modified by candle sooting for (**b**) 20 s, (**c**) 60 s, and (**d**) 120 s.

3.4. Microbial Colonization

Figure 8a–d shows SEM images of the surfaces of the unmodified carbon cloth and those modified by 20, 60, and 120 s of candle sooting, respectively, after immersion in the chambers of the MFCs for 24 h. Figure 8a shows that some microorganisms colonized on the surface of the unmodified carbon cloth. In addition, some biofilm segments formed between the carbon fibers. Figure 8b–d show more abundant microorganisms and biofilms formed on the surfaces of the candle-soot-modified carbon cloths. Figure 8 demonstrates that the surface modification by candle soot could effectively accelerate the propagation of microorganisms and formation of biofilms onto the surfaces of the carbon cloths.

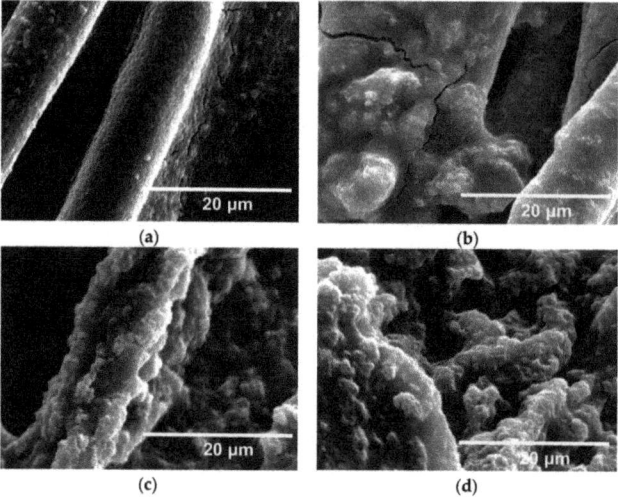

Figure 8. SEM images of the surfaces of the (**a**) unmodified carbon cloth and those modified by (**b**) 20, (**c**) 60, and (**d**) 120 s of candle sooting after immersion in the chambers of the MFCs for 24 h.

3.5. Electrochemical Performance

Figure 9a shows the LSV results and power density response curves of MFCs configured with the unmodified and candle-soot-modified carbon cloth electrodes. Figure 9a indicates that the highest power densities of MFCs configured with the electrode of unmodified carbon cloth and with those modified by 20, 60, and 120 s of candle sooting were ca. 10.2 ± 0.2, 13.6 ± 0.9, 19.8 ± 0.2, and 17.6 ± 0.8 mW·m^{-2}, respectively. This implied that the power-generating efficiencies of MFCs could be effectively enhanced by the candle soot modification. To compare internal resistance figures of different MFCs, EIS analysis on MFCs configured with the unmodified and candle-soot-modified carbon cloth electrodes were implemented (Figure 9b). As shown in Figure 9b, each MFC exhibits a single capacitive loop, which was fitted by the constant-phase-element (CPE) circuit model. The circuit comprises a CPE in parallel with a charge-transfer resistance (RCT), as shown in Figure 9c; the impedance of the CPE can be calculated as: $Z_{CPE} = \frac{1}{T(j\omega)^\varphi}$ [32]. The Z-View® software (ZMAN™2.3) was adopted for fitting the impedance of the CPE; φ is denoted as CPE-P and T is denoted as CPE-T. Table 1 presents the calculated values of RS, CPE-T, CPE-P, and RCT of the MFCs configured with the untreated and candle-soot-modified carbon cloth electrodes. The RCT values of the MFCs configured with the unmodified carbon cloth electrode and with those modified by 20, 60, and 120 s of candle sooting were 2342, 1875, 619, and 1138 Ω, respectively. As RCT corresponds to the resistance of the electrochemical reaction on the electrode [33], Figure 9b demonstrates that the candle soot effectively improved the charge transfer efficiencies of the MFCs. The optimal duration of candle soot to minimize electron transfer resistance was ca. 60 s.

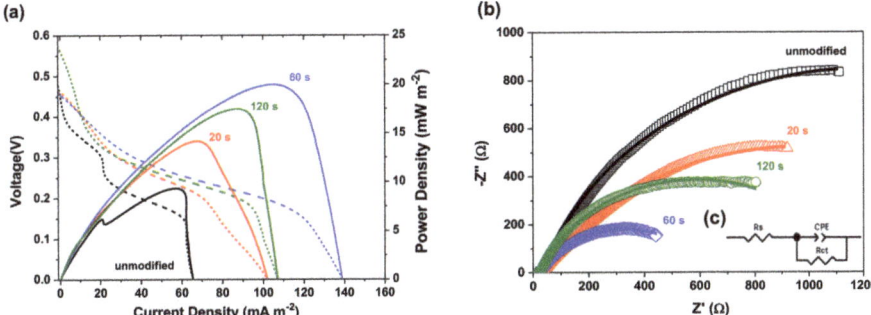

Figure 9. (a) Power density response curves and (b) EIS results of the MFCs configured with the unmodified and candle-soot-modified carbon cloths; (c) Equivalent circuit model.

Table 1. R_S, CPE-T, CPE-P, and R_{CT} of the MFCs configured with the unmodified and 20-, 60-, and 120-s candle-soot-modified carbon cloth electrodes.

Modified Sample	R_S (Ω)	CPE-T	CPE-P	R_{CT} (Ω)
unmodified	27.21	0.0075	0.7985	2342
20 s of candle sooting	25.73	0.0054	0.6458	1875
60 s of candle sooting	22.06	0.0091	0.6665	619
120 s of candle sooting	20.82	0.0070	0.7580	1138

4. Discussion

According to the electrochemical results in Figure 9, evidently the MFCs configured with the candle-soot-modified carbon cloth electrodes exhibited higher power densities and lower total internal resistances than those of the MFC with the unmodified carbon cloth electrode. This was very likely attributed to the sponge-like porous structure of the deposited candle soot that effectively increased the surface areas of the electrodes and facilitated the microbial colonization (Figures 2 and 8). The more promising electron transfer capabilities of modified MFCs can also be attributed to the hydrophilic surfaces of the candle-soot-modified carbon cloths. According to the XPS results in Figures 6 and 7, only sp^3 C–C and small number of C–O functional groups were observed on the surface of the unmodified carbon cloth, whereas abundant sp^2 C–C, C–O, C=O, and C–OH functional groups were observed on the surfaces of the candle-soot-modified carbon cloths. The abundant C–O, C=O, and C–OH functional groups led to the super-hydrophilic surfaces of the candle-soot-modified carbon cloths with more electroactive characteristics for electron transfer in power generation. As bacteria were more likely to attach and propagate onto the hydrophilic surface [34], the candle-soot-modified carbon cloths would favor the microbial growth as demonstrated in Figure 5. In addition, the carboxyl functional groups on the surfaces of the candle-soot-modified carbon cloths facilitated the transfer of electrons from the attached bacteria to solid electrodes. This was possibly due to the hydrogen bonding with the membrane-bound peptide bonds in bacterial cytochromes associated with the intracellular electron transfer chain [17]. Furthermore, the conductive nature of sp^2 C–C in the candle soot was beneficial to the power generation efficiencies of MFCs as the transfer of electrons from the aqueous-phase media to the solid-phase electrodes in the MFCs was not impeded by the deposited candle soot.

Although the candle-soot-modified MFCs exhibited better electrochemical characteristics than that of the unmodified MFC, the appropriate residence time of candle soot still significantly affected the electrochemical performances of the MFCs. Compared to those of the other candle-soot-modified MFCs, the MFC configured with the 20-s candle-soot-modified carbon cloth electrode exhibited the lowest power density of 13.6 ± 0.9 mW·m^{-2} and highest RCT of 1875 Ω. This could be attributed to the insufficient candle soot residence time, leading to a partially hydrophilic surface of the carbon cloth. On the contrary, the MFC configured with the 60-s candle-soot-modified carbon cloth electrode had the highest power density of 19.8 ± 0.2 mW·m^{-2} and lowest RCT of 619 Ω. This was likely attributed to the nearly complete formation of the highly hydrophilic surface of the 60-s candle-soot-modified carbon cloth. In addition, the toxicity was possibly not significantly increased after modification, thus it was more favorable for microbial colonization and biofilm formation. Although the 120-s candle-soot-modified carbon cloth exhibited comparable surface characteristics to those of the 60-s candle-soot-modified carbon cloth, the maximum power density of the MFC configured with the 120-s candle-soot-modified carbon cloth electrode (17.6 ± 0.8 mW·m^{-2}) was slightly lower than that of the MFC configured with the 60-s candle-soot-modified carbon cloth electrode. This unexpected result could be explained by the fact that the conductivity of the candle soot film was normally deteriorated when the film was thicker than the optimal threshold thickness for maximal electron transport efficiency. That is, candle soot with coating in less layers would significantly increase hydrophilicity to reduce electron transfer resistance for power generating augmentation in MFCs. However, a dense coating of candle soot would result in increased resistance across multiple layers as an electron transfer barrier for power generation. Similar results have been obtained for other nanostructured carbonaceous or flame-soot nanoparticle films [35,36].

Extensive studies have been performed to improve the electrochemical performances of MFCs by modifying the surface properties of the anode electrodes. Table 2 lists the comparison of the chemicals and instruments used in these anode modifications [14–25,30]. As shown in Table 2, most of these techniques involve expensive instruments, complex and time-consuming processes, or potentially toxic chemicals. That was why this study intentionally used some procedure-simple and cost-effective alternatives for the surface modification of cloth electrodes with practicability. Although the power outputs achieved from the MFC with candle-soot modified electrodes are relatively low compared to the ones reported in most of the works of Table 2, this study simply focused on the applicability of using candle soot as a possible means to modify electrode characteristics for the enhancement of power generation in MFCs. Such modification was just a first-step treatability assessment and did not cover the overall optimization of the MFC system. As a matter of fact, several factors (e.g., microbial characteristics, MFC bioreactor operation strategy, biofilm development, bacterial community structure, solid-solid, solid-liquid interfacial electron-transfer resistance, exogenous electron shuttles) were inevitably still required to be explored for system optimization. However, the influences of individual factors after the modification were still worthy to be uncovered for the follow-up practicability. The carbon cloth-electrode was regularly used as control/reference to compare with our new and/or novel methods of modification and/or literature data. Recently, Singh et al. [30] were the first to deposit candle soot on ultrafine stainless steel wire disks as the anode and cathode electrodes and successfully enhanced the electrochemical properties of double-chamber MFCs. Their findings indicated that the synthesis of the carbon-nanoparticle-based electrodes by candle soot was simple, cost-effective, reproducible, and scalable, and the fabricated MFC could produce a high amount of bioenergy. In our study, we directly deposited candle soot on the surface of the carbon cloth electrode, rather than on a stainless-steel disk. Noticeably, the candle soot modification could effectively facilitate bacterial colonization and biofilm formation on the surface of the carbon cloth, increasing the extracellular electron transfer efficiency and the power generation capabilities of MFCs. However, it should be noted that the MFCs could exhibit the optimized efficiencies only when the carbon cloth was modified using an appropriate candle soot residence time for maximal electron transfer capacities to be expressed. That is, the super-hydrophilicities of the candle-soot-modified carbon cloths typically deteriorated with longer time of exposure. Therefore, the candle-soot-modified carbon cloth electrodes should be fabricated into MFCs as soon as possible after the modification is completed to exhibit an optimal performance. Besides, the candle-soot-modified carbon cloth electrodes are suitable for large-scale MFC applications because of their low cost and easily prepared. Nevertheless, the modification method for a large-scale anode should be carefully designed and controlled to obtain a homogeneous surface.

Table 2. Comparison of the chemicals and instruments used in various anode modifications.

Methods	Anode Materials	Chemicals and Instruments Used in Modifications	Performance	Ref.
NH_3 gas treatment	Carbon cloth	Ammonia	Maximum power density of 1970 mW·m^{-2}	[14]
Acid soaking and heating	Carbon fiber brush	Ammonium Peroxydisulfate, Concentrated Sulfuric Acid, Muffle Furnace	Maximum power density of 1370 mW·m^{-2}	[15]
Oxidized anode and modified by AQDS	Graphite plate	Anthraquinone-1,6-disulfonic Acid, Perchloric Acid, 1,4-Naphthoquinone, Ethanol	Increase kinetic activity of a factor of 218	[16]
Electrochemical treatment for 12 h	Graphite felt	H_2SO_4	Increase current production of 39.5%	[17]
Atmospheric pressure plasma jets	Carbon cloth	Atmospheric Pressure Plasma Jets	Increase maximum power density from 2.38 to 7.56 mW·m^{-2}	[18]
Coating carbon nanotube	Carbon cloth	Multiwall Carbon Nanotubes, Ethanol	Maximum power density of 65 mW·m^{-2}	[19]
Coating Ferric Oxide	Carbon paper	Ferric Citrate, Acetate	Increase maximum power density from 2 to 40 mW·m^{-2}	[20]
Sputtering Au nanoparticles	Carbon paper	Au Nanoparticles, Electron Beam Physical Vapor Deposition (EBPVD) Machine	1.22–1.88-Fold increase in power density	[21]
Coating goethite nanowhiskers	Carbon paper	$Fe(NO_3)_3 \cdot 9H_2O$, Teflon-lined Stainless Steel Autoclave, KOH	60% increase in current density	[22]
Coating NiO nanoflaky array	Carbon cloth	H_2SO_4, Nickel Chloride, $CO(NH_2)_2$, Hexadecyl Trimethyl Ammonium Bromide, Teflon-lined Stainless Steel Autoclave	3-Fold increase in power density	[23]
Atmospheric-pressure plasma jet processed reduced graphene oxides	Carbon cloth	Atmospheric Pressure Plasma Jets, Reduced Graphene Oxide, Terpineol, Ethanol, Ethyl Cellulose	Increase maximum power density from 6.02 to 10.80 mW·m^{-2}	[24]
Tungsten carbide	Graphite foil	Tungsten Carbide, Yellow Tungsten Acid, Oxalic Acid, NH_4Cl, Tube Furnace	Achieve current density of 8.8 mA·m^{-2}	[25]
Coating candle soot on ultrafine stainless steel wire disks	Ultrafine stainless steel wire disk	HCl, Acetone, Ethanol, Candle, Hydraulic Press	Produced a high OCP (0.68 V), limiting current density (7135 mA/m^2) and power generation (1650 mW/m^2)	[30]
Coating candle soot on carbon cloths	Carbon cloth	Candle	Increase maximum power density from 10.2 to 19.8 mW·m^{-2}	This study

5. Conclusions

This study demonstrated that the use of candle soot is an effective and economic method for the surface modification of carbon cloth electrodes to improve the electrochemical performances of MFCs. The SEM results showed that the carbon fibers in carbon cloths were not fractured or attenuated during the candle sooting. The deposited soot particle films exhibited sponge-like structures, providing larger surface areas for bacterial adhesion. The wettability measurements revealed that a residence time of only 60 s was inevitably required to alter the hydrophobic surfaces of the carbon cloths to super-hydrophilic. XPS results showed that abundant sp^2 C–C, C–O, and C=O functional groups existed on the surfaces of the candle-soot-modified carbon cloths. The C–O and C=O functional groups were responsible for the super-hydrophilicity of the non-toxic surfaces of the candle-soot-modified carbon cloths. The carboxyl and sp^2 C–C functional groups favored the transfer of electrons from the attached bacteria to the electrodes. The electrochemical measurements demonstrated that the MFC configured with the 60-s candle-soot-modified carbon cloth electrode exhibited the highest power density of 19.8 ± 0.2 mW·m^{-2} and lowest total internal resistance of 619 Ω, among the considered MFCs. Therefore, the use of candle soot is a rapid, economic, and simple method for the surface modification of carbon cloth electrodes. However, after the candle soot modification is completed, a postponement in the MFC construction may lead to the deterioration in the super-hydrophilicity of the candle-soot-modified carbon cloth.

Author Contributions: Conceptualization, B.-Y.C.; Data Curation, Y.-T.T.; Funding Acquisition, B.-Y.C. and S.-H.C.; Investigation, B.-Y.C., Y.-T.T. and S.-H.C.; Methodology, Y.-T.T. and S.-H.C.; Project Administration, S.-H.C.; Supervision, S.-H.C.; Writing–Original Draft Preparation, S.-H.C.; Writing–Review & Editing, B.-Y.C.

Funding: This research was funded by the Ministry of Science and Technology, Taiwan (MOST 106-2221-E-197-020-MY3 and 107-2621-M-197-001 and MOST 107-2221-E197-004).

Conflicts of Interest: The authors declare no conflict of interest.

References

1. Logan, B.E.; Hamelers, B.; Rozendal, R.; Schröder, U.; Keller, J.; Freguia, S.; Aelterman, P.; Verstraete, W.; Rabaey, K. Microbial fuel cells: Methodology and technology. *Environ. Sci. Technol.* **2006**, *40*, 5181–5192. [CrossRef] [PubMed]
2. Pant, D.; Van Bogaert, G.; Diels, L.; Vanbroekhoven, K. A review of the substrates used in microbial fuel cells (MFCs) for sustainable energy production. *Bioresour. Technol.* **2010**, *10*, 1533–1543. [CrossRef]
3. Pandey, P.; Shinde, V.N.; Deopurkar, R.L.; Kale, S.P.; Patil, S.A.; Pant, D. Recent advances in the use of different substrates in microbial fuel cells toward wastewater treatment and simultaneous energy recovery. *Appl. Energy* **2016**, *168*, 706–723. [CrossRef]
4. Santoro, C.; Arbizzani, C.; Erable, B.; Ieropoulos, I. Microbial fuel cells: From fundamentals to applications. A review. *J. Power Sources* **2017**, *356*, 225–244. [CrossRef] [PubMed]
5. Grattieri, M.; Minteer, S.D. Self-Powered Biosensors. *ACS Sens.* **2018**, *3*, 44–53. [CrossRef]
6. Logan, B.E.; Regab, J.M. Microbial fuel cells—Challenges and applications. *Environ. Sci. Technol.* **2006**, *40*, 5172–5180. [CrossRef] [PubMed]
7. Li, W.W.; Yu, H.Q.; He, Z. Towards sustainable wastewater treatment by using microbial fuel cells-centered technologies. *Energy Environ. Sci.* **2014**, 911–924. [CrossRef]
8. Yu, F.; Wang, C.; Ma, J. Applications of graphene-modified electrodes in microbial fuel cells. *Materials* **2016**, *9*, 807. [CrossRef]
9. Chaudhuri, S.K.; Lovely, D.R. Electricity generation by direct oxidation of glucose in mediator less microbial fuel cells. *Nat. Biotechnol.* **2003**, *21*, 1229–1232. [CrossRef]
10. Park, D.H.; Zeikus, J.G. Improved fuel cell and electrode designs for producing electricity from microbial degradation. *Biotechnol. Bioeng.* **2003**, *81*, 348–355. [CrossRef]
11. Ishii, S.; Watanabe, K.; Yabuki, S.; Logan, B.E.; Sekiguchi, Y. Comparison of electrode reduction activities of geobacter sulfurreducens and an enriched consortium in an air-cathode microbial fuel cell. *Appl. Environ. Microbiol.* **2008**, *74*, 7348–7355. [CrossRef] [PubMed]
12. Wei, J.; Huang, X. Recent progress in electrodes for microbial fuel cell. *Bioresour. Technol.* **2011**, *102*, 9335–9344. [CrossRef] [PubMed]
13. Kalathil, S.; Pant, D. Nanotechnology to rescue bacterial bidirectional extracellular electron transfer in bioelectrochemical systems. *RSC Adv.* **2016**, *6*, 30582–30597. [CrossRef]
14. Cheng, S.A.; Logan, B.E. Ammonia treatment of carbon cloth anodes to enhance power generation of microbial fuel cells. *Electrochem. Commun.* **2007**, *9*, 492–496. [CrossRef]
15. Feng, Y.; Yang, Q.; Wang, X.; Logan, B.E. Treatment of carbon fiber brush anodes for improving power generation in air–cathode microbial fuel cells. *J. Power Sources* **2010**, *195*, 1841–1844. [CrossRef]
16. Lowy, D.A.; Tender, L.M. Harvesting energy from the marine sediment-water interface III—Kinetic activity of quinone- and antimony-based anode materials. *J. Power Sources* **2008**, *185*, 70–75. [CrossRef]
17. Tang, X.; Guo, K.; Li, H.; Du, Z.; Tian, J. Electrochemical treatment of graphite to enhance electron transfer from bacteria to electrodes. *Bioresour. Technol.* **2011**, *102*, 3558–3560. [CrossRef]

18. Chang, S.H.; Liou, J.S.; Liu, J.L.; Chiu, Y.F.; Xu, C.H.; Chen, B.Y.; Chen, J.Z. Feasibility study of surface-modified carbon cloth electrodes using atmospheric pressure plasma jets for microbial fuel cells. *J. Power Sources* **2016**, *336*, 99–106. [CrossRef]
19. Tsai, H.Y.; Wu, C.C.; Lee, C.Y.; Shih, E.P. Microbial fuel cell performance of multiwall carbon nanotubes on carbon cloth as electrodes. *J. Power Sources* **2009**, *194*, 199–205. [CrossRef]
20. Kim, J.R.; Min, B.; Logan, B.E. Evaluation of procedures to acclimate a microbial fuel cell for electricity production. *Appl. Microbiol. Biotechnol.* **2005**, *68*, 23–30. [CrossRef]
21. Alatraktchi, F.A.; Zhang, Y.; Angelidaki, I. Nanomodification of the electrodes in microbial fuel cell: Impact of nanoparticle density on electricity production and microbial community. *Appl. Energy* **2014**, *116*, 216–222. [CrossRef]
22. Wang, L.; Su, L.; Chen, H.; Yin, T.; Lin, Z.; Lin, X.; Yuan, C.; Fu, D. Carbon paper electrode modified by goethite nanowhiskers promotes bacterial extracellular electron transfer. *Mater. Lett.* **2015**, *141*, 311–314. [CrossRef]
23. Qiao, Y.; Wu, X.S.; Li, C.M. Interfacial electron transfer of Shewanella putrefaciens enhanced by nanoflaky nickel oxide array in microbial fuel cells. *J. Power Sources* **2014**, *266*, 226–231. [CrossRef]
24. Chang, S.H.; Huang, B.Y.; Wan, T.H.; Chen, J.Z.; Chen, B.Y. Surface modification of carbon cloth anodes for microbial fuel cells using atmospheric-pressure plasma jet processed reduced graphene oxides. *RSC Adv.* **2017**, *7*, 56433–56439. [CrossRef]
25. Rosenbaum, M.; Zhao, F.; Quaas, M.; Wulff, H.; Schröder, U.; Scholz, F. Evaluation of catalytic properties of tungsten carbide for the anode of microbial fuel cells. *Appl. Catal. B* **2007**, *74*, 261–269. [CrossRef]
26. Wei, Z.; Yan, K.; Chen, H.; Yi, Y.; Zhang, T.; Long, X.; Li, J.; Zhang, L.; Wang, J.; Yang, S. Cost-efficient clamping solar cells using candle soot for hole extraction from ambipolar perovskites. *Energ. Environ. Sci.* **2014**, *7*, 3326–3333. [CrossRef]
27. Kakunuri, M.; Sharma, C.S. Candle soot derived fractal-like carbon nanoparticles network as high-rate lithium ion battery anode material. *Electrochim. Acta* **2015**, *180*, 353–359. [CrossRef]
28. Khalakhan, I.; Fiala, R.; Lavková, J.; Kúš, P.; Ostroverkh, A.; Václavů, M.; Vorokhta, M.; Matolínová, I.; Matolín, V. Candle soot as efficient support for proton exchange membrane fuel cell catalyst. *Fuel Cells* **2016**, *16*, 652–655. [CrossRef]
29. Liang, C.J.; Liao, J.D.; Li, A.J.; Chen, C.; Lin, H.Y.; Wang, X.J.; Xu, Y.H. Relationship between wettabilities and chemical compositions of candle soots. *Fuel* **2014**, *128*, 422–427. [CrossRef]
30. Singh, S.; Bairagi, P.K.; Verma, N. Candle soot-derived carbon nanoparticles: An inexpensive and efficient electrode for microbial fuel cells. *Electrochim. Acta* **2018**, *264*, 119–127. [CrossRef]
31. Chen, B.Y.; Zhang, M.M.; Chang, C.T.; Ding, Y.; Lin, K.L.; Chiou, C.S.; Hsueh, C.C.; Xu, H. Assessment upon azo dye decolorization and bioelectricity generation by *Proteus hauseri*. *Bioresour. Technol.* **2010**, *101*, 4737–4741. [CrossRef] [PubMed]
32. Jorcin, J.B.; Orazem, M.E.; Pebere, N.; Tribollet, B. CPE analysis by local electrochemical impedance spectroscopy. *Electrochim. Acta* **2006**, *51*, 1473–1479. [CrossRef]
33. Qiao, Y.; Li, C.M.; Bao, S.J.; Bao, Q.L. Carbon nanotube/polyaniline composite as anode material for microbial fuel cells. *J. Power Sources* **2007**, *170*, 790–794. [CrossRef]
34. Han, T.H.; Sawant, S.Y.; Hwang, S.J.; Cho, M.H. Three-dimensional, highly porous N-doped carbon foam as microorganism propitious, efficient anode for high performance microbial fuel cell. *RSC Adv.* **2016**, *6*, 25799–25807. [CrossRef]
35. Bruzzi, M.; Piseri, P.; Miglio, S.; Bongiorno, G.; Barborini, E.; Ducati, C.; Robertson, J.; Milani, P. Electrical conduction in nanostructured carbon and carbon-metal films grown by supersonic cluster beam deposition. *Eur. Phys. J. B* **2003**, *36*, 3–13. [CrossRef]
36. De Falco, G.; Commodo, M.; Barra, M.; Chiarella, F.; D'Anna, A.; Aloisio, A.; Cassinese, A.; Minutolo, P. Electrical characterization of flame-soot nanoparticle thin films. *Synth. Met.* **2017**, *229*, 89–99. [CrossRef]

© 2018 by the authors. Licensee MDPI, Basel, Switzerland. This article is an open access article distributed under the terms and conditions of the Creative Commons Attribution (CC BY) license (http://creativecommons.org/licenses/by/4.0/).

Article

Synthesis of Core-Shell MgO Alloy Nanoparticles for Steelmaking

Jinglong Qu [1], Shufeng Yang [2,3,*], Hao Guo [2,3], Jingshe Li [2,3] and Tiantian Wang [2,3]

1. Central Iron and Steel Research Institute, Beijing 100081, China; 13810256459@139.com
2. School of Metallurgical and Ecological Engineering, University of Science and Technology Beijing, Beijing 100083, China; guohaoustb@sina.com (H.G.); lijingshe@ustb.edu.cn (J.L.); wangtiantian@xs.ustb.edu.cn (T.W.)
3. Beijing Key Laboratory of Special Melting and Preparation of High-End Metal Materials, Beijing 100083, China
* Correspondence: yangshufeng@ustb.edu.cn

Abstract: In this present study, we aimed to reduce the wetting angle of nanoparticles (NPs) in molten steel and thus, increase their utilization ratio in steel. In order to achieve this, a two-step process was used to synthesize a core-shell AlTi-MgO@C NP structure for steelmaking through a dopamine polymerization process, which used an ammonium persulfate oxidant and high-temperature carbonization. The NP surface characterization was tested by scanning electron microscopy and field emission transmission electron microscopy, while the hydrodynamic NP size was measured by dynamic light scattering. The results showed that a carbon coating that had a thickness of 10 nm covered the NP surface, with the dispersion and stability of the particles in the aqueous solution having improved after the coating. The contact angle of the surface-treated NP was less than that of the uncoated NP in high-temperature molten steel and the corresponding wetting energy was smaller, which indicated improved wettability.

Keywords: MgO nanoparticle; dopamine polymerization; carbonization process; core shell structure; contact angle

1. Introduction

The surface modification of materials plays a vital role in modern chemistry, biology and the science of materials science. This has been widely applied in the fields of science, engineering and technology [1,2]. There are numerous and varied methods for modification, including Langmuir–Blodgett deposition, monolayer self-assembly and layer-by-layer self-assembly. This can be chosen according to optimal compatibility with the specific material requirements [3–6]. However, these methods lack efficacy on a broad range of material surfaces.

In the science of surface coatings, carbon coating technology has been widely used due to its chemical stability. Glucose has been used as a carbon source, but its specific capacity is limited due to a relatively high carbon content. Oleic acid [7], citric acid [8], ethylenediaminetetraacetic acid [9] and other organic compounds have been used as a substitute for glucose to provide a carbon source but obtaining a relatively uniform carbon layer remains a challenge.

Dopamine can be utilized to form polydopamine (PDA) in alkaline pH conditions to create oxide-induced self-polymerization [10]. After this, the PDA can become adhered to the surface of a solid material through an amino- or mercapto-nucleophilic reaction. This widely-used method exhibits strong adhesion properties, easily produces a functional modification of the material surface and is unaffected by the size and shape of the material. The advantages of dopamine-encapsulated materials are linked to the simple reaction conditions and the homogeneous coating [11]. In addition, the thickness of the dopamine layer can be controlled by changing the initial concentration and polymerization time of the dopamine. This is important especially considering that thickness control is very important for material design and surface modification [12]. Due to the excellent and versatile coating capabilities of dopamine, it can be used as a precursor of carbon to form a uniform and continuous coating on the NP surface [13,14]. After this, by controlling the heat treatment atmosphere, the coating of PDA on the material can be converted into a carbon layer through high-temperature carbonization [15].

The purity of molten steel plays an important role in determining the properties of the steel and the distribution of second-phase nanoparticle (NP) dispersions in the steel can be used as an external method to improve steel purity [16]. This is accomplished because NPs can alter inclusion characteristics so that the precipitation of molten steel in the cooling process becomes finer and can reach sizes on even the nanoscale, avoiding the adverse effects of large inclusions in the steel [17]. However, due to the large surface properties of the NPs, they easily agglomerate in the high-temperature steel and reduce the yield.

Using these dispersant and nanoscale materials, some groups have mechanically hot-pressed the NPs mixed with the metal alloy powder in experiments, with some having added the NPs to the molten steel. The addition of the NPs was found to have a significant effect [18,19], although the NPs that entered into the welding pool were found to gather and float to the surface of the steel liquid. Thus, the second-phase particle yield in oxide metallurgy technology is a key problem that must be solved before this production can be applied for real-life applications [20]. In this study, NP surfaces were modified by a chemical method. The characterization showed that the treated NPs had good stability and dispersibility, with successful preparation of NPs with a core-shell structure. The surface coating method was beneficial in reducing the contact angle between the NP and the steel liquid and in improving the yield of the nanoparticles in the steel liquid.

2. Materials and Methods

2.1. Materials

The dopamine hydrochloride and ammonium persulfate used in this present study were purchased from Sigma (Beijing, China), while the Al, Ti and MgO NPs were purchased from Chao Wei Co., Ltd. (Shanghai, China). All other chemicals used were of at least analytical reagent grade and were obtained from Sinopharm Chemical Reagent Co., Ltd. (Beijing, China). Aqueous solutions were prepared using deionized water (Milli-Q system).

Our first aim was to improve the dispersion of the MgO NPs in liquid steel. Firstly, the AlTi (diameter of 50–70 nm) NP alloy was manufactured by mechanically mixing the individual NPs with a weight ratio of Al and Ti NP of 7:3. This NP mixture was called the AlTi alloy NP. The AlTi (Al with 70 wt %; Ti with 30 wt %) NP alloy and the MgO NPs were mixed in a planetary ball mill (IKN Mechanical Equipment Co., Ltd., Berlin, Germany) with a mass fraction of the MgO to AlTi alloy of 5:1. During mixing, the planetary ball mill was operated at 6500 rpm/min for 4 h at a low temperature and under low oxygen conditions, which aims to avoid the risk of high temperatures resulting from particle collisions.

2.2. Characterization

The morphologies and chemical compositions of the samples were measured using field emission scanning electron microscopy (SEM, Hitachi, Tokyo, Japan) and field emission transmission electron microscopy (TEM, Hitachi, Tokyo, Japan). X-ray diffraction spectra (XRD, PANalytical Co., Ltd., Amsterdam, The Netherlands) of the samples were obtained at a scan rate of 1°/min. The hydrodynamic sizes of NPs were measured by dynamic light scattering (DLS, Malvern, Worcestershire, UK) in deionized water at room temperature.

2.3. Preparation of AlTi-MgO@PDA nanoparticles (NPs)

To coat the NPs with PDA, 300 mg of dopamine hydrochloride and 300 mg of AlTi-MgO NPs were added to a phosphate-buffered solution (0.1 M, pH of 7.0) with magnetic stirring for 2 h. After this, 170 mg of ammonium persulfate was added into the mixture, which was stirred for another 1 h. This produced AlTi-MgO@PDA NPs, which were obtained after several rounds of centrifugation and washing with distilled water and ethanol. All reactions were conducted at room temperature. The process of self-polymerization is shown in Figure 1.

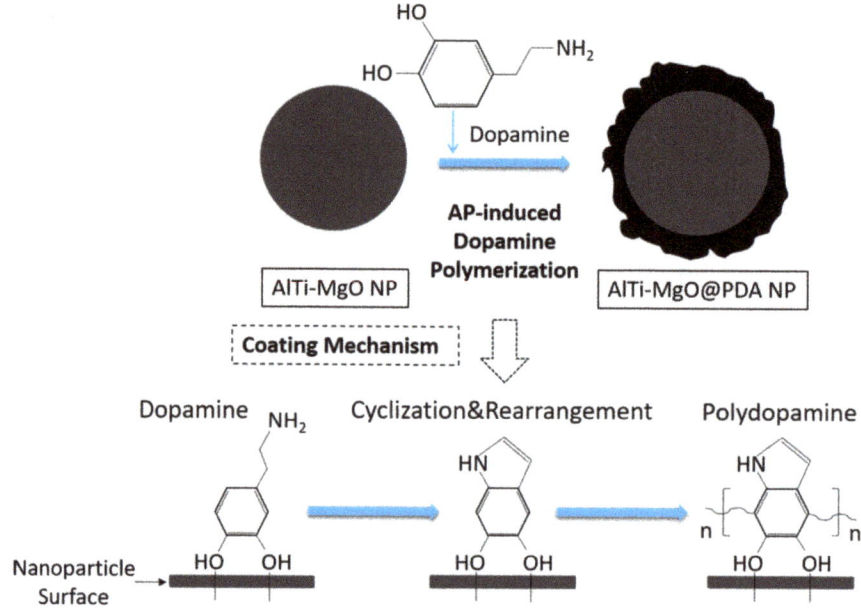

Figure 1. Schematic diagram of the formation of the PDA film on AlTi-MgO nanoparticles (NPs).

2.4. Preparation of AlTi-MgO@C NPs

To carbonize the PDA film, the AlTi-MgO@PDA NPs were placed into a tubular furnace, which was maintained at 550 °C for 4 h in a nitrogen environment. After the heat treatment, the AlTi-MgO@C NPs were soaked in a 0.24 M HCl solution for 30 min, before being rinsed several times with deionized water. Figure 2 shows the process of the high-temperature carbonization of NPs.

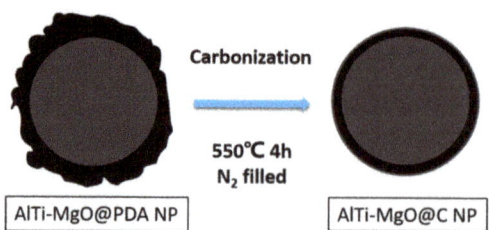

Figure 2. Schematic diagram of the formation of the carbon film after heat treatment.

3. Results

3.1. NP Characteristics

Figure 3 shows the typical SEM and TEM images of the AlTi-MgO NP cores for steelmaking. After the pre-dispersion ball-milling treatment, the NPs were shown to be spherical with a smooth surface that exhibits no coating. Figure 3C shows the XRD pattern of the AlTi-MgO NPs, which depicts many narrow peaks. The narrowness of the peaks indicated that the NPs have good crystallinity after the pre-dispersion treatment. The corresponding phases were identified as Al and Ti as individual substances and MgO oxide. Furthermore, the MgO NPs easily combined with the water vapor in the air to form $Mg(OH)_2$.

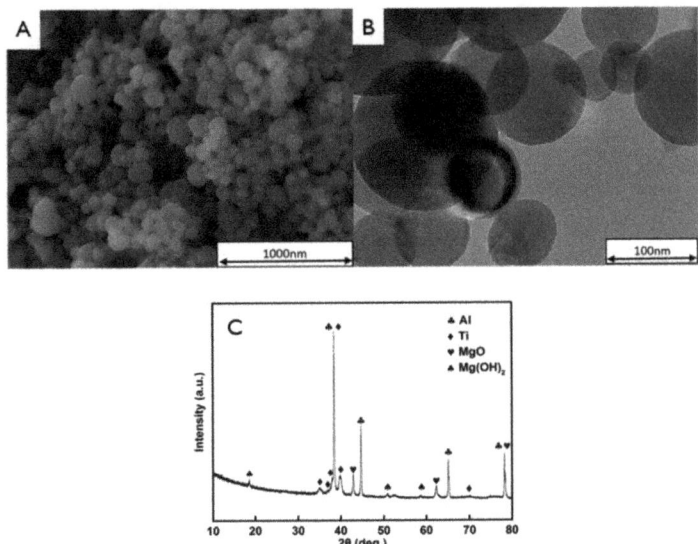

Figure 3. Analysis of the bare AlTi-MgO NPs, showing the: (**A**) Scanning electron microscopy (SEM) image; (**B**) Transmission electron microscopy (TEM) image; and (**C**) X-ray diffraction spectra (XRD) pattern.

Figure 4 shows the typical SEM and TEM images of the AlTi-MgO@PDA NPs for steelmaking. The SEM image (Figure 4A) shows that after the process of dopamine self-polymerization induced by the ammonium persulfate oxidant, the surface of the NP becomes rough, which is the typical morphology of a dopamine self-polymerization covering. The shell layer of these surface-treated NPs had a thickness of approximately 7–18 nm, which was measured from the comparison of the TEM images with the original NP diameters (Figure 3B). Figure 4C shows the XRD pattern of the AlTi-MgO@PDA NPs. After heating at 550 °C for 4 h in a nitrogen environment, a proportion of the Al and Ti was retained as individual substances and the mass fraction of the MgO NPs was relatively small. The corresponding metal complex oxide phases with Al and Ti elements were easily formed at this high temperature: $MgTiO_3$, $MgTi_2O_5$ and $MgAl_2O_4$ phases. At the same time, the amorphous carbon and graphitic carbon phases appeared with a broadened peak, which indicated that a C-AlTi-MgO composite phase was formed after the carbonization process.

Figure 5 shows the TEM images of AlTi-MgO@C NPs after the carbonization process. During this process, the structure of the surface organic substance breaks down and the loss of these hybrid atoms leads to a contraction of the PDA film [21]. The high-resolution TEM images are shown in Figure 5A (insets), which demonstrated that the outer layer of the particle is coated with a carbon layer. No obvious lattice fringes can be observed in the outer layer of the carbon shell, while the inner layer of the coating exhibits an inter-crystalline spacing of $d_{(002)}$ = 0.34 nm, which corresponds to the graphitic carbon crystal. After dispersion and carbonization, no obvious agglomeration phenomenon was observed and the particles were well dispersed in the aqueous solution. Figure 5B shows the XRD pattern of the AlTi-MgO@C NPs after hot HCl acid pickling for 30 min, which exhibits peak shapes that corresponded to amorphous carbon and graphite. Furthermore, the latter peak showed good crystallinity. According to these results, AlTi-MgO@C NP were synthesized after the surface treatment.

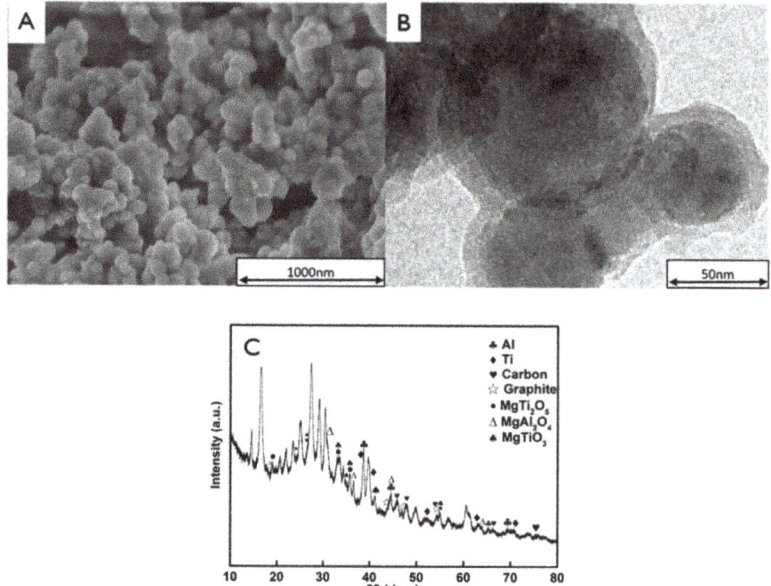

Figure 4. Analysis of the AlTi-MgO NPs coated with PDA, showing the: (**A**) SEM image; (**B**) TEM image; and (**C**) XRD pattern.

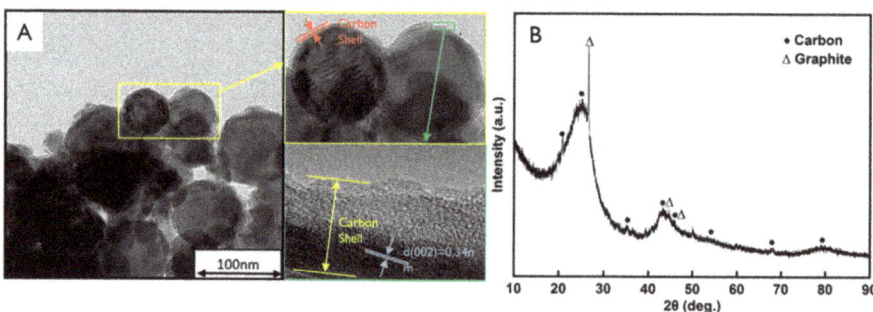

Figure 5. (**A**) High-resolution TEM images of the AlTi-MgO@C NP and (**B**) XRD pattern of the AlTi-MgO NP after HCl acid pickling for 30 min.

Figure 6 shows the size distribution of the AlTi-MgO, AlTi-MgO@PDA core-shell and AlTi-MgO@C core-shell NP samples. The dynamic light scattering (DLS) measurement detected one particle size in each aqueous solution. The average size of the AlTi-MgO NPs was approximately 50 nm, while the AlTi-MgO@PDA NP size increased to 70 nm after the surface modification. This reflects a combination of the core diameter (50 nm) and the shell (20 nm) thicknesses found in the TEM images. The carbonization process broke down the structure of the PDA film, which causes the layer to become thinner. As a result, the mean hydrodynamic size of the AlTi-MgO@C NPs was reduced to 60 nm.

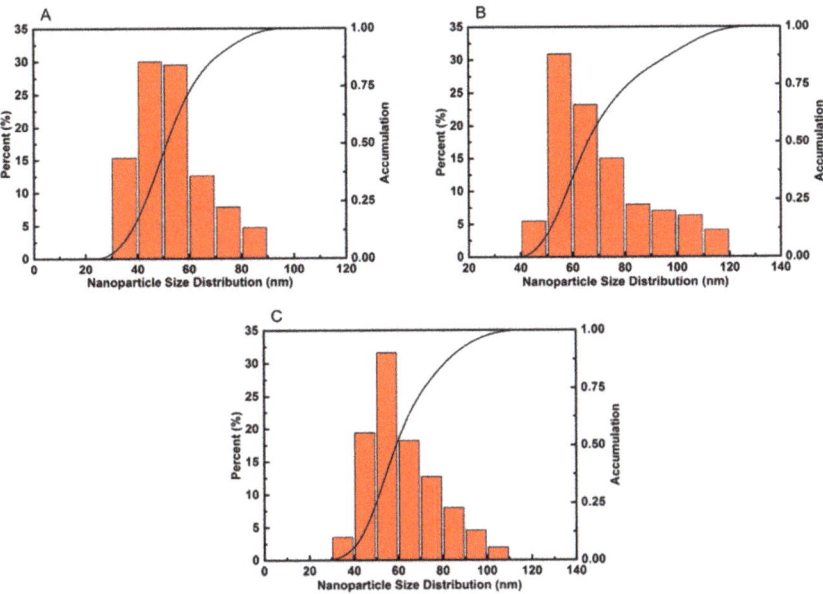

Figure 6. Hydrodynamic sizes of: (**A**) AlTi-MgO NPs; (**B**) AlTi-MgO@PDA NPs; and (**C**) AlTi-MgO@C NPs in the aqueous solution.

3.2. Surface Treatment and Physical Properties of the NPs in Steel

Dai et al. reported that polydopamine can act as a template for functionalizing a coated material surface and will form a carbon structure after subsequent heat treatment [15]. The carbonized NPs were found to exist in a stable form in an appropriate solvent, exhibiting good monodispersity and reduction in the aggregation of the NPs [15,22,23]. In the subsequent process of high-temperature steelmaking, the insignificant amount of carbon found in the coating of the surface-treated NPs exhibited little or no pollution in the molten steel [24].

The NPs could be regarded as the heterogenous nuclei that inhibits the movement of the grain boundary. Furthermore, the dispersive and fine inclusions can induce intragranular acicular ferrite. Both these processes can be used to improve the properties of materials. However, NPs can easily become agglomerated due to the effect of size and intense Brownian motion collisions. This agglomeration in the molten steel was closely linked to the wettability of the NPs in the molten steel and thus, can significantly affect the utilization ratio of the NPs in the steel.

After the NPs were added to the molten steel, they passed through the air–steel interface and were stirred by the Ar gas flow and by the force exerted by the furnace. After the NPs entered the liquid steel, they evolved into nucleating agents and denaturants of the inclusions in the steelmaking process. The agglomerating nature of the NPs has a significant influence on the distribution and characteristics of the inclusions in the steel. In the simulation of inclusions in a metallurgic reactor, the Stokes formula typically considers the kinematic viscosity, the equivalent radius of the inclusions and the difference in the density of the fluid and the inclusion. Furthermore, Wang [25] assumed that in liquid steel, NPs with a particle size of 50 nm floated up by only about 100 nm in 60 min and thus, the movement of NPs in the vertical direction can be ignored. Therefore, we can conclude that the NPs immersed into the molten steel will not float to the air–steel interface. The wettability of the NP in high-temperature molten steel is an important factor affecting its utilization ratio in liquid steel. The wetting energy is typically used to assess the wettability of an NP in liquid steel, which can be calculated as follows:

$$W_i = \gamma_{LG} \times \cos\theta, \tag{1}$$

where W_i is the wetting work of the ith NP; γ_{LG} is the surface tension of the liquid steel; and θ is the contact angle, which is shown in the diagram in Figure 7. Equation (1) demonstrates that the wettability of the NPs in the molten steel is determined by the contact angle (θ) as minimizing the contact angle improves the wettability. A contact angle that is greater than 90° indicates that the NPs are not wetted with the molten steel as the resulting strong hydrophobic effect would cause the NPs to strongly repel the surrounding medium.

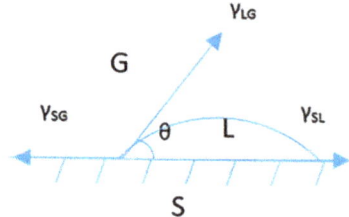

Figure 7. Contact angle diagram, showing the gas (G), liquid (L) and solid (S) regions, with definitions of the contact angle (θ) and the solid–gas (γ_{SG}), liquid–gas (γ_{LG}) and solid–liquid (γ_{SL}) surface tensions.

Depending on the type of inclusions present in the steel, it is assumed that the added NPs will form corresponding oxides after they react with the dissolved oxygen in the molten steel. The contact angles of Al_2O_3, MgO and pyrographite with liquid steel were found to be 100–141° [26], 130° [27] and 50° [28], respectively, under a helium gas environment. This signifies that the contact angle of the AlTi-MgO@C NPs coated with a carbon layer is much lower than that of the AlTi-MgO NPs. Thus, the AlTi-MgO@C NPs possess superior wettability in the high-temperature molten steel and a greater number of effective particles will enter into the molten steel, creating an increased NP utilization ratio. Although the contact angle of these AlTi-MgO@C NPs have been shown in theory, more experimental evidence is needed to prove the wettability of these AlTi-MgO@C NPs in the molten steel.

4. Conclusions

1. The AlTi-MgO alloy NPs were selected as the pre-dispersed medium in this study, with the average AlTi-MgO NPs size being 50 nm after pre-dispersion. A new type of core-shell structure that was comprised of AlTi-MgO@C NPs was successfully synthesized by dopamine self-polymerization under alkaline conditions and a high-temperature carbonization process, which can be used for steelmaking.
2. After the surface treatment, the size and composition of the NPs exhibited good dispersion of the NPs in the aqueous solution, with the width of the shell layer being about 10 nm. After immersion in hot HCl acid for several hours, the NP coating contained only amorphous and graphitic carbon, which was confirmed by XRD pattern analysis and the high-resolution TEM images of the surface-treated NPs. These results confirmed the presence of a carbon layer on the surface of the treated NPs.
3. Due to the different compositions of the NP coatings, the theoretical results show that after the surface treatment, the AlTi-MgO NPs have a smaller contact angle in the high-temperature liquid steel compared to the original AlTi-MgO NPs coated with PDA, indicating the superior

wettability of the AlTi-MgO@C NPs. Furthermore, this suggests that surface treatment will greatly increase the NP utilization ratio.

Author Contributions: Jinglong Qu conceived and designed the experiments and interpreted the data; Shufeng Yang and Jingshe Li wrote the paper; Shufeng Yang and Hao Guo analyzed the data and collected the literatures; Hao Guo and Tiantian Wang performed the experiments.

Acknowledgments: This research is supported by the National Science Foundation of China (No. 51574190).

Conflicts of Interest: The authors declare no conflict of interest.

References

1. Gelde, L.; Cuevas, A.L.; Martínez de Yuso, M.V.; Benavente, J.; Vega, V.; González, A.S.; Prida, V.M.; Hernando, B. Influence of TiO_2-coating layer on nanoporous alumina membranes by ALD technique. *Coatings* **2018**, *8*, 60. [CrossRef]
2. Vázquez Martínez, J.M.; Salguero Gómez, J.; Batista Ponce, M.; Botana Pedemonte, F.J. Effects of laser processing parameters on texturized layer development and surface features of Ti_6Al_4V alloy samples. *Coatings* **2017**, *8*, 6. [CrossRef]
3. Murugan, P.; Krishnamurthy, M.; Jaisankar, S.N.; Samanta, D.; Baran Mandal, A. Controlled decoration of the surface with macromolecules: Polymerization on a self-assembled monolayer (SAM). *Chem. Soc. Rev.* **2015**, *44*, 3212–3243. [CrossRef] [PubMed]
4. Jiang, J.H.; Zhu, L.P.; Zhu, L.J.; Zhu, B.K.; Xu, Y.Y. Surface characteristics of a self-polymerized dopamine coating deposited on hydrophobic polymer films. *Langmuir* **2011**, *27*, 14180–14187. [CrossRef] [PubMed]
5. Wei, Q.; Zhang, F.L.; Li, J.; Li, B.J.; Zhao, C.S. Oxidant-induced dopamine polymerization for multifunctional coatings. *Polym. Chem.* **2010**, *1*, 1430–1433. [CrossRef]
6. Ball, V.; Del Frari, D.; Michel, M.; Buehler, M.J.; Toniazzo, V.; Singh, M.K.; Gracio, J.; Ruch, D. Deposition mechanism and properties of thin polydopamine films for high added value applications in surface science at the nanoscale. *BioNanoScience* **2012**, *2*, 16–34. [CrossRef]
7. Wang, S.B.; Min, Y.L.; Yu, S.H. Synthesis and magnetic properties of uniform hematite nanocubes. *J. Phys. Chem. C* **2007**, *111*, 3551–3554. [CrossRef]
8. Yu, X.L.; Deng, J.J.; Zhan, C.Z.; Lv, R.; Huang, Z.H.; Kang, F.Y. A high-power lithium-ion hybrid electrochemical capacitor based on citrate-derived electrodes. *Electrochim. Acta* **2017**, *228*, 76–81. [CrossRef]
9. Lei, C.; Han, F.; Li, D.; Li, W.C.; Sun, Q.; Zhang, X.Q.; Lu, A.H. Dopamine as the coating agent and carbon precursor for the fabrication of N-doped carbon coated Fe_3O_4 composites as superior lithium ion anodes. *Nanoscale* **2013**, *5*, 1168–1175. [CrossRef] [PubMed]
10. Yue, Q.; Wang, M.H.; Sun, Z.K.; Wang, C.; Wang, C.; Deng, Y.H.; Zhao, D.Y. A versatile ethanol-mediated polymerization of dopamine for efficient surface modification and the construction of functional core-shell nanostructures. *J. Mater. Chem. B* **2013**, *1*, 6085–6093. [CrossRef]
11. Liu, Y.L.; Ai, K.L.; Lu, L.H. Polydopamine and its derivative materials: Synthesis and promising applications in energy, environmental, and biomedical fields. *Chem. Rev.* **2014**, *114*, 5057–5115. [CrossRef] [PubMed]
12. Kasemset, S.; Lee, A.; Miller, D.J.; Freeman, B.D.; Sharma, M.M. Effect of polydopamine deposition conditions on fouling resistance, physical properties, and permeation properties of reverse osmosis membranes in oil/water separation. *J. Membr. Sci.* **2013**, *425*, 208–216. [CrossRef]
13. Lee, H.; Dellatore, S.M.; Miller, W.M.; Messersmith, P.B. Mussel-inspired surface chemistry for multifunctional coatings. *Science* **2007**, *31*, 426–430. [CrossRef] [PubMed]
14. Kang, S.M.; You, I.; Cho, W.K.; Shon, H.K.; Lee, T.G.; Choi, I.S.; Karp, J.M.; Lee, H. One-step modification of superhydrophobic surfaces by a mussel-inspired polymer coating. *Angew. Chem. Int. Ed.* **2010**, *49*, 9401–9404. [CrossRef] [PubMed]
15. Liu, R.; Mahurin, S.M.; Li, C.; Unocic, R.R.; Idrobo, J.C.; Gao, H.J.; Pennycook, S.J.; Dai, S. Dopamine as a carbon source: The controlled synthesis of hollow carbon spheres and yolk-structured carbon nanocomposites. *Angew. Chem. Int. Ed.* **2011**, *50*, 6799–6802. [CrossRef] [PubMed]
16. Gao, X.Z.; Yang, S.F.; Li, J.S.; Yang, Y.D.; Chattopadhyay, K. Inclusion characteristics and microstructure properties under different cooling conditions in steel with nanoparticles addition. *Ironmak. Steelmak.* **2017**, 1–7. [CrossRef]

17. Zener, C. Private Communication to CS Smith. *Trans. Am. Inst. Metall. Eng.* **1949**, *175*, 15–17.
18. Gao, X.Z.; Yang, S.F.; Li, J.S.; Liao, H.; Gao, W.; Wu, T. Addition of MgO nanoparticles to carbon structural steel and the effect on inclusion characteristics and microstructure. *Metall. Mater. Trans. B* **2016**, *47*, 1124–1136. [CrossRef]
19. Gao, X.Z.; Yang, S.F.; Li, J.S.; Yang, Y.D.; Chattopadhyay, K.; Mclean, A. Effects of MgO nanoparticle additions on the structure and mechanical properties of continuously cast steel billets. *Metall. Mater. Trans. A* **2016**, *47*, 461–470. [CrossRef]
20. Gao, X.Z.; Yang, S.F.; Li, J.S.; Ma, A.; Chattopadhyay, K. Improvement of utilization ratio of nanoparticles in steel and its influence on acicular ferrite formation. *Steel Res. Int.* **2017**, *88*. [CrossRef]
21. Gong, W.; Chen, W.S.; He, J.P.; Tong, Y.; Liu, C.; Su, L.; Gao, B.; Yang, H.K.; Zhang, Y.; Zhang, X.J. Substrate-independent and large-area synthesis of carbon nanotube thin films using ZnO nanorods as template and dopamine as carbon precursor. *Carbon* **2015**, *83*, 275–281. [CrossRef]
22. Liang, Y.R.; Liu, H.; Li, Z.H.; Fu, R.W.; Wu, D.C. In situ polydopamine coating-directed synthesis of nitrogen-doped ordered nanoporous carbons with superior performance in supercapacitors. *J. Mater. Chem. A* **2013**, *1*, 15207–15211. [CrossRef]
23. Kong, J.H.; Yee, W.A.; Wei, Y.F.; Yang, L.P.; Anq, J.M.; Phua, S.L.; Wong, S.Y.; Zhou, R.; Dong, Y.L.; Li, X.; et al. Silicon nanoparticles encapsulated in hollow graphitized carbon nanofibers for lithium ion battery anodes. *Nanoscale* **2013**, *5*, 2967–2973. [CrossRef] [PubMed]
24. Mu, W.Z.; Mao, H.H.; Jönsson, P.G.; Nakajima, K.J. Effect of carbon content on the potency of the intragranular ferrite formation. *Steel Res. Int.* **2016**, *87*, 311–319. [CrossRef]
25. Wang, G.C.; Wang, T.M.; Li, S.N.; Fang, K.M. Study on the process of adding Al$_2$O$_3$ nano-powder to molten pure iron. *Chin. J. Eng.* **2007**, *29*, 578–581. (In Chinese)
26. Zhao, L.Y.; Sahajwalla, V. Interfacial phenomena during wetting of graphite/alumina mixtures by liquid iron. *ISIJ Int.* **2003**, *43*, 1–6. [CrossRef]
27. Humenik, M.; Kingery, W.D. Metal-ceramic interactions: III, surface tension and wettability of metal-ceramic systems. *J. Am. Ceram. Soc.* **1954**, *37*, 18–23. [CrossRef]
28. Kostikov, V.I.; Maurakh, M.A.; Nozhkina, A.V. Wetting of diamond and graphite by liquid iron-titanium alloys. *Powder Metall. Met. Ceram.* **1971**, *10*, 62–64. [CrossRef]

© 2018 by the authors. Licensee MDPI, Basel, Switzerland. This article is an open access article distributed under the terms and conditions of the Creative Commons Attribution (CC BY) license (http://creativecommons.org/licenses/by/4.0/).

Communication

Modulating the Partitioning of Microparticles in a Polyethylene Glycol (PEG)-Dextran (DEX) Aqueous Biphasic System by Surface Modification

Chang Kyu Byun *, Minkyung Kim and Daehee Kim

Department of Applied Chemistry, Daejeon University, 62, Daehak-ro, Dong-gu, Daejeon 34520, Korea; kmink2000@naver.com (M.K.); kimdh9022@naver.com (D.K.)
* Correspondence: byunck@dju.kr; Tel.: +82-42-280-4421

Received: 1 January 2018; Accepted: 11 February 2018; Published: 26 February 2018

Abstract: Aqueous two-phase systems (ATPSs) or aqueous biphasic systems are useful for biological separation/preparation and cell micropatterning. Specifically, aqueous two-phase systems (ATPSs) are not harmful to cells or biomaterials; therefore, they have been used to partition and isolate these materials from others. In this study, we suggest chemically modifying the surface of target materials (micro/nanoparticles, for example) with polymers, such as polyethylene glycol and dextran, which are the same polymer solutes as those in the ATPS. As a simple model, we chemically coated polyethylene glycol or dextran to the surface of polystyrene magnetic particles and observed selective partitioning of the surface modified particles to the phase in which the same polymer solutes are dominant. This approach follows the principle "like dissolves like" and can be expanded to other aqueous biphasic or multiphasic systems while consuming fewer chemicals than the conventional modulation of hydrophobicities of solute polymers to control partitioning in aqueous biphasic or multiphasic systems.

Keywords: aqueous two-phase system (ATPS); microparticle; partitioning; surface modification; polyethylene glycol (PEG); dextran (DEX)

---

1. Introduction

The surface of micro/nanoparticles dramatically affects their physical and chemical properties [1]. The partitioning of micro/nanoparticles for separation and analysis has been studied for various purposes such as improving sensitivity and efficiency for material collection [2] and sensing methods [1,3]. For example, partitioning has been applied to actuate interactions of a model bacterial system via the magnetic movement of bacterial microcolonies. The introduction of an aqueous biphasic system with dextran-conjugated microparticles enabled successful partitioning *E. coli* into sub-microliter droplets [4]. Although this method could two-dimensionally locate target colonies in a biological interaction study for the first time, the fundamental mechanisms underlying the phenomenon were not fully investigated or explained in detail within a short period of time. Dextran was chemically conjugated to the surface of polystyrene (PS)-coated microparticles so that the attachment of dextran onto the particle surface might increase the positive interaction or affinity of the microparticle to the lower layer of a polyethylene glycol (PEG)-dextran (DEX) aqueous biphasic system, which is usually called the "like dissolves like" rule. As a result, the DEX-conjugated microparticles showed the opposite preference for partition behavior compared to the unmodified PS microparticles. However, the reason underlying the partition inversion was not fully explained, although the partitioning to lower DEX-rich layer was achieved overall. There was a possibility that the reason the PS particles moved to the lower DEX-rich layer was not mainly the affinity increase of the DEX-conjugated particles but simply the increase in mass due to the surface attachment of the DEX-polymer to the particle.

In addition, using bovine serum albumin (BSA) for reaction processes [4] could interfere with this phenomenon and could increase the hydrophilicity of the particle, so the partition inversion might not have been due to the attachment of DEX but mainly due to that of BSA. Therefore, another important assumption naturally arises, that is., if the "like dissolve like" rule works for the surface modification using dextran to partition microparticles, the same rule would apply to attach PEG to the surface, and the resultant PEG-modified microparticles must show different partition behavior from the DEX-modified microparticles. Our motive and the challenge of studying partition was informed by these points of view. Recently, a prerequisite infrared (IR) study was attempted and published as a first step in solving the above questions [5]. Here, we aim to investigate the partition effect of surface modifying polystyrene (PS)-coated magnetic microparticles in a PEG-DEX aqueous two-phase systems (ATPS). We aim to answer the questions mentioned above to better explain the useful partition behavior that we found in previous research [4]. In addition, future directions for this study will also be suggested with a review of related papers and the limitations of the present study.

2. Materials and Methods

2.1. Materials

Bovine serum albumin (BSA, heat shock fraction, protease free, pH 7, >98%), methoxy polyethylene glycol amine (mPEG) (molecular weight (MW): 10,000), and amino-dextran (aDEX, MW: 10,000) were each purchased from Sigma-Aldrich (St. Louis, MO, USA), Creative PEGWorks (Chapel Hill, NC, USA), and Molecular Probes, Life Technologies (Eugene, OR, USA), respectively. Other chemicals were obtained from Fisher Scientific (Fair Lawn, NJ, USA). Deionized (DI) water was obtained using a water purifier device (Human Corporation, Seoul, Korea).

2.2. Preparation of Polymer-Bound Magnetic Particles

The surface modification of Dynabead® M-280 which were tosyl-activated (Life Technologies, Oslo, Norway) by amino-dextran, has been described previously [3,4]. The same approach was attempted for mPEG and BSA. Amino-dextran was reacted with Dynabead® M-280, which was tosyl activated by the following procedures. Briefly, 50 µL of a vortexed Dynabead® suspension was transferred into a 1 mL plastic tube, centrifuged, and placed on a DynaMag™-2 (Life Technologies, Oslo, Norway) magnet holder to remove the storage solution. The beads were additionally rinsed with 1 mL of buffer A (0.3 g boric acid/50 mL water, and the pH was adjusted to 9.5 with 5 M NaOH). To these beads were added approximately 3 mg of amino-dextran in 150 µL of buffer A and 100 µL of buffer C (2 g ammonium sulfate dissolved in 50 mL of buffer A). The amount of amino-dextran (aDEX) used to modify the surface of the microbeads was determined for the following reason. The tosyl-activated Dynabead® M-280 manual lists the active chemical functionality as 0.1–0.2 mmol/g. Therefore, 50 µL of a 30 mg/mL Dynabead® suspension contains 1.5 mg of Dynabeads® and $0.13–0.27 \times 10^{-3}$ mmol of tosyl groups. Amino-dextran, which has one to two amino groups per molecule, will react with the tosyl group at a 1:1 ratio, and thus, $0.13–0.27 \times 10^{-3}$ mmol of amino-dextran is required, which corresponds to 1.3–2.7 mg. The mixture was incubated on a roller at 37 °C overnight (12–18 h). After centrifuging, the dextran-conjugated beads were collected magnetically. Then, the beads were stored in a refrigerator or resuspended in 100 µL of DI water or the PEG-rich/DEX-rich layer of the ATPS (10% PEG/5% DEX or 7.5% PEG/7.5% DEX). The reaction scheme of amino-dextran with tosyl-activated Dynabeads®, was shown in Figure 1.

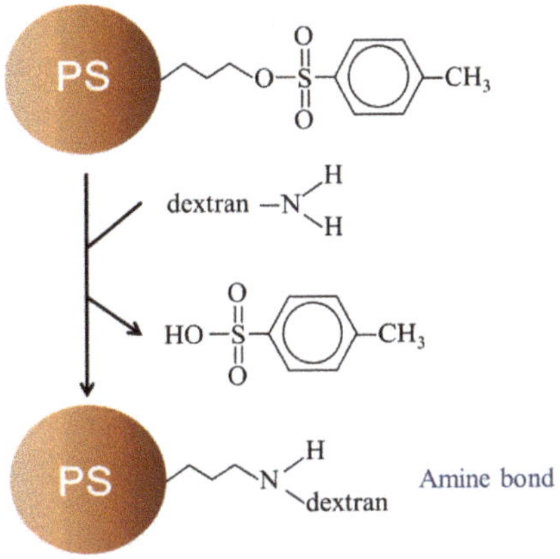

Figure 1. Reaction scheme between amino-dextran and tosyl-activated magnetic particles (Dynabeads®), which are coated with polystyrene (PS).

2.3. Preparation of the ATPS Solutions

First, 10% PEG and 5% DEX ATPS solutions were prepared by mixing 20% polyethylene glycol (PEG, MW: 8000, Fisher Scientific, Co., Fair Lawn, NJ, USA) and 10% dextran (DEX T20, MW: 20,000, Pharmacosmos, Holbaek, Denmark) solutions at a 1:1 ratio (500 µL of each PEG and DEX solutions were mixed for 1 mL). The 7.5% PEG and 7.5% DEX ATPS solutions were prepared analogously. The polymers were dissolved in DI water.

2.4. Partition Observation of Polymer-Bound Magnetic Particles

A sample (50 µL) of the previously prepared suspension of DEX-, PEG-, or BSA-conjugated magnetic beads (Section 2.2) was added to 1 mL of the ATPS solution (Section 2.3) in a 1.5 mL centrifugal plastic tube and briefly vortexed. Images were recorded after the two separated layers formed in 5 min. The procedures (Figure 2) were performed three times, each time by a different person.

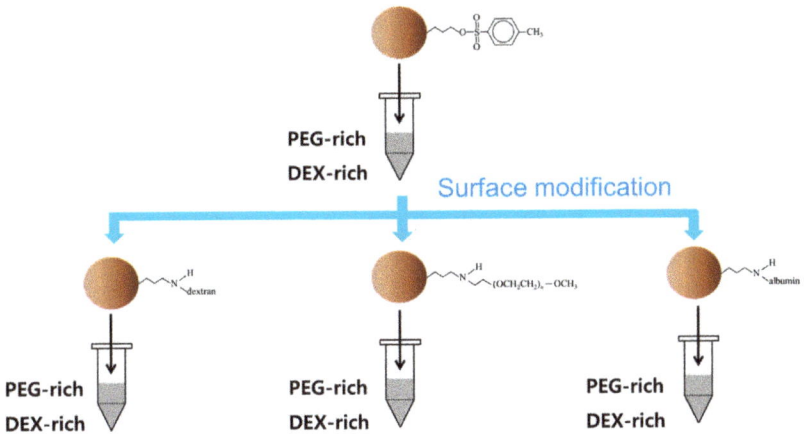

Figure 2. Partition observation in the aqueous two-phase system (ATPS) with three different surface-modified PS microparticles and one unmodified tosyl-activated Dynabead®. Amino-dextran (aDEX), methoxy polyethylene glycol amine (mPEG), and bovine serum albumin (BSA) were used for comparison with the unmodified case.

3. Results

3.1. Theoretical Considerations

The fundamentals of aqueous two-phase systems (ATPSs) have been studied and described after Albertsson's broad research and review [3,6]. The PEG-DEX binodal curve (Figure 3) that was applied to our study shows the concentration relationship between the two water-soluble polymers, which form two phases or one phase. Below the curve, the solution cannot form two separate layers and exists as a monophase solution. If the two polymer concentrations are above the binodal curve, the aqueous solution separates into layers of two phases, similar to water and oil. One of the most important characteristics of ATPSs is that both layers contain both water-soluble polymers. In other words, the two polymers are dissolved in both separated layers, but the concentrations (or the ratio) of the polymers are different in each layer. The interfacial tension of ATPSs is known to be orders of magnitude lower than that of water-oil systems [3]. The upper PEG-rich layer is known to be more hydrophobic, whereas the bottom or lower DEX-rich layer is known to be more hydrophilic. This difference arises naturally because dextran is more hydrophilic than PEG, and PEG is more hydrophobic than dextran, even though PEG is a highly water-soluble polymer.

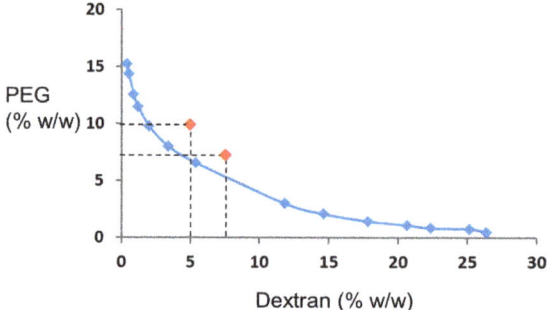

Figure 3. Phase diagram of the dextran (DEX)-polyethylene glycol (PEG) system, assuming that the compositions are approximately the same as those in [3]. The blue dots and the binodal curve are reconstructed from the data of the top and bottom phase table in ref. [3]. The average molecular weight (MW) of their polymers were 6000 (PEG) and 23,000 (DEX), which are similar to those of our system (PEG 8000–DEX 20,000). The red dots indicate the total system compositions that were attempted to investigate the partitioning of surface-modified microbeads in our study.

In fact, hydrophobicity/hydrophilicity is quite a relative and abstractive concept, and the hydrophobic ladder that describes the relative hydrophobicities of water-soluble polymers frequently used in aqueous biphasic or multiphasic systems are found in Albertsson's book [3]. In this case, the lower layer has a higher density than the upper layer, which explains why the aqueous biphasic or multiphasic systems can be formed with or even without centrifugation. The formation and maintenance of aqueous two-phase systems are known to be more dependent on entropic change than on the enthalpic change of the polymer dissolution process [3], which explains why the binodal curve changes with temperature.

The distribution of particles in an aqueous two-phase system (ATPS) was also described based on Albertsson's work [3]. Figure 4 shows three different cases of partition in the ATPS: (Figure 4A) a particle is partitioned in the upper layer (or top phase); (Figure 4B) a particle is partitioned in the lower layer (or bottom phase); and (Figure 4C) a particle is located between the two layers. The partition coefficient K can be expressed by the following equation:

$$K = \frac{C_1}{C_2} = e^{-A(\gamma_{P1} - \gamma_{P2})/kT} \qquad (1)$$

where C_1 and C_2 are the particle concentrations of upper and lower layers, respectively, A is the surface area of a particle, γ_{P1} is the interfacial tension between the particles and the upper layer, γ_{P2} is the interfacial tension between the particles and the lower layer, k is the Boltzmann coefficient, and T is the temperature in degrees Kelvin. If we assume that A is the constant for our surface modification, K will be greatly affected by the surface characteristics of the particles and the relative affinity between the particles and the upper/lower layers. If the interfacial tension between the particles and the upper layer (γ_{P1}) is smaller than the interfacial tension between the particles and the lower layer (γ_{P2}), then K will be greater than e^0, which is one (number 1), and the particles will tend to locate in the upper layer (Figure 4A). The interfacial tension between the particles and the solution layer will be smaller if the surface materials/chemicals of the particles are the same as the materials/chemicals in solution. Therefore, this strategy, which modifies the particle surface with the same solute polymer of the upper or lower layer, is expected to work for selective partitioning in the ATPS. If γ_{P1} is larger than γ_{P2}, K will be less than one, and the particles will tend to locate in the lower layer (Figure 4B). If γ_{P1} is similar or equal to γ_{P2}, K value will be approximately one, and the particles will tend to locate at the interface between upper and lower layer (Figure 4C).

More recent information about the theoretical and experimental estimation of binodal curves [7–9] and particle distribution in ATPSs [10] can be found elsewhere.

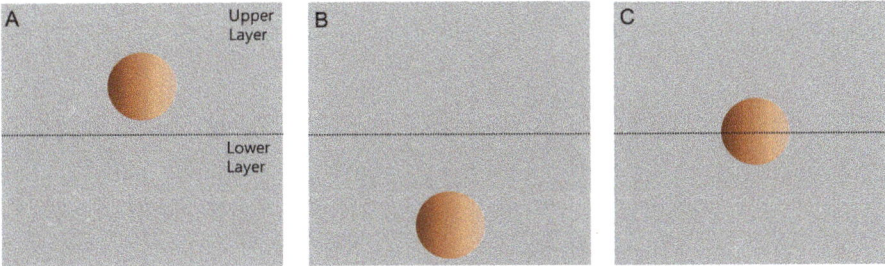

Figure 4. Distribution of microparticles in PEG-DEX ATPS. (**A**) microparticle that prefers the upper PEG-rich layer; (**B**) microparticle that is partitioned in the DEX-rich lower layer; (**C**) the microparticle that locates at the interface between two layers. The pictures are drawn conceptually and not to scale.

3.2. Partitioning of Unmodified PS Microbeads in PEG-DEX ATPS

The unmodified tosyl-activated PS microparticles called Dynabeads® are partitioned in the PEG-rich layer, as seen in Figure 5A. IR spectral investigation of the tosyl-activated Dynabeads® [5] showed that the tosyl functional group exists on the Dynabead surface, and both polystyrene and the tosyl group on the Dynabead M-280 surface are hydrophobic, which is the driving force of partition to upper PEG-rich layer.

3.3. Partitioning of Amino-Dextran (aDEX)-Modified Microbeads in PEG-DEX ATPS

Amino-dextran (aDEX) was used for the surface modification to partition Dynabeads® into the DEX-rich layer. Its molecular weight (MW 10,000) was chosen to be similar to that of the DEX solute in the ATPS for the best interactions between aDEX and DEX molecules. The modified particles were partitioned into the DEX-rich layer (Figure 5B), as in our previous studies [4]. The IR spectrum of the DEX-modified microspheres [5] proved that there were chemical changes after the reaction with amino-dextran (see Figure S1). We tested the microspheres without the BSA reaction steps, which are normally used to replace unreacted tosyl residues on the microbeads. Thus, amino-dextran obviously works to create "intimacy" in the DEX-rich layer, which contains more DEX than PEG in the aqueous solution, and this hydrophilicity is the driving force of partition for low DEX-rich layer.

3.4. Partitioning of Methoxy-Polyethylene Glycol Amine (mPEG)-Modified Microbeads in PEG-DEX ATPS

If partition inversion is required for microparticles in the ATPS (in other words, if some particles must be partitioned into the top layer, not the bottom layer) in applications such as immunoassays, modifying the particle surface with PEG will work if the same "like dissolves like" phenomenon occurs for PEG. We found various types of PEG analogs from a vendor (see Experiments) and chose methoxy-PEG-amine, which has an amine functional group on one end, while the other end is protected by the methoxy group. In addition, it has a similar average MW to that of the previously used aDEX (MW 10,000). The IR spectra of the PEG-modified Dynabeads®, established that the methoxy-PEG-amine (mPEG) modification was successful according to the IR peak of the amino bond after the reaction [5] (see Figure S1). The PEG-modified Dynabead® was partitioned into the PEG-rich upper layer (Figure 5C), similar to the unmodified tosyl-activated Dynabead® (Figure 5A). The relative hydrophobic characteristic of PEG compared to DEX, is the driving force for partitioning to upper PEG-rich layer.

3.5. Partitioning of Bovine Serum Albumin (BSA)-Modified Microbeads in PEG-DEX ATPS

BSA (MW ~66.5 kDa) is used to remove possible unreacted tosyl groups from the surface of the Dynabeads®, as explained in our previous papers [4,5], but the use of BSA might interfere with partitioning because it is a blood protein that is well known to be hydrophilic, which is the driving force of partitioning to the lower DEX-rich layer. The partitioning effect was tested using BSA as a modifier. A similar amount (3–6 mg of BSA) was added instead of aDEX or mPEG for the modification reaction because BSA has multiple amine groups per molecule. Thus, we did not need to add a one-to-one molar ratio of BSA for the tosyl group substitution reaction. Figure 5D shows that the BSA-modified Dynabeads® were partitioned into the DEX-rich layer. In addition, the IR spectrum for the modified particles revealed the attachment of BSA on the modified microparticles [5] (also see Figure S1).

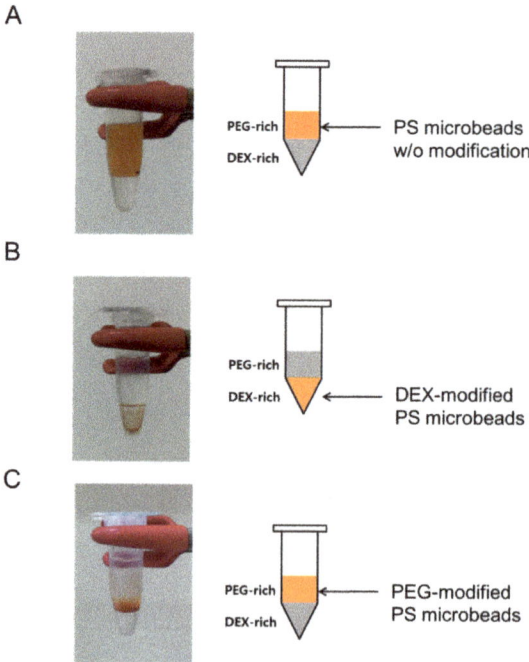

Figure 5. *Cont.*

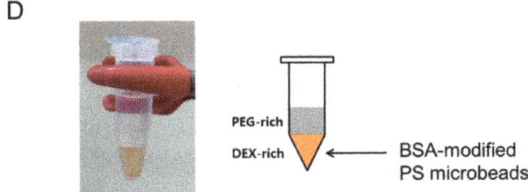

Figure 5. Side-view pictures and results of the partitioning of microparticles with different surfaces. (**A**) unreacted, tosyl-activated Dynabead®; (**B**) microparticles modified with amino-dextran (MW 10,000); (**C**) microparticles modified with methoxy-(polyethylene)glycol-amine (MW 10,000) and (**D**) microparticles modified with bovine serum albumin (MW ~66,500). The composition of the ATPS was 10% (w/w) PEG–5% (w/w) DEX. Other conditions are described in the Experimental section.

4. Discussion

Controlling partitioning of microparticles is of importance for collecting target material from mixtures which are very similar in size and features. Jauregi et al. [11] could separate ampicillin and phenylglycine crystal mixture by selective interfacial partitioning in a water/alkanol biphasic system. Recently, Cakmak and Keating [12] reported that natural clays, which work as reaction catalysts, partitioned differently by their kind in PEG/DEX ATPS and showed slight increase of reaction rate in ATPS compared to that in buffer. Also, Deng et al. [13] reported that PEG/DEX ATPS could separate Merrifield resins and N-methylimidazolium grafted Merrifield resins by controlling the ATPS compositions, graft ratio and different anions of resins.

Surface-modified microparticles were demonstrated to be useful for selectively partitioning particles in an ATPS. The surface modification, not only with DEX but also with PEG, triggered the partitioning of the modified microparticles into the specific layer (phase) in the ATPS, that is, the layer that contained more of the same polymer as a solute in the ATPS. In addition, our results showed that not only the density [14] but also the affinity and hydrophobicity/hydrophilicity must be considered for the partitioning of microparticles. Compared to the conventional approach reported by Albertsson [3], which achieved partitioning in an ATPS by selecting hydrophobic/hydrophilic polymer solutes from the library or hydrophobic ladder, our approach has the benefit of requiring and consuming much less polymer (milligrams) to obtain the desired partitioning in the ATPS.

Characterization of surface chemical modification about micro/nanoparticles has been reviewed by many researchers [1,15,16]. We chose FT-IR microscopy [5] mainly because of its simplicity and minimal amount of sample needed, compared to other known methods. Among them, zeta potential measurement [12,17], vibrating sample magnetometry (VSM) [18], X-ray photoelectron spectroscopy (XPS) [17,19], and thermogravimetry [15] can be considered as other tools for confirmation of surface modification, which may provide more quantitative information on the degree of modification.

Several topics may be considered in further investigation. For example, we tested polymers with similar average MWs to that of the ATPS polymer solutes for optimal interaction. However, not only the type of polymers/chemicals that are attached to the particle surfaces but also other parameters, such as the average molecular weights of the polymers or the mixing ratio of polymers, may affect the partitioning of the polymer–modified particle. We tested a 1:1 ratio of aDEX/mPEG for the modification, and the particles were observed to locate narrowly at the interface of the PEG-DEX ATPS layers (data not shown). Granick et al. [20] devoted several studies to the conformation and assembly of micro-sized particles. Similar approaches may possibly achieve partitioning in PEG-DEX ATPS or other ATPSs with our surface modification strategy.

Supplementary Materials: The following are available online at http://www.mdpi.com/2079-6412/8/3/85/s1, Figure S1. Infrared spectra of (a) PS microparticle (blue), DEX-modified particle (red), and amino-dextran (green); (b) PS microparticle (blue), PEG-modified particle (red), and methoxy-PEG-amine (green); (c) PS microparticle (blue), BSA-modified particle (red), and BSA (green); see reference [5] for peak assignments and conditions.

Acknowledgments: This research was supported by the Daejeon University fund (201501410001). The authors are thankful to Samkeun Lee and In-Ho Lee, Department of Applied Chemistry, Daejeon University, for their advice on this research and their kind permission to use their instruments.

Author Contributions: Chang Kyu Byun, Minkyung Kim and Daehee Kim conceived and designed the experiments; Chang Kyu Byun, Minkyung Kim and Daehee Kim performed the experiments; Chang Kyu Byun contributed reagents/materials and wrote the paper; Minkyung Kim and Daehee Kim contributed equally.

Conflicts of Interest: The authors declare no conflict of interest. The founding sponsors had no role in the design of the study; in the collection, analyses, or interpretation of data; in the writing of the manuscript, and in the decision to publish the results.

References

1. Lin, P.-C.; Lin, S.; Wang, P.C.; Sridhar, R. Techniques for physicochemical characterization of nanomaterials. *Biotechnol. Adv.* **2014**, *32*, 711–726. [CrossRef] [PubMed]
2. Chen, L.; Holmes, M.; Schaefer, E.; Mulchandani, A.; Ge, X. Highly active spore biocatalyst by self-assembly of co-expressed anchoring scaffoldin and multimeric enzyme. *Biotechnol. Bioeng.* **2018**, *115*, 557–564. [CrossRef] [PubMed]
3. Albertsson, P.-A. *Partition of Cell Particles and Macromolecules*, 3rd ed.; John Wiley & Sons: New York, NY, USA, 1986.
4. Byun, C.K.; Hwang, H.; Choi, W.S.; Yaguchi, T.; Park, J.; Kim, D.; Mitchell, R.J.; Kim, T.; Cho, Y.-K.; Takayama, S. Productive chemical interaction between a bacterial microcolony couple is enhanced by periodic relocation. *J. Am. Chem. Soc.* **2013**, *135*, 2242–2247. [CrossRef] [PubMed]
5. Byun, C.K. Investigation of chemical modification on tosyl-activated polystyrene microsphere magnetic particle surface by infrared microscopy. *Anal. Sci. Technol.* **2016**, *29*, 225–233. [CrossRef]
6. Iqbal, M.; Tao, Y.; Xie, S.; Zhu, Y.; Chen, D.; Wang, X.; Huang, L.; Peng, D.; Satter, A.; Shabbir, M.A.B.; et al. Aqueous two-phase system (ATPS): An overview and advances in its applications. *Biol. Proced. Online* **2016**, *18*, 18. [CrossRef] [PubMed]
7. Croll, T.; Munro, P.D.; Winzor, D.J.; Trau, M.; Nielsen, L.K. Quantitative prediction of phase diagrams for polymer partitioning in aqueous two-phase systems. *J. Polym. Sci. Part B Polym. Phys.* **2003**, *41*, 437–443. [CrossRef]
8. Amrhein, S.; Schwab, M.-L.; Hoffmann, M.; Hubbuch, J. Characterization of aqueous two phase systems by combining lab-on-a-chip technology with robotic liquid handling stations. *J. Chromatogr. A* **2014**, *1367*, 68–77. [CrossRef] [PubMed]
9. Atefi, E.; Fyffe, D.; Kaylan, K.B.; Tavana, H. Characterization of aqueous two-phase systems from volume and density measurements. *J. Chem. Eng. Data* **2016**, *61*, 1531–1539. [CrossRef]
10. Asenjo, J.A.; Andrews, B.A. Aqueous two-phase systems for protein separation: A perspective. *J. Chromatogr. A* **2011**, *1218*, 8826–8835. [CrossRef] [PubMed]
11. Jauregi, P.; Hoeben, M.A.; van der Lans, R.G.J.; Kwant, G.; van der Wielen, L.A.M. Selective separation of physically near-identical microparticle mixtures by interfacial partitioning. *Ind. Eng. Chem. Res.* **2001**, *40*, 5815–5821. [CrossRef]
12. Cakmak, F.P.; Keating, C.D. Combining catalytic microparticles with droplets formed by phase coexistence: adsorption and activity of natural clays at the aqueous/aqueous interface. *Sci. Rep.* **2017**, *7*, 3215. [CrossRef] [PubMed]
13. Deng, Y.F.; Zhang, D.L.; Zhu, L.L.; Long, T.; Chen, J. Applying aqueous biphasic systems for partitioning N-methylimidazolium grafted Merrifield resin microparticles. *Solv. Extr. Ion Exch.* **2010**, *28*, 653–664. [CrossRef]
14. Mace, C.R.; Akbulut, O.; Kumar, A.A.; Shapiro, N.D.; Derda, R.; Patton, M.R.; Whitesides, G.M. Aqueous multiphase systems of polymers and surfactants provide self-assembling step-gradients in density. *J. Am. Chem. Soc.* **2012**, *134*, 9094–9097. [CrossRef] [PubMed]

15. Mohamad, N.R.; Marzuki, N.H.C.; Buang, N.A.; Huyop, F.; Wahab, R.A. An overview of technologies for immobilization of enzymes and surface analysis techniques for immobilized enzymes. *Biotechnol. Biotechnol. Equip.* **2015**, *29*, 205–220. [CrossRef] [PubMed]
16. Wang, L.S.; Hong, R.Y. Synthesis, surface modification and characterization of nanoparticles. In *Advances in Nanocomposites—Synthesis, Characterization and Industrial Applications*; Reddy, B., Ed.; InTech: Rijeka, Croatia, 2011.
17. Yu, Y.; Addai-Mensah, J.; Losic, D. Surface modification of diatomaceous earth silica microparticles with functional silanes for metal ions sorption. *Chemeca* **2010**, *38*, 379–388.
18. Bhayani, K.R.; Kale, S.N.; Arora, S.; Rajagopal, R.; Mamgain, H.; Kaul-Ghanckar, R.; Kundaliya, D.C.; Kulkarni, S.D.; Parsricha, R.; Dhole, S.D.; et al. Protein and polymer immobilized $La_{0.7}Sr_{0.3}MnO_3$ nanoparticles for possible biomedical applications. *Nanotechnology* **2007**, *18*, 345101. [CrossRef]
19. Yang, C.; Guan, Y.; Xing, J.; Liu, H. Surface functionalization and characterization of magnetic polystyrene microbeads. *Langmuir* **2008**, *24*, 9006–9010. [CrossRef] [PubMed]
20. Zhang, J.; Luijten, E.; Granick, S. Toward design rules of directional Janus colloidal assembly. *Annu. Rev. Phys. Chem.* **2015**, *66*, 581–600. [CrossRef] [PubMed]

© 2018 by the authors. Licensee MDPI, Basel, Switzerland. This article is an open access article distributed under the terms and conditions of the Creative Commons Attribution (CC BY) license (http://creativecommons.org/licenses/by/4.0/).

Review

Progress in Wear Resistant Materials for Total Hip Arthroplasty

Rohit Khanna [1],*, Joo L. Ong [2], Ebru Oral [3,4] and Roger J. Narayan [5]

1. Department of Mechanical Engineering, The University of Texas at San Antonio, San Antonio, TX 78249, USA
2. Department of Biomedical Engineering, The University of Texas at San Antonio, San Antonio, TX 78249, USA; Anson.Ong@utsa.edu
3. Harris Orthopaedic Laboratory, Massachusetts General Hospital, Boston, MA 02114, USA; EORAL@mgh.harvard.edu
4. Department of Orthopaedic Surgery, Harvard Medical School, Boston, MA 02115, USA
5. Joint Department of Biomedical Engineering, University of North Carolina and North Carolina State University, Raleigh, NC 27695, USA; Roger_Narayan@unc.edu
* Correspondence: rkhannaJP@gmail.com; Tel.: +1-210-458-5524

Received: 20 June 2017; Accepted: 5 July 2017; Published: 9 July 2017

Abstract: Current trends in total hip arthroplasty (THA) are to develop novel artificial hip joints with high wear resistance and mechanical reliability with a potential to last for at least 25–30 years for both young and old active patients. Currently used artificial hip joints are mainly composed of femoral head of monolithic alumina or alumina-zirconia composites articulating against cross-linked polyethylene liner of acetabular cup or Co-Cr alloy in a self-mated configuration. However, the possibility of fracture of ceramics or its composites, PE wear debris-induced osteolysis, and hypersensitivity issue due to metal ion release cannot be eliminated. In some cases, thin ultra-hard diamond-based, TiN coatings on Ti-6A-4V or thin zirconia layer on the Zr-Nb alloy have been fabricated to develop high wear resistant bearing surfaces. However, these coatings showed poor adhesion in tribological testing. To provide high wear resistance and mechanical reliability to femoral head, a new kind of ceramic/metal artificial hip joint hybrid was recently proposed in which 10–15 µm thick dense layer of pure α-alumina was formed onto Ti-6Al-4V alloy by deposition of Al metal layer by cold spraying or cold metal transfer methods with 1–2 µm thick Al_3Ti reaction layer formed at their interface to improve adhesion. An optimal micro-arc oxidation treatment transformed Al to dense α-alumina layer, which showed high Vickers hardness 1900 HV and good adhesion to the substrate. Further tribological and cytotoxicity analyses of these hybrids will determine their efficacy for potential use in THA.

Keywords: wear resistance; artificial hip joints; UHMWPE; ceramics; coatings; alumina/Ti alloy hybrid; cold spraying; cold metal transfer; micro-arc oxidation; total hip arthroplasty

1. Introduction

Every year more than one million hip and knee replacements are performed worldwide, which provide relief from pain, improved mobility, and better quality of life for patients. There will be an estimated 3.48 million knee replacements needed by 2030, an increase of approximately 673% as compared to the current number of procedures [1]. Artificial hip replacements are estimated to increase by 174% up from its current total procedures performed to 572,000 procedures per year in 2030. Current developments in the field of artificial hip joints are focused on the use of (a) short stems for minimal invasive surgery [2,3]; (b) new Ti alloys with better mechanical strength and biocompatibility than conventional Ti alloys [4–7]; (c) calcium phosphate coatings [8–12] or simple alkali and heat treatments on the stem to provide bioactivity [13–17]; and (d) materials that impart better wear resistance and mechanical reliability on the articulating surfaces to minimize polyethylene wear debris-induced osteolysis [18–27]. Periprosthetic osteolysis is the primary cause of hip implant failure, which is the result of activation of an innate immune response caused by wear of bearing materials in total hip prostheses. Taken up by macrophages and multi-nucleated giant cells, the presence of wear debris particles may cause the release of cytokines, thereby resulting in inflammation that further activates osteoclasts at the interface between bone and implant and eventually leading to implant loosening and failure. One of the primary strategies to avoid implant revision is to eliminate the release of wear debris by improving the wear resistance and mechanical reliability of bearing couples that would ultimately enhance the longevity of the implant. Despite progressive attempts over last few decades, currently available hip replacement joints last only about 10–15 years. This relatively short implant life span is problematic for patients under 60 years; about 44% of this population demands a joint implant life expectancy of up to 20–25 years [28]. The demand for reliable hip joints is relatively high for young and active patients to maintain their work and life style comfortably. In this regard, this review is mainly focused on describing the evolution of materials and technologies developed to impart high wear resistance and reliability to bearing surfaces of artificial hip joints.

2. History of Development of Artificial Hip Joints

During 1955–1965, metal-on-metal (MoM) bearings were fabricated using large ball diameters (32 mm, 35 mm, and 41.5 mm). However, the use of MoM bearings declined in the 1970s for some years after Sir John Charnley introduced a total hip replacement (THR) device based on metal-on-polymer (MoP) composed of a small metal ball and a cemented polyethylene (PE) cup in 1963 [29]. Although this new tribocouple device received widespread acceptance, it was revealed over 6–8 years of clinical studies that implants with PE cups failed mainly due to osteolysis, a result of a destructive reaction by the body in the presence of PE wear debris [30]. Anticipating the issue of "polyethylene disease" or "cement disease", Pierre Boutin, a French surgeon, began implementing the use of alumina ceramic-on-ceramic (CoC) hip implants in clinical trials in the 1970s [31]. Since then, CoC devices have continuously been used in THA. These developments also made ceramic-on-polyethylene (CoP) combinations as competitive bearing alternative along with MoM and CoC over 1963–1973 period. Currently, the most frequently used artificial hip joints are composed of an acetabular cup, femoral head and stem, all of which are typically made of cast Cobalt-Chromium (Co-Cr) alloy. The lining of the acetabular cup is made of ultra-high molecular weight polyethylene (UHMWPE). Alternative materials are dense alumina ceramic and titanium−6% aluminum−4% vanadium (Ti-6Al-4V) alloys for the femoral head and the stem, respectively.

3. Ultra-High Molecular Weight Polyethylene

UHMWPE wear debris-mediated osteolysis is widely known as one of the most formidable challenges in hip arthroplasty [32,33]. To improve its wear resistance, UHMWPE is commonly exposed to irradiation at doses higher than that required for sterilization (40–100 kGy) to form cross-linked PE (XLPE) [34]. However, improvement of UHMWPE's wear resistance comes at the cost of reducing its ultimate tensile strength as a result of reduced ductility [35]. In addition, many of its mechanical properties are sensitive to oxidation [36,37], which is mainly due to the reactions of residual free radicals caused by irradiation and trapped in the material for prolonged periods of time [38,39]. Oxidative changes gradually decrease the polymer's molecular weight and affect its clinical performance through structural changes [40]. To overcome these problems, XLPE is stabilized by post re-melting or annealing and by the addition of antioxidants such as vitamin E to eliminate, reduce, or stabilize free radicals, with the intention of increased wear resistance as well as reduced potential for oxidation as compared to conventional XLPEs [18,19,41]. While some highly cross-linked UHMWPEs show higher oxidation *in vivo* than others [42], radiation cross-linking of UHMWPE has tremendously decreased (87%) the incidence of wear particle-induced osteolysis in the first ten years of clinical use [43], especially in hips where wear damage is prominent.

In addition to state-of the art technology incorporating antioxidants for long-term stability [44–46], a recent application of a surface treatment on the XLPE articulating surface has improved the wear resistance by covering its surface with a thin layer (100–200 nm) of chemically-bonded Poly(2-methacryloyloxyethyl phosphorylcholine) (PMPC), which is formed by photo-induced graft polymerization; this material creates a super-lubricious layer that mimics articular cartilage [20]. A recent hip simulator study reported that MPC polymer grafted on the XLPE surface dramatically reduced the wear up to 70 million cycles [47]. Wear debris induced osteolysis and inflammatory responses may be suppressed by this new design of PMPC-grafted XLPE, which is attributed to the bio-inertness of bio-inspired MPC polymers [47]. Although this kind of artificial hip joint has been implanted in more than 20,000 patients since 2011 in Japan [47], it is a new technology whose success and impact will be determined in the longer term. Analysis of retrieved components and radiographic follow-up of patients [48] will continue to inform us on the effects of wear on future clinical performance.

4. Metallic Materials

Metallic materials such as cobalt-chromium-molybdenum (Co-Cr-Mo) and Ti alloys are commonly used in THA due to their superior mechanical strength, fracture toughness and ductility. Typically, CoCrMo alloys and Ti alloys are used as stem of artificial hip joints. Co-Cr alloy used as head in MoM artificial hip joint model has demonstrated mechanical reliability due to high mechanical strength and fracture toughness, but it tends to produce metal debris over the long-term, releasing metal ions that cause inflammation and blackening of periprosthetic tissue termed as metallosis. Incidents of metal sensitivity were reported to affect 10%–15% of the general population [49–51] and were higher among patients with failed joints. Ti-6Al-4V alloy is the most commonly used alloy for stem and acetabular cementless components of THR due to its low density, high mechanical strength, excellent corrosion resistance, and biocompatibility with bone [52]. Compared with Co-Cr alloys, the elastic modulus of 110 GPa for Ti alloy offers a more physiologically uniform stress distribution at the bone-implant interface. However, the release of potentially harmful metal ions such as aluminum (Al) and vanadium (V) from the Ti alloy has been reported to be associated with long term health problems such as peripheral neuropathy, osteomalacia and Alzheimer's disease [53,54].

Over the last few decades, vanadium-free Ti implants such as $\alpha + \beta$ Ti-6Al-7Nb alloy (ISO 5832-11), near β alloys such as Ti-13Nb-13Zr alloy (ASTM F1713-96), Ti-12Mo-6Zr-2Fe alloy (ASTM F1813-97) with improved biocompatibility have been developed by incorporating biocompatible elements such as Ta, Zr or Nb [4–7]. Metastable β type Ti-15Mo-5Zr-3Al and $\alpha + \beta$ type Ti-6Al-2Nb-1Ta alloys have been clinically developed for both cemented and cementless types of artificial hip joints. A new Ti-15Zr-4Nb-4Ta alloy with excellent mechanical properties and biocompatibility has been developed for artificial hip joint application [7]. To overcome stress shielding effects, recent research on Ti alloy development is focused on controlling processing to reduce the elastic modulus while retaining high biocompatibility and mechanical strength. In this regard, a metastable β type Ti alloy such as Ti-35Nb-7Zr-5Ta alloys with an elastic modulus of 55 GPa has been developed [55]. Nevertheless, Ti alloys are not used for manufacturing of femoral head due to their poor wear resistance. To overcome this issue, thermal oxidation treatments have been applied to Ti alloy to form oxides to improve surface hardness [56,57], laser surface treatments to alter its surface microstructures [58] and/or friction stir processing to change its metallic properties through localized plastic deformation [59]. However, these treatments do not provide superior wear resistance, which limit their use as the bearing surfaces of artificial hip joints.

5. Ceramics

5.1. Alumina

Alumina ceramics are most widely used in THA as femoral heads due to their bio-inertness, high wear resistance, and chemical durability. In terms of design, the surface finish of materials used for manufacturing of femoral heads or cups is an important requirement. Advances in alumina processing have revealed that an excellent surface finish can be achieved using high purity alumina with <0.5 wt % magnesium oxide as a sintering aid and compaction using hot-isostatic pressing before sintering to obtain a dense microstructure of fine-grained pure α-alumina with improved mechanical properties. As per ASTM F603, medical grade alumina for orthopedic implant use should have a bulk density of >3.94 gm/cc, grain size of <4.5 µm, and flexural strength of >400 MPa. Compared with MoP, artificial hip joints with CoP couple have shown 25%–30% reduction in wear rate in hip simulator and clinical studies [60,61]. Using an alumina-on-alumina couple, wear in THA was the lowest (0.01–0.1 mm^3/million cycles) as compared with MoP, CoP or MoM combinations [61,62]. Since its inception by Boutin in the 1970s, more than 2.5 million femoral heads of alumina and 100,000 liners have been implanted worldwide. Although incidences of fracture of alumina ceramics in THA are rare (0.14% reported in the USA in the mid-1990s), the uncertainty about risk of brittle fracture, rim chipping, squeaking [63] (audible noise) and stripe wear [64,65] in artificial hip joints with alumina CoC couple are mainly due to improper positioning [66] and cannot be eliminated. Squeaking in THA articulations with alumina CoC have been reported especially in some young patients, aged forty or under, despite high survivorship up to 97.4% at ten years [67,68]. Squeaking can be associated with impingement of femoral neck on the rim of the ceramic cup, due to difference in diameters of femoral heads, edge loading effect due to improper positioning during surgery or could also be due to micro-separation between femoral head and liner of cup, all of which increase friction in case of CoC bearing couples and hip dislocation and potentially lead to to revision of surgery. Current trends to increase the diameters of ceramic balls to 32 mm, 36 mm, or larger tend to make a positive effect. Nevertheless, the risk of catastrophic fracture of alumina cannot be eliminated.

5.2. Zirconia

To improve the mechanical reliability of the head, partially stabilized zirconia and yttria-stabilized tetragonal zirconia polycrystal (Y-TZP) ceramic with almost double fracture toughness of 6–10 MPa $\sqrt{m}$ and flexural strength of 1000 MPa than alumina, were introduced as alternatives in the 1980s. Improved mechanical properties in zirconia are attributed to stress-induced transformation toughening mechanism. Briefly, as crack propagates, zirconia undergoes phase transformation from tetragonal to monoclinic in leading to 3%–4% volume expansion in the zirconia grains, resulting in compressive stress to slow or arrest the crack, thereby increasing the toughness. Despite some good clinical results with seven-year [69] and ten-year [70] follow up, several controversial hip simulator and clinical reports have been published over 15 years of its clinical use in patients [71,72]. Unfortunately, a worldwide recall concluded in 2001 led to a sudden halt of manufacturing and clinical use of zirconia after many zirconia heads manufactured by St. Gobain were fractured *in vivo* due to change in sintering procedure. Even though transformation toughening led to high strength and toughness of designed Y-TZP heads, the spontaneous transformation to monoclinic polymorph in the presence of body fluid results in roughening and micro-crack formation on the surface of the head, thereby enhancing wear and eventual fracture as reported in *in vivo* studies [72]. This phenomenon is often referred as low-temperature hydrothermal aging. Close examinations of more than 100 retrieved Zr femoral heads revealed extensive cratering in addition to considerable monoclinic phase [71]. Further, eight and nine-year follow up revealed up to 85% monoclinic phase and dramatically higher surface roughness of up to 250 nm on the Prozyr heads retrieved from Sydney [73,74]. These critical issues almost restricted the use of Zr femoral heads in the US and EU markets. Although zirconia is commonly used in dentistry due to its aesthetics and high mechanical properties, its microstructure and processing steps must be carefully examined before use. Hydrothermal aging is the Achilles' heel of zirconia, limiting its longevity.

5.3. Silicon Nitride

Silicon nitride is a non-oxide ceramic material with high strength and toughness and has been used as bearings, turbine blades for more than 50 years. In the medical field, it has been used in cervical spacer and spinal fusion devices [75] since 2008, with few adverse reports among 25,000 implanted spinal cages [76]. Silicon nitride has been recently considered as a bearing material in artificial hip and knee replacements due to its high biocompatibility, moderate Vickers hardness of 12–13 GPa, Young's modulus of 300 GPa, high fracture toughness of 10–12 MPa $\sqrt{m}$ and flexural strength of 1 GPa, with a typical grain size of 0.6 µm after alloying with small amounts of yttria and alumina [77]. The exceptional strength and toughness of silicon nitride are derived from its asymmetric needle-like interlocking grains surrounded by thin glassy phase at the grain boundary, which dissipates energy as crack propagates [78]. Silicon nitride has also been used as wear resistant coating by PVD or CVD methods [79,80]. However, coatings are non-stoichiometric, not fully crystalline, and do not resemble similar strength and toughness as polycrystalline materials, which limited their use as coatings for implants. Bulk silicon nitrides in articulation against itself and Co-Cr materials have shown wear rates that are comparable to those of alumina-on-alumina couple, which are the lowest among currently used bearing couples [81]. A recent hip simulator study indicated that self-mated silicon nitride couples have comparable wear performance to that of self-mated alumina up to three million cycles; however, some self-mated silicon nitride couples showed increased wear at the end of five million cycles as compared to alumina CoC [82]. This transition in wear regime from fluid film to lubrication was attributed to the formation and disruption of tribochemical film composed of a gelatinous silicic acid. Silicon nitrides articulating against PE or XLPE revealed similar wear performance as that of CoCr or alumina heads. There is some discrepancy between the hip simulator and *in vivo* studies for some femoral heads, further long-term clinical studies of retrieved heads of silicon nitride and hip simulator studies by others might be necessary before confirming the potential use of silicon nitride as bearing material for hip replacements.

5.4. Alumina-Zirconia Composites

Despite the long clinical history of alumina and zirconia in THR, both materials pose potential drawbacks. Attempts to overcome these material's weaknesses by combining alumina's hardness with zirconia's toughness led to the development of zirconia-toughened alumina (ZTA), which was first commercialized by CeramTec under the trade name of BIOLOX® Delta in around 2000. ZTA is an alumina matrix composite containing 75% fine grained alumina of 0.5–0.6 µm in diameter and 25% Y-TZP with a grain size of 1 µm or smaller to obtain a flexural strength of 1200 MPa and a fracture toughness of 6.5 MPa $\sqrt{m}$. The flexural strength and fracture toughness of ZTA are significantly higher than that of alumina while retaining superior chemical and hydrothermal stability like alumina. Superior mechanical strength and toughness of ZTA is attributed to the stress-induced transformation toughening mechanism offered by an optimal amount of fine grained zirconia dispersed throughout the ZTA microstructure. Additional toughening to alumina is provided by crack deflection at the boundaries of platelet-like alumina grains with magneto plumbite structure that are formed by solid solutions with a small amount of SrO and Cr_2O_3 additives during high temperature sintering.

This unique combination of transformation toughening and crack deflection mechanisms provides exceptional durability to BIOLOX® delta, which is not achieved by any other ceramic material used in the body. Addition of alumina to zirconia slows down the kinetics of hydrothermal ageing, which is a potential advantage over monolithic zirconia. Laboratory hip simulator wear tests on ZTA-on-ZTA couple revealed that the wear performance of ZTA CoC was better than that of alumina CoC [83]. Moreover, more strip wear was revealed on alumina CoC as compared with ZTA CoC tested under same tribological parameters. However, up to 30% monoclinic zirconia was detected on the surface of

ZTA, which shows that transformation from tetragonal to monoclinic zirconia happened during the articulation between ZTA cup and the ball [84]. *In vivo* wear results are more realistic as compared to hip wear simulator results. Even though BIOLOX delta components are clinically used for more than 12 years and have been implanted in more than six million patients, according to CeramTec company web information, longer term clinical reports will determine the efficacy of this new ceramic composite for long lasting hip prostheses.

Alumina toughened zirconia (ATZ) is another alumina-zirconia composite bearing material containing 80% Y-TZP and 20% alumina, which was developed and manufactured as Ceramys® by Mathys Orthopedics [85] in Germany in 2010. ATZ Ceramys is composed of 61% tetragonal zirconia, 17% cubic zirconia, 1% monoclinic zirconia and remaining alumina which form a fine-grained alumina dispersed in the zirconia matrix with average grain size of 0.4 µm for both zirconia and alumina. ATZ revealed a high flexural strength of more than 1200 MPa, a moderate hardness of 1500 HV, and a fracture toughness of 7.4 MPa $\sqrt{m}$ [23]. In addition, ATZ is reported to have improved wear performance with a wear rate of 0.06 mm^3/million cycles as compared with the 0.74 mm^3/million cycles for monolithic alumina heads in CoC configuration [24]. In addition to the critical hydrothermal aging effect of zirconia, there remain major concerns with the variation in manufacturing steps of zirconia-based materials by different manufacturers, which needs to be carefully assessed every time before implantation in the human body. Nevertheless, even composite ceramics are brittle and liable to fracture.

6. Ultra-Hard Coatings on Metals

To maintain the active lifestyle of patients requiring hip replacements, reliable designs of artificial hip joints with high wear resistance and mechanical reliability are needed to extend their current life span from 10–15 years to 25–30 years, which otherwise will require multiple revision surgeries that will burden healthcare expenditure all over the world. The dire need to produce more durable artificial joints is made explicit by the fact that hip and knee replacement surgeries have increased by 20% during the last five years [86]. Trends for revision surgeries are becoming more common among young and active patients aged between 45 and 64 who will require functional artificial hip for at least 30 years [86] to maintain their active lifestyle. While Co-Cr alloy in self-mated configuration or the alloy heads sliding against PE or XLPE are frequently used in THA, over 50% artificial hip joints fail mainly due to osteolysis-mediated aseptic loosening in addition to metal ion allergies over a long-term period [87]. A frequently used alternative hybrid approach is to coat the metallic alloys with ultra-hard and biocompatible surface layers such as diamond-like carbon (DLC; 5000 HV) [88] or titanium nitride (TiN 2100 HV) [89]. This approach ensures that the original properties of high strength metallic substrate are retained while: (a) supporting a bearing surface; and (b) avoiding the release of toxic metal ions from the underlying the Ti alloy substrate. The release of wear debris from TiN-coated Ti alloy after delamination enhanced wear by third-body wear mechanisms *in vivo* [89]. DLC-coated Ti-6Al-4V alloy showed numerous pits, local delamination, and crevice corrosion while articulating against PE and showed only 54% survival rate after 8.5 years of clinical study [90]. To improve the adhesion between the coating and the substrate, interlayers of Ti, tantalum (Ta), or CrC have been used to minimize the residual stresses at the interface. Recently, a DLC layer with a varying thickness of multi-interlayers of Cr-Cr$_2$N was reported to increase the fracture strength or adhesion of the DLC coating [91]. However, the hardness of these multilayer coatings (800–1000 HV) was almost half of that of sintered alumina (1800–2000 HV) [91]. In addition, there could be concerns about inflammation due to release of Cr over the long term.

Another method is to deposit pure diamond on the metal heads. In this regard, coatings of ultra-nanocrystalline diamond (UND) with a grain size of 3–100 nm have been applied directly to Ti and Co-Cr alloys using microwave plasma CVD [92,93]. UND coatings possess high hardness (56–80 GPa) and low surface roughness, RMS value < 10 nm, that is shown to provide high wear resistance to third-body wear particles [94]. These coatings have been applied to Ti and its alloys;

good corrosion resistance, adhesion, and toughness have been reported in addition to comparable tribological performance to those of MoM, MoP and CoC couples, which makes them suitable for THA application [95]. However, large compressive stresses are retained within the UND coatings due to impurities at their grain boundaries, which affected their adhesion with the substrate [96]. Further improvements in processing parameters are needed to avoid the risk of delamination of the UND coatings, which will govern their clinical use. In a nutshell, further improvements are needed in these coating technologies to meet high wear resistance, mechanical reliability and adhesion requirements for long lasting total hip prostheses.

7. Hybrid Design of Oxide Ceramic Layer on Metal: Oxinium™

A promising approach of a dense oxide layer on metal substrate combines high mechanical strength and toughness of underlying metal with advantages of an oxide layer with bio-inertness, high corrosion and wear resistance as an articulating surface. If native TiO_2 layer on the surface of high mechanical strength Ti alloys was at least a few microns thick and wear resistant for long lasting hip prostheses or bulk ceramics were not inherently brittle, there could have been no interventions to deposit alternative wear resistant materials on metals over the last few decades. These drawbacks have spurred alternative explorations of bio-inert oxide ceramic layers on metallic substrates, adopting a hybrid approach by wedding the superior wear resistance of ceramics with the superior toughness of metals.

A layer of monoclinic zirconia 5 μm thick formed on the surface of Zr-2.5% Nb alloy by heat treatment of the alloy at 500 °C [25] has been successfully applied to a femoral head commercialized as Oxinium™ (OxZr; Smith & Nephew, Memphis, TN, USA). OxZr is not a coating, but a surface transformed layer formed by oxygen diffusion hardening treatment, which is expected to provide improved resistance under load bearing operations. Compared to monolithic zirconia, thin zirconia layer on the Zr-Nb alloy avoids catastrophic failure of the implant. *In vitro* hip simulator studies have shown remarkable reductions in wear of OxZr by 45% articulating against XLPE for total hip and knee arthroplasty as compared with Co-Cr alloys [26,97]. Although OxZr heads are being routinely used for young and active patients, some clinical studies have reported extensive damage of OxZr heads articulating against XLPE [98,99], indicating poor mechanical reliability of the joint. Moreover, OxZr head has only 5 μm thick layer of zirconia formed on the Zr-Nb alloy, which is not expected to last up to 25–30 years to meet the demands of durable artificial hip joints that are needed for patients aged 40 or younger. OxZr is still in clinical use; receipients of OxZr implants will need to undergo regular check-up to ensure the good condition of implants in their bodies. From the materials design perspective, it is intriguing to reveal the reasons for the good adhesion between OxZr layer and the metal substrate given that they are incompatible materials due to significant differences in their physical, chemical and mechanical properties. However, this information is not clarified in the literature yet despite more than eight years of clinical use of OxZr heads. The ideal design of a strong interface between ceramic and metal is not revealed yet and is a fundamental gap in the knowledge. Once this gap is filled, it would lead to new knowledge in terms of methodologies that are needed for designing reliable artificial hips with an extended lifespan of more than 25–30 years.

8. Rationale for Design of New Kind of Ceramic/Metal Hybrid Artificial Joint

With increasing aging population and demands for active and young patients aged 40 and under, the life span of artificial joints is expected to be more than 25–30 years. To meet these demands, the focus is to design and develop new kinds of artificial joints with high wear resistance and mechanical reliability.

8.1. Selection and Design of Materials

Materials used to design future generation durable femoral head or cup made of ceramic/metal hybrids should meet demands of high wear resistance at bearing surface, supported by a metallic material that can sustain high toughness. These demands have not been met by any of the currently used materials for artificial joints including Oxinium™, as discussed in Section 7. To meet these demands, a promising ceramic/metal hybrid design was recently proposed [100–103] in which better wear resistance might be obtained if zirconia is replaced with the alumina as an articulating surface because the latter has higher hardness. Mechanical reliability of articulating surface might increase if Zr-Nb alloy substrate is replaced by Ti-6Al-4V alloy because the latter has higher mechanical strength and fracture toughness than that of the former or Co-Cr alloy, commonly used as the head in THA. Furthermore, alumina layer formed on the tough Ti alloy is expected to offer better mechanical reliability as compared with monolithic alumina that is liable to catastrophic fracture. In addition, alumina shows better long-term *in vivo* stability than zirconia. Oxinium™ is composed of only 5 µm thick layer of monolithic zirconia on the Zr-Nb alloy. Therefore, formation of dense α-alumina layer with thickness of 5 µm or more is expected to last longer than Oxinium™. In addition to requirement for high wear resistance, a good adhesion between alumina layer and the Ti alloy is critical for reliability of alumina/Ti alloy hybrid for efficient load transfer mechanism during operations of artificial hip.

It is technically challenging to fabricate dense and pure α-alumina layer on the Ti alloy given that alumina and the Ti alloy are very incompatible with each other due to significant differences in their coefficient of thermal expansion ($\alpha_{Alumina}$ = 7.5 × 10^{-6}/K and $\alpha_{Ti-6Al-4V}$ = 9.1–9.8 × 10^{-6}/K) and mechanical properties (Hardness, $H_{Alumina}$ = 2100 HV, $H_{Ti\ alloy}$ = 350 HV; fracture toughness, $K_{Ic,Alumina}$ = 3–4 MPa √m, $K_{Ic,Ti\ alloy}$ = 75 MPa √m), which could result in the residual stresses at the interface between them. High residual stresses could lead to cracks at the interface and subsequent delamination of alumina layer. Although the ideal design of strong interface between ceramic and metal is not fully revealed yet, it was recently proposed that formation of an intermediate AlTi type of bonding layer with metallurgical bond-like characteristics might tend to minimize the residual stresses at alumina-Ti interface and could also enhance the mechanical reliability of the joint as well [100–103]. The cross-sectional view of newly proposed alumina/Ti alloy hybrid (ATH) artificial hip joint model is shown schematically in Figure 1a. This kind of designed hybrid could be used for both the cup and head in alumina-on-alumina type of contact configuration, as shown in Figure 1b. The rationale behind this configuration is that alumina CoC combination has shown lowest wear rate (0.01 mm^3/million cycles) among the currently used materials in THA. Hence, it is expected that wear of ATH-on-ATH could be similar as well. Compared with monolithic alumina, ATH design could offer better mechanical reliability to artificial hip since bearing surface is supported by tough Ti alloy. The proposed ATH head could also be used in articulation against conventional PE or XLPE liner.

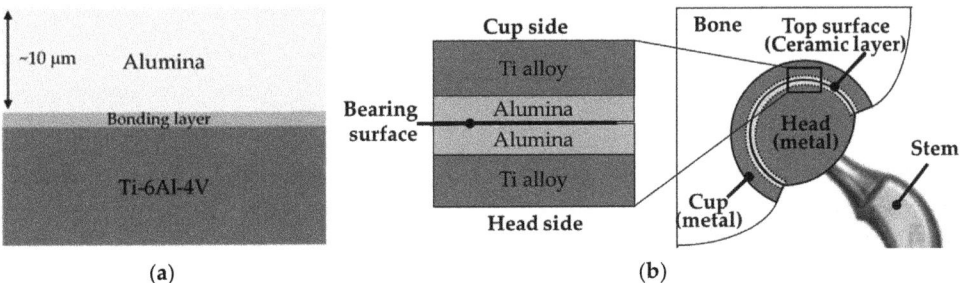

Figure 1. Cross-sectional views of alumina/Ti alloy (ATH) hybrid model (**a**), and ATH-on-ATH configuration of the artificial hip joint (**b**). Reproduced from [102] with permission from Elsevier.

8.2. Rationale for New Artificial Hip Joint Design

Advanced implant designs are focused on reducing wear of bearing surfaces by increasing diameter of heads. Despite being successful in this regard, all MoM, MoP, CoC couples share common issues of edge loading or impingement of taper joint of the stem with the rim of the cup, both of which result in the release of a large amount of wear debris and thereby leading to high revision rates. Edge loading of the cup is reported in case contact patch between ball and cup extends to the rim of the cup, resulting in an increase in local stresses and more wear at the rim. It could also be the loss of lubrication due to lack of synovial fluid between contacting surfaces [104]. Almost 70% of the MoP implants have been reported to be associated with impingement related rim damage [105]. For CoC implants, incidences of wear of rim have been reported to enhance third body wear, leading to catastrophic fracture of the ceramic part [106]. Evidence of metal wear debris has been reported for both MoM and CoC implant designs, which indicates the wear of taper joint and neck-cup impingement [107,108]. To overcome the major issue of wear of taper joint, an alternative design is proposed in which a single part for both stem and head could be made of the Ti alloy, as shown in Figure 2a. The head portion of Ti alloy could be selectively covered with alumina layer, as shown schematically in Figure 2b. The stem can be directly bonded to the bone after its porous part of Ti metal is subjected to alkali-heat treatment, as demonstrated by our previous research groups [13,15,109]. This surface treatment technology is currently marketed as AHFIX® by KYOCERA Medical. Clinical reports have demonstrated 98% success rate after ten years of implantation [109].

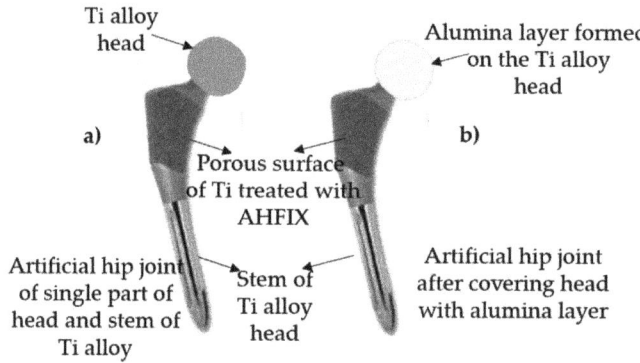

Figure 2. New design of artificial hip joint made of single part of head and stem of Ti alloy before (**a**) and after alumina layer formed on the head part (**b**).

8.3. Science and Technology of Novel Artificial Hip Joint: A Layer of Dense α-Alumina on the Ti Alloy

Conventionally, an alumina layer is fabricated on the metal substrates by plasma spraying of pure α-alumina powders. However, γ-alumina is mostly detected in plasma-sprayed alumina coatings with porous microstructures, leading to poor hardness of 1000–1200 HV, which makes them unsuitable for wear resistance applications. Although anodic oxidation of Al metal [110] or thermal treatments of Al metal layer on Ti alloy [100] have also been attempted to produce α-alumina, a dense and pure α-alumina layer cannot be fabricated using these methods. One of the promising processing strategy to form alumina layer on the Ti alloy includes the deposition of dense Al metal layer onto the Ti alloy and subsequent oxidation of Al.

8.3.1. Formation of Al Layer on the Ti Alloy

In order to form 5–10 μm thick dense α-alumina layers, at least a couple of microns thick layer of Al metal is needed to form on the Ti alloy, which is not yet possible with conventional Al deposition techniques such as PVD [111], arc ion plating [112], and sputtering [113]. Ideally, a method of formation of a thick layer of Al on the Ti alloy substrate should be chosen with minimal or no chemical reaction with the substrate during Al layer formation, which otherwise might lead to formation of thick layer of AlTi intermetallic compounds containing several cracks or pores [114] and leading to delamination of Al layer. To address these needs, cold spraying (CS) technique is used to deposit thick layers of Al metal on soft metals such as Al alloys [115,116] at temperature much lower than the melting point of Al (T_m = 660 °C). CS technique allows the deposition of a thick metal coating on a metallic substrate by accelerating small particles (<50 µm) to high velocities (500–1500 m/san) by a supersonic jet of compressed gas at a temperature below the melting point of the material to be deposited. In the field of biomaterials, for the first time, dense and thick layers of pure Al with thickness from 250 to 1000 µm were formed on Ti-6Al-4V alloy by CS of Al particles in N_2 or He gas atmosphere at varying pressures of 1–3 MPa and at a temperature of 380 °C [100].

Cold sprayed Al layers (CS-Al) on the Ti alloy formed dense microstructures both in He or N_2 gas atmosphere at an optimal pressure of 3 MPa and a temperature of 380 °C. However, CS-Al layers formed in He atmosphere shows larger cracks or gaps at the interface as a result of higher critical velocities of Al particles (972.5 m/s) [100]. CS-Al layers formed in N_2 gas atmosphere also showed poor adhesion as shown by gaps at the interface as shown in Figure 3a. Nevertheless, these samples showed relatively better adhesion than those treated in He atmosphere which could be due to lower critical velocity of 552.1 m/s in N_2 gas atmosphere. These kinds of gaps could arise due to significant difference in materials properties of Al and Ti alloy or massive plastic deformation within CS-Al layer that might have been caused by ballistic impacts during CS, thereby resulting in accumulation of residual stresses at the interface and ultimately leading to cracks at the interface [101].

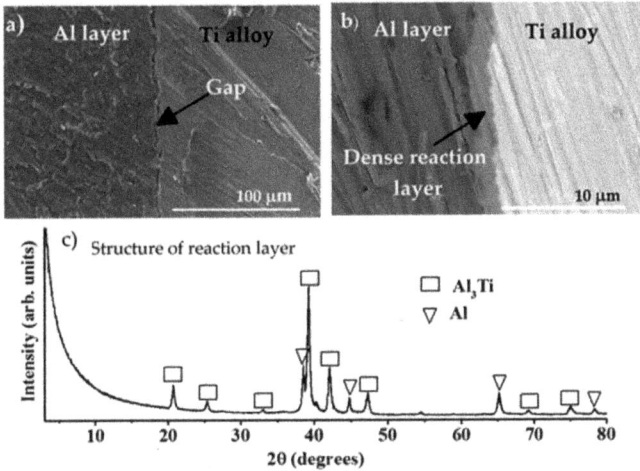

Figure 3. SEM image of Al layer formed on the Ti alloy (**a**), BSE image after heat treatment (**b**), and TF-XRD of the reaction layer (**c**). Reproduced [101] from with permission from Elsevier

8.3.2. Formation of Reaction Layer at Alumina-Ti Alloy Interface

The formation of a reaction layer of AlTi type of intermetallic compound at the interface between CS-Al layer and the Ti alloy substrate could be a promising solution to improve the adhesion and minimize the residual stresses at their interface. In this regard, a heat treatment of CS-Al layer/Ti alloy at 640 °C for 1 h in air or Ar gas atmosphere formed a dense and crack-free reaction layer 2 µm thick at the interface as shown in backscattered electron image (BSE), Figure 3b. This reaction layer could be formed by solid state diffusion of Al into Ti, according to Equation (1), since Al has lower melting point (T_m = 660 °C) as compared with the Ti alloy.

$$\text{Ti (s)} + 3 \text{ Al (s)} \rightarrow \text{Al}_3\text{Ti (s)} \tag{1}$$

Thin film XRD (TF-XRD) revealed that the reaction layer is composed of an Al_3Ti intermetallic compound, as shown in Figure 3c. When CS-Al/Ti is heated at 640 °C for more than 3 h or at temperatures more than 850 °C, a several tens of micrometers thick reaction layer containing a mixture of the AlTi type of intermetallic compounds with several pores or cracks is formed due to differences in densities of thick Ti–Al layer and compacted CS-Al from which it is formed. As a result, thick reaction layers are not be suitable for a load bearing application of artificial hip.

Even though the reaction layer of Al_3Ti layer improves adhesion between Al and Ti, Al_3Ti is itself a brittle intermetallic compound and might be prone to fracture. To reveal the mechanical reliability of Al_3Ti reaction layers, high mechanical load of 500 N was applied by compressing steel balls of 3 mm diameter directly onto the surface of reaction layers. Several large cracks along the periphery of spherical indentation on the thick reaction layer in the SEM image, Figure 4a, suggests a typical signature of brittle fracture which indicate poor mechanical deformability of thick layer, almost like a bulk ceramic. The thin reaction layer of Al_3Ti on the Ti alloy did not reveal any cracks in the SEM image, Figure 4b, suggesting its good mechanical deformability. Sharp Vickers indentations also showed better mechanical reliability of thin reaction layer [101]. Overall, about 2 µm thick reaction layer seems to provide a good interface between alumina and the Ti alloy hybrid.

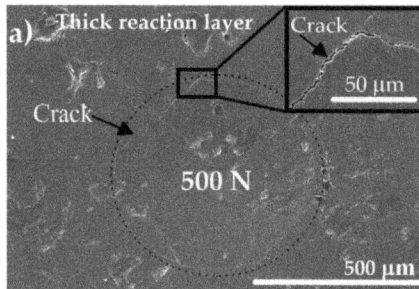 

Figure 4. SEM images showing the impressions of ball indentations on the surface of thick after applying a load of 500 N (**a**), and thin reaction layers after applying a load of 500 N (**b**). Reproduced from [101] with permission from Elsevier.

8.3.3. Formation of Alumina Layer on the Ti Alloy

Heat treatment of Al-coated Ti alloy at or above melting point of Al, could result in uncontrolled melting of Al and diffusing into Ti to form an undesirable thick layer of AlTi type of intermetallics along with only small scales of transition alumina. Dense α-alumina starts forming at temperatures above 1000 °C. However, such high temperatures cannot be applied to Al coated onto Ti-6A-4V alloy since an α → β Ti transformation at 995 °C results in a decrease in the mechanical strength of the alloy, which defeats the purpose of using high strength Ti alloy for the designed hybrid. Attempts were made to form alumina in two steps: firstly by forming a thick layer of Al_3Ti by heating it at 640 °C for 12 h in air and subsequent heat treatment at 850 °C for 96 h [100] according to the Equation (2).

$$2TiAl_3 \text{ (s)} + 3/2\, O_2 \text{ (g)} \rightarrow 2TiAl_2 \text{ (s)} + Al_2O_3 \text{ (s)} \tag{2}$$

Thin film XRD revealed that mainly Al_3Ti phase is detected after first heat treatment in Figure 5a and mainly, Al_2Ti phase is detected after subsequent heat treatment in Figure 5b.

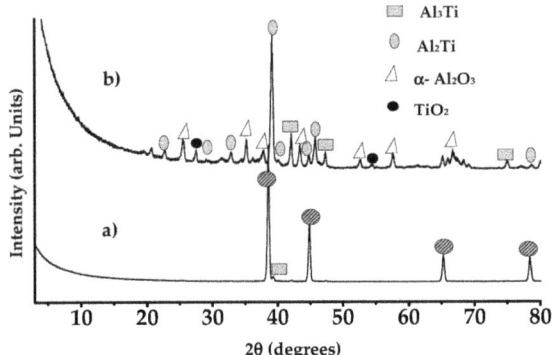

Figure 5. TF-XRD of Al layer on the Ti alloy after heat treatment at 640 °C for 12 h (**a**) and subsequent heat treatment at 850 °C for 96 h (**b**) both in an air atmosphere. Reproduced from [100] with permission from Trans Tech.

Clearly, dense α-alumina cannot be formed on the Ti alloy by using conventional methods such as heat treatments or plasma spraying of alumina. An alternative approach was recently proposed in which CS-Al layer was transformed to a dense alumina layer by micro-arc oxidation (MAO) treatment. MAO is a plasma-assisted electrochemical surface treatment carried out under 400–700 V in a dilute alkaline medium to produce ultra-hard oxide layers on the surface of soft metals such as Al, Mg or Zr for structural applications [117–121]. However, ultra-hard oxide layers formed on the soft metals do not provide the mechanical support for load bearing application of artificial hip joint. Thus, a duplex coating is required to form a α-alumina layer on the Ti alloy. To meet these goals, a promising processing strategy was proposed and demonstrated in which a dense CS-Al layer was formed onto the Ti alloy by cold spraying, followed by heat treatment at 640 °C for 1 h in air to form about 2 µm thick reaction layer to improve adhesion and finally, an optimal MAO treatment of the CS-Al layer to transform it into alumina layer, as shown schematically in Figure 6.

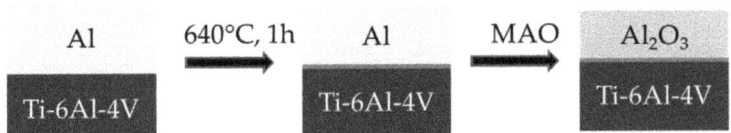

Figure 6. Schematic of deposition of Al layer, followed by heat treatment to form a thin reaction layer at the interface and micro-arc oxidation (MAO) to form alumina.

Compared with MAO treatments of Al in unipolar DC mode, bipolar pulse power supply with electrical parameters of pulse frequency, duty cycle and current density offer control over the electrochemical processes in MAO treatment. Despite many studies performed so far on MAO of Al [117–121], it is not revealed whether a α-alumina layer of at least 10 microns with dense microstructures and high Vickers hardness can be formed. In a quest to reveal this, effects of MAO electrical parameters on phase, purity, thickness and hardness of oxide layers formed by MAO of CS-Al layers/Ti alloy were systematically studied recently to determine the optimal process to form at least 10–20 µm thick layer of pure α-alumina with high Vickers hardness [101,102].

Typically, micro-arc oxidation treatment of Al layer forms a phase and density gradient oxide layer in which outer porous region is composed of mainly γ-alumina and with increasing depth, a mixed α and γ-alumina denser regions are formed in which amount of γ-alumina decreases and that of α-alumina increases and an innermost dense regions containing pure α-alumina is formed as shown schematically in Figure 7a. Corresponding cross-sectional BSE image of oxide layer formed by MAO of CS-Al/Ti after treatment time of 60 min is shown in Figure 7b. Due to poor thermal conductivity, alumina absorbs heat input to undergo phase transition and as a result, innermost region absorbs maximum heat input to transform to pure α-alumina. Outer surface regions contains mainly γ-alumina and several pores and cracks due to entrapped Si-oxides. Outer surface in contact with electrolytic solution remains relatively cooler than inner regions and hence, does not absorb sufficient heat from discharges [117–121]. As a result, mainly γ-alumina is detected on the outer surface in MAO coatings, as shown in TF-XRD of outer regions of oxide layers in Figure 7c,d. Typically, thin oxide layers (30–40 µm thick) contain γ-alumina as the major phase, which may be due to lack of sufficient heat input to undergo phase transition from γ → α alumina. Thick oxide layers 90–100 µm or more, absorbs more heat input from intense discharges, which is sufficient to trigger γ → α alumina phase transition. Outer regions of oxide layer contains several large sized pores and cracks and contain a mixture of α and γ-alumina, which are undesirable due to inferior load bearing support for artificial hip joint application. Moreover, commercial alumina of femoral head is composed of α-alumina phase and has shown long-term *in vivo* stability. From application point of view, the outer regions of oxide layer need to be carefully abraded to reveal an underlying dense α-alumina as shown by TF-XRD of abraded oxide layer in Figure 7e. Subsequent microscopic observations revealed 30 µm α-alumina formed by MAO treatment of Al layer for 60 min [101].

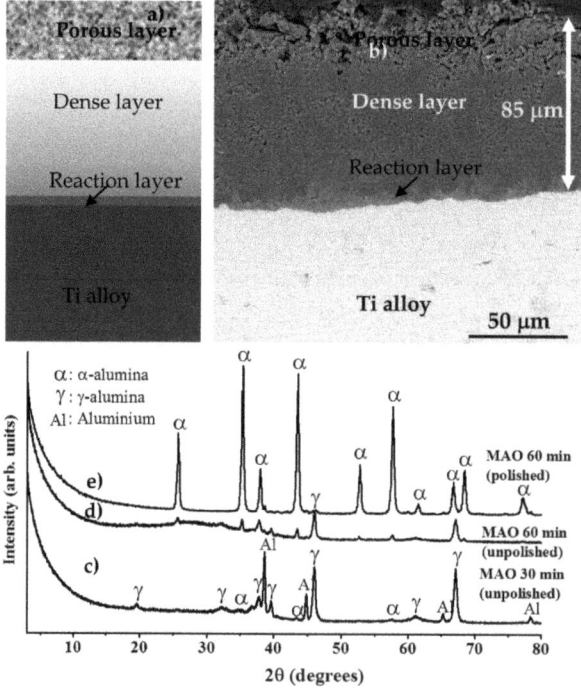

Figure 7. Schematic view of section of designed alumina/Ti alloy model (**a**); corresponding BSE image of alumina layer on the Ti alloy (**b**); and TF-XRDs of CS-Al layer after MAO for 30 min (**c**); 60 min (**d**); and after polishing (**e**). Reproduced from [101] with permission from Elsevier.

Close microscopic observations revealed several small micro-cracks formed within the 30 µm thick, pure α-alumina layer as shown in Figure 8a, which showed Vickers hardness of about 1700 HV; this value is closer to that of commercial alumina but still needs improvement. To increase the hardness of α-alumina, application of higher frequency in bipolar pulse mode is commonly known to increase thickness of oxide layer which could form a thicker α-alumina layer. These hypotheses were recently corroborated by revealing that thickness of α-alumina layer increased from 30 to 45 µm by increasing pulse frequency from 500 to 1000 Hz, both at a duty cycle of 40%. However, with increasing frequency from 500 to 1000 Hz, mean Vickers hardness decreased from 1658 to 1520 HV since the mean size of micro-cracks increased from 2 to 3 µm [102], as shown in Figure 8b.

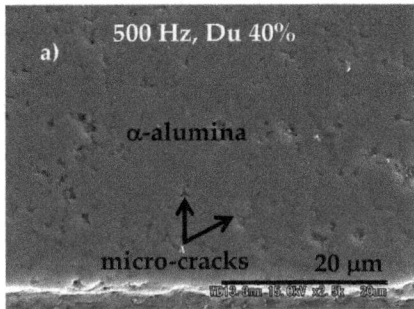 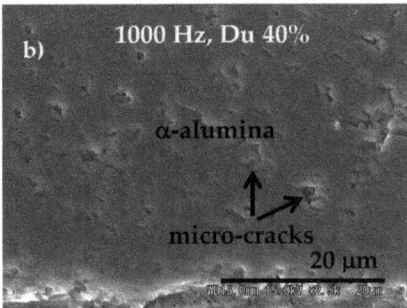

Figure 8. SEM image of cross-section of alumina layer on the Ti alloy formed by MAO of CS-Al layer at pulse frequency of 500 Hz (**a**) and 1000 Hz (**b**) both at duty cycle of 40% for 60 min. Reproduced from [102] with permission from Elsevier.

Another possible way to improve the Vickers hardness of α-alumina is to increase the duty cycle process parameter of MAO treatment. Increasing the duty cycle results in longer working time in a single pulse that imparts large discharge energy within a pulse that might improve sinterability of alumina. It was revealed that increasing duty cycle from 40% to 60% increased Vickers hardness of α-alumina layer from 1658 to 1900 HV, which was reported to be due to decrease in mean micro-crack size from 2 to 1 µm, while the amount or thickness of the layer remained the same [102]. Pure and dense α-alumina formed under optimal MAO parameters of pulse frequency 500 Hz, duty cycle 60% and treatment time of 60 min is typically shown in SEM cross-sectional image (Figure 9).

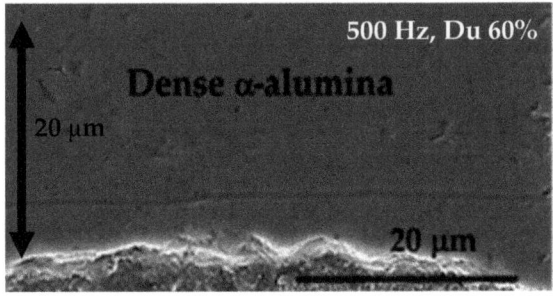

Figure 9. SEM image of cross-section of alumina layer on the Ti alloy formed by MAO of CS-Al layer at pulse frequency of 500 Hz and at duty cycle of 60% for 60 min. Reproduced from [102] with permission from [102].

These findings clearly demarcated the key role of micro-cracks in the densification of the α-alumina layer. Micro-cracks are inherent micro-structural feature in MAO coatings and are difficult to eliminate due to thermal/residual stresses that are generated because of differential cooling rates within graded layer of alumina. To obtain pore or crack free α-alumina, post-laser surface treatments can be applied carefully using Nd-YAG laser.

8.3.4. Adhesion of α-Alumina Layer with the Ti Alloy

In order to realize the potential of ATH for wear resistance applications, the α-alumina layer should have good adhesion to the Ti alloy substrate, which could also ensure good reliability of the joint. Ideally, reaction layer of Al_3Ti between CS-Al layer and the Ti alloy should not be oxidized by MAO treatment. Nevertheless, reaction layer was partially oxidized, which increased its thickness from 2 to 5 µm, as shown in BSE image, Figure 10. To evaluate the adhesion of α-alumina layer, Vickers indentations were made directly at the interfaces. No cracks within or around the impression of indentation were revealed in BSE image, Figure 10, which suggests good adhesion between the alumina layer and the Ti alloy. Grazing angle TF-XRD of the polished α-alumina layer revealed that the phase of reaction layer was changed from Al_3Ti to Al_2TiO_5 [102]. Usually, thermal expansion coefficients of oxides are less than metals or intermetallics; hence, the formation of Al_2TiO_5 might improve the adhesion strength of the joint. However, the thickness of oxidized reaction layer is very critical; if reaction layer is too much oxidized to a thickness of 15–20 µm by increasing MAO treatment time, several cracks are formed at the interface, suggesting poor adhesion [102]. Further quantitative measure of adhesion strength is needed to evaluate its potential for artificial joint applications.

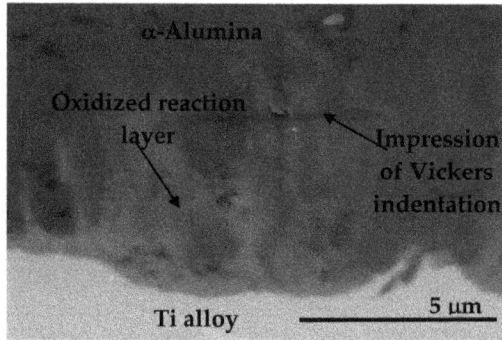

Figure 10. BSE image show sectional view of alumina layer-reaction layer-Ti alloy and impression of Vickers indentation at alumina-Ti interface. Reproduced from [102] with permission from Elsevier.

8.3.5. Dense Alumina Layer on Ti Alloy by Cold Metal Transfer and MAO Methods

Evidently, reaction layer at the interface is needed to improve adhesion between alumina and the Ti alloy, which is generated by heat treatment of cold sprayed Al layer on the Ti alloy [100–102]. If a similar reaction layer at the interface can be formed during Al layer deposition itself, it could eliminate the additional heat treatment step, which could reduce the cost of manufacturing of alumina/Ti alloy hybrids. To explore this possibility, weld cladding method has been commonly used to deposit metal coatings on the substrates using gas metal arc welding (GMAW) or laser cladding. However, composition of deposited metal coating might significantly change due to mixing with the metal substrate during GMAW process, which inadvertently increases the thickness of intermetallics formed at the interface between the coating and the substrate in the case of Al/Fe; similar change might be expected in Al/Ti system, which is undesirable. The thickness of intermetallic reaction layer should be less than 5 µm, which otherwise can result in cracks within the reaction layer due to residual stresses within it.

A promising alternative is to use a low heat input Al deposition method that could result in minimal diffusion of the substrate material into the Al coating to form a thin intermetallic layer. In this regard, cold metal transfer (CMT) method has been shown to form 2–4 µm thick intermetallic layers in joints of Al and Zn-coated steels. CMT method was used for the first time to deposit about 3 mm thick Al layer on the Ti alloy after careful process optimization [103]. Cross-sectional analysis of deposited layer of Al by CMT method (CMT-Al) revealed 1–1.5 µm thick reaction layer in Figure 11a with gradient compositions indicated by gradual EDS element profiles of Al and Ti in Figure 11b, which suggests formation of AlTi type intermetallic compounds with Al_3Ti as major intermetallic compound, as confirmed by TF-XRD measurements [103]. Optimized MAO process parameters applied to CMT-Al layer on the Ti alloy revealed almost dense α-alumina layer on the Ti alloy as shown in Figure 11c, while retaining the nascent gradient reaction layer at the interface as shown by EDS elemental line profies of Al and Ti in Figure 11d.

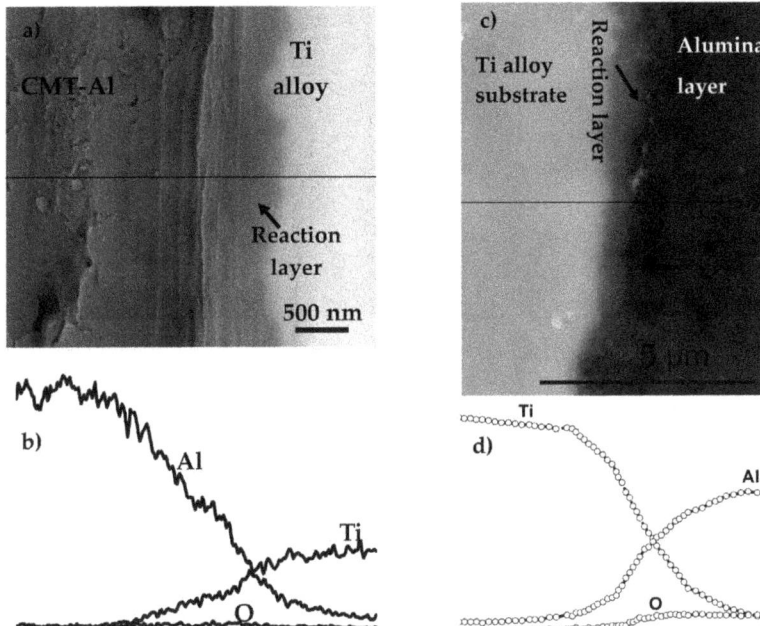

Figure 11. BSE image and line profile of cross-section of CMT-Al: Before MAO treatment (**a,b**); and after MAO treatment (**c,d**). Reproduced from [103] with permission from Cambridge Core.

This kind of gradient reaction layer could offer better mechanical reliability to the joint between alumina and the Ti alloy by minimizing the residual stresses at their interfaces and avoiding steep changes in physical or mechanical properties from top ceramic layer to the Ti alloy substrate. This is the major issue that has restricted the use of alumina/Ti hybrid combination for applications in many other R&D fields over many decades. Close transmission electron microscopy (TEM) analysis of the structures of this gradient reaction layer is further needed, which could provide the guidelines for designing future generation ceramic-metal hybrid based prosthetic devices with high mechanical reliability.

9. Summary and Conclusions

Comparing to bulk alumina used for the femoral head, studies suggested that dense α-alumina layers with high Vickers hardness and better mechanical reliability can be successfully fabricated onto the Ti alloy using a combination of cold spray or cold metal transfer and micro-arc oxidation methods. A schematic of cross-section of designed alumina/Ti alloy hybrid is shown in Figure 12.

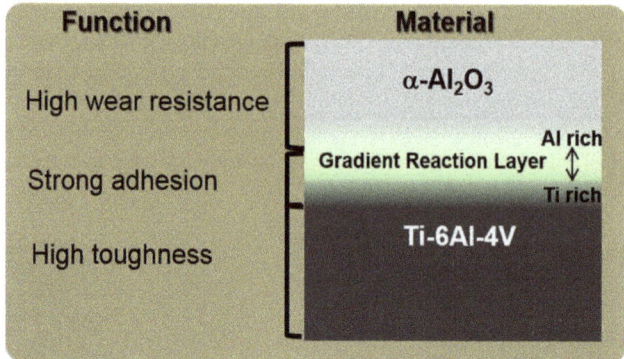

Figure 12. Schematic of cross-sectional view of a α-alumina layer on the Ti alloy with a gradient reaction layer at their interface. Reproduced from [103] with permission from Cambridge Core.

This designed alumina/Ti alloy could be a potential candidate for future generation of artificial hip joint prostheses. These investigations also suggest that the combination of cold spray or cold metal transfer and micro-arc oxidation methods will be effective in forming an adherent layer of dense α-alumina and is expected to enable advances in the field of artificial hip joints. The designed ceramic/metal hybrid technology is novel in a broad sense since it can be suitably applied to other medical implants such as knee joint and as a dense ceramic coating on Ti metal for dental implants. Before considering these hybrids for commercial applications, it is necessary to evaluate the cytotoxicity, inflammatory responses, and chemical analyses of wear debris of alumina layer on the Ti alloy subjected to tribological testing as well as TEM analysis of the interfaces between the alumina layer-reaction layer-Ti alloy and corrosion testing of the reaction layer. The outcomes of these investigations will provide paths forward for the possibility of extending the research methodologies to develop a femoral head of dense α-alumina layer on the Ti alloy.

Conflicts of Interest: The authors declare no conflict of interest.

References

1. Kurtz, S.; Ong, K.; Lau, E.; Mowat, F.; Halpern, M. Projections of primary and revision hip and knee arthroplasty in the United States from 2005 to 2030. *J. Bone Jt. Surg. Am.* **2007**, *89*, 780–785. [CrossRef]
2. Renkawitz, T.; Santori, F.S.; Grifka, J.; Valverde, C.; Morlock, M.M.; Learmonth, I.D. A new short uncemented, proximally fixed anatomic femoral implant with a prominent lateral flare: Design rationals and study design of an international clinical trial. *BMC Musculoskelet. Disord.* **2008**, *9*, 147. [CrossRef] [PubMed]
3. Steens, W.; Boettner, F.; Bader, R.; Skripitz, R.; Schneeberger, A. Bone mineral density after implantation of a femoral neck hip prosthesis-a prospective 5 year follow-up. *BMC Musculoskelet. Disord.* **2015**, *16*, 192. [CrossRef] [PubMed]
4. Miura, K.; Yamada, N.; Hanada, S.; Jung, T.K.; Itoi, E. The bone tissue compatibility of a new Ti-Nb-Sn alloy with a low Young's modulus. *Acta Biomater.* **2011**, *7*, 2320–2326. [CrossRef] [PubMed]
5. Guo, S.; Bao, Z.Z.; Meng, Q.K.; Hu, L.; Zhao, X.Q. A novel metastable Ti-25Nb-2Mo-4Sn alloy with high strength and low Young's modulus. *Metall. Mater. Trans. A Phys. Metall. Mater. Sci.* **2012**, *43*, 3447–3451. [CrossRef]
6. Niinomi, M.; Hattori, T.; Morikawa, K.; Kasuga, T.; Suzuki, A.; Fukui, H.; Niwa, S. Development of low rigidity beta-type titanium alloy for biomedical applications. *Mater. Trans.* **2002**, *43*, 2970–2977. [CrossRef]
7. Okazaki, Y. A new Ti-15Zr-4Nb-4Ta alloy for medical applications. *Curr. Opin. Solid State Mater. Sci.* **2001**, *5*, 45–53. [CrossRef]
8. Bai, X.; Sandukas, S.; Appleford, M.R.; Ong, J.L.; Rabiei, A. Deposition and investigation of functionally graded calcium phosphate coatings on titanium. *Acta Biomater.* **2009**, *5*, 3563–3572. [CrossRef] [PubMed]

9. Bai, X.; Sandukas, S.; Appleford, M.; Ong, J.L.; Rabiei, A. Antibacterial effect and cytotoxicity of Ag-doped functionally graded hydroxyapatite coatings. *J. Biomed. Mater. Res. Part B Appl. Biomater.* **2012**, *100*, 553–561. [CrossRef] [PubMed]
10. Chen, W.; Liu, Y.; Courtney, H.S.; Bettenga, M.; Agrawal, C.M.; Bumgardner, J.D.; Ong, J.L. In vitro anti-bacterial and biological properties of magnetron co-sputtered silver-containing hydroxyapatite coating. *Biomaterials* **2006**, *27*, 5512–5517. [CrossRef] [PubMed]
11. Ong, J.L.; Lucas, L.C.; Lacefield, W.R.; Rigney, E.D. Structure solubility and bond strength of thin calcium-phosphate coatings produced by ion-beam sputter deposition. *Biomaterials* **1992**, *13*, 249–254. [CrossRef]
12. Yang, Y.Z.; Kim, K.H.; Ong, J.L. Review on calcium phosphate coatings produced using a sputtering process—an alternative to plasma spraying. *Biomaterials* **2005**, *26*, 327–337. [CrossRef] [PubMed]
13. Kim, H.M.; Miyaji, F.; Kokubo, T.; Nakamura, T. Preparation of bioactive Ti and its alloys via simple chemical surface treatment. *J. Biomed. Mater. Res.* **1996**, *32*, 409–417. [CrossRef]
14. Kim, H.M.; Miyaji, F.; Kokubo, T.; Nishiguchi, S.; Nakamura, T. Graded surface structure of bioactive titanium prepared by chemical treatment. *J. Biomed. Mater. Res.* **1999**, *45*, 100–107. [CrossRef]
15. Kim, H.M.; Takadama, H.; Miyaji, F.; Kokubo, T.; Nishiguchi, S.; Nakamura, T. Formation of bioactive functionally graded structure on Ti-6Al-4V alloy by chemical surface treatment. *J. Mater. Sci. Mater. Med.* **2000**, *11*, 555–559. [CrossRef] [PubMed]
16. Kizuki, T.; Takadama, H.; Matsushita, T.; Nakamura, T.; Kokubo, T. Preparation of bioactive Ti metal surface enriched with calcium ions by chemical treatment. *Acta Biomater.* **2010**, *6*, 2836–2842. [CrossRef] [PubMed]
17. Kokubo, T.; Pattanayak, D.K.; Yamaguchi, S.; Takadama, H.; Matsushita, T.; Kawai, T.; Takemoto, M.; Fujibayashi, S.; Nakamura, T. Positively charged bioactive Ti metal prepared by simple chemical and heat treatments. *J. R. Soc. Interface* **2010**, *7*, S503–S513. [CrossRef] [PubMed]
18. Oral, E.; Christensen, S.D.; Malhi, A.S.; Wannomae, K.K.; Muratoglu, O.K. Wear resistance and mechanical properties of highly cross-linked, ultrahigh-molecular weight polyethylene doped with vitamin E. *J. Arthroplast.* **2006**, *21*, 580–591. [CrossRef] [PubMed]
19. Oral, E.; Muratoglu, O.K. Vitamin E diffused, highly crosslinked UHMWPE: A review. *Int. Orthop.* **2011**, *35*, 215–223. [CrossRef] [PubMed]
20. Moro, T.; Kawaguchi, H.; Ishihara, K.; Kyomoto, M.; Karita, T.; Ito, H.; Nakamura, K.; Takatori, Y. Wear resistance of artificial hip joints with poly(2-methacryloyloxyethyl phosphorylcholine) grafted polyethylene: Comparisons with the effect of polyethylene cross-linking and ceramic femoral heads. *Biomaterials* **2009**, *30*, 2995–3001. [CrossRef] [PubMed]
21. Kyomoto, M.; Moro, T.; Iwasaki, Y.; Miyaji, F.; Kawaguchi, H.; Takatori, Y.; Nakamura, K.; Ishihara, K. Superlubricious surface mimicking articular cartilage by grafting poly(2-methacryloyloxyethyl phosphorylcholine) on orthopaedic metal bearings. *J. Biomed. Mater. Res. Part A* **2009**, *91*, 730–741. [CrossRef] [PubMed]
22. Clarke, I.C.; Manaka, M.; Green, D.D.; Williams, P.; Pezzotti, G.; Kim, Y.H.; Ries, M.; Sugano, N.; Sedel, L.; Delauney, C.; et al. Current status of zirconia used in total hip implants. *J. Bone Jt. Surg. Am.* **2003**, *85*, 73–84. [CrossRef]
23. Begand, S.; Oberbach, T.; Glien, W. Investigations of the mechanical properties of an alumina toughened zirconia ceramic for an application in joint prostheses. *Key Eng. Mater.* **2005**, *284*, 1019–1022. [CrossRef]
24. Al-Hajjar, M.; Jennings, L.M.; Begand, S.; Oberbach, T.; Delfosse, D.; Fisher, J. Wear of novel ceramic-on-ceramic bearings under adverse and clinically relevant hip simulator conditions. *J. Biomed. Mater. Res. Part B Appl. Biomater.* **2013**, *101*, 1456–1462. [CrossRef] [PubMed]
25. Hobbs, L.W.; Rosen, V.B.; Mangin, S.P.; Treska, M.; Hunter, G. Oxidation microstructures and interfaces in the oxidized zirconium knee. *Int. J. Appl. Ceram. Technol.* **2005**, *2*, 221–246. [CrossRef]
26. Good, V.; Ries, M.; Barrack, R.L.; Widding, K.; Hunter, G.; Heuer, D. Reduced wear with oxidized zirconium femoral heads. *J. Bone Jt. Surg. Am.* **2003**, *85*, 105–110. [CrossRef]
27. Burger, W.; Richter, H.G. High strength and toughness alumina matrix composites by transformation toughening and 'in situ' platelet reinforcement (ZPTA)—The new generation of bioceramics. *Key Eng. Mater.* **2000**, *192–195*, 545–548. [CrossRef]
28. Brown, A.S. Hip New World. *ASME Mech. Eng.* **2006**, *128*, 28–33.

29. Charnley, J.; Kamangar, A.; Longfield, M.D. The optimum size of prosthetic heads in relation to wear of plastic sockets in total replacement of hip. *Med. Biol. Eng. Comput.* **1969**, *7*, 31–39. [CrossRef]
30. Amstutz, H.C.; Campbell, P.; Kossovsky, N.; Clarke, I.C. Mechanism and clinical significance of wear debris-induced osteolysis. *Clin. Orthop. Relat. Res.* **1992**, *276*, 7–18. [CrossRef]
31. Boutin, P. Total arthroplasty of the hip by fritted alumina prosthesis. Experimental study and 1st clinical applications. *Orthop. Traumatol. Surg. Res.* **2014**, *100*, 15–21. [CrossRef] [PubMed]
32. Harris, W.H. The prolem is osteolysis. *Clin. Orthop. Relat. Res.* **1995**, *311*, 46–53.
33. Kim, Y.H.; Kim, J.S.; Park, J.W.; Joo, J.H. Periacetabular osteolysis is the problem in contemporary total hip arthroplasty in young patients. *J. Arthroplast.* **2012**, *27*, 74–81. [CrossRef] [PubMed]
34. Kurtz, S. Compendium of highly crosslinked UHMWPEs. In *UHMWPE Biomaterials Handbook*; Kurtz, S., Ed.; Elsevier: New York, NY, USA, 2009; pp. 291–308.
35. Oral, E.; Malhi, A.S.; Muratoglu, O.K. Mechanisms of decrease in fatigue crack propagation resistance in irradiated and melted UHMWPE. *Biomaterials* **2006**, *27*, 917–925. [CrossRef] [PubMed]
36. Reinitz, S.D.; Currier, B.H.; Van Citters, D.W.; Levine, R.A.; Collier, J.P. Oxidation and other property changes of retrieved sequentially annealed UHMWPE acetabular and tibial bearings. *J. Biomed. Mater. Res. Part B Appl. Biomater.* **2015**, *103*, 578–586. [CrossRef] [PubMed]
37. Oral, E.; Neils, A.L.; Doshi, B.N.; Fu, J.; Muratoglu, O.K. Effects of simulated oxidation on the *in vitro* wear and mechanical properties of irradiated and melted highly crosslinked UHMWPE. *J. Biomed. Mater. Res. Part B Appl. Biomater.* **2016**, *104*, 316–322. [CrossRef] [PubMed]
38. Bhateja, S.K.; Duerst, R.W.; Aus, E.B.; Andrews, E.H. Free radicals trapped in polyethylene crystals. *J. Macromol. Sci. Part B Phys.* **1995**, *34*, 263–272. [CrossRef]
39. Jahan, M.S.; King, M.C.; Haggard, W.O.; Sevo, K.L.; Parr, J.E. A study of long-lived free radicals in gamma-irradiated medical grade polyethylene. *Radiat. Phys. Chem.* **2001**, *62*, 141–144. [CrossRef]
40. Lewis, G. Properties of crosslinked ultra-high-molecular-weight polyethylene. *Biomaterials* **2001**, *22*, 371–401. [CrossRef]
41. Dumbleton, J.H.; D'Antonio, J.A.; Manley, M.T.; Capello, W.N.; Wang, A. The basis for a second-generation highly cross-linked UHMWPE. *Clin. Orthop. Relat. Res.* **2006**, *453*, 265–271. [CrossRef] [PubMed]
42. Currier, B.H.; Van Citters, D.W.; Currier, J.H.; Carlson, E.M.; Tibbo, M.E.; Collier, J.P. In vivo oxidation in retrieved highly crosslinked tibial inserts. *J. Biomed. Mater. Res. Part B Appl. Biomater.* **2013**, *101*, 441–448. [PubMed]
43. Kurtz, S.M.; Gawel, H.A.; Patel, J.D. History and systematic review of wear and osteolysis outcomes for first-generation highly crosslinked polyethylene. *Clin. Orthop. Relat. Res.* **2011**, *469*, 2262–2277. [CrossRef] [PubMed]
44. Oral, E.; Neils, A.L.; Wannomae, K.K.; Muratoglu, O.K. Novel active stabilization technology in highly crosslinked UHMWPEs for superior stability. *Radiat. Phys. Chem.* **2014**, *105*, 6–11. [CrossRef]
45. Green, J.M.; Hallab, N.J.; Liao, Y.S.; Narayan, V.; Schwarz, E.M.; Xie, C. Anti-oxidation treatment of ultra high molecular weight polyethylene components to decrease periprosthetic osteolysis: Evaluation of osteolytic and osteogenic properties of wear debris particles in a murine calvaria model. *Curr. Rheumatol. Rep.* **2013**, *15*, 325. [CrossRef] [PubMed]
46. Sillesen, N.H.; Greene, M.E.; Nebergall, A.K.; Nielsen, P.T.; Laursen, M.B.; Troelsen, A.; Malchau, H. Three year RSA evaluation of vitamin E diffused highly cross-linked polyethylene liners and cup stability. *J. Arthroplast.* **2015**, *30*, 1260–1264. [CrossRef] [PubMed]
47. Ishihara, K. Highly lubricated polymer interfaces for advanced artificial hip joints through biomimetic design. *Polym. J.* **2015**, *47*, 585–597. [CrossRef]
48. Bragdon, C.R.; Doerner, M.; Martell, J.; Jarrett, B.; Palm, H.; Malchau, H. The 2012 John Charnley Award: Clinical multicenter studies of the wear performance of highly crosslinked remelted polyethylene in THA. *Clin. Orthop. Relat. Res.* **2013**, *471*, 393–402. [CrossRef] [PubMed]
49. Skipor, A.K.; Campbell, P.A.; Patterson, L.M.; Anstutz, H.C.; Schmalzried, T.P.; Jacobs, J.J. Serum and urine metal levels in patients with metal-on-metal surface arthroplasty. *J. Mater. Sci. Mater. Med.* **2002**, *13*, 1227–1234. [CrossRef] [PubMed]
50. Catelas, I.; Wimmer, M.A. New insights into wear and biological effects of metal-on-metal bearings. *J. Bone Jt. Surg. Am.* **2011**, *93*, 76–83. [CrossRef] [PubMed]

51. Basketter, D.A.; Briaticovangosa, G.; Kaestner, W.; Lally, C.; Bontinck, W.J. Nickel, cobalt and chromium in consumer products: A role in allergic contact-dermatitis? *Contact Dermat.* **1993**, *28*, 15–25. [CrossRef]
52. Head, W.C.; Bauk, D.J.; Emerson, R.H. Titanium as the material of choice for cementless femoral components in total hip arthroplasty. *Clin. Orthop. Relat. Res.* **1995**, *311*, 85–90.
53. Walker, P.R.; Leblanc, J.; Sikorska, M. Effects of aluminum and other cations on the structure of brain and liver chromatin. *Biochemistry* **1989**, *28*, 3911–3915. [CrossRef] [PubMed]
54. Rao, S.; Ushida, T.; Tateishi, T.; Okazaki, Y.; Asao, S. Effect of Ti, Al, and V ions on the relative growth rate of fibroblasts (L929) and osteoblasts (MC3T3-E1) cells. *Bio-Med. Mater. Eng.* **1996**, *6*, 79–86.
55. Long, M.; Crooks, R.; Rack, H.J. High-cycle fatigue performance of solution-treated metastable-beta titanium alloys. *Acta Mater.* **1999**, *47*, 661–669. [CrossRef]
56. Guleryuz, H.; Cimenoglu, H. Surface modification of a Ti-6Al-4V alloy by thermal oxidation. *Surf. Coat. Technol.* **2005**, *192*, 164–170. [CrossRef]
57. Guleryuz, H.; Cimenoglu, H. Effect of thermal oxidation on corrosion and corrosion-wear behaviour of a Ti–6Al–4V alloy. *Biomaterials* **2004**, *25*, 3325–3333. [CrossRef] [PubMed]
58. Singh, R.; Kurella, A.; Dahotre, N.B. Laser surface modification of Ti–6Al–4V: Wear and corrosion characterization in simulated biofluid. *J. Biomater Appl.* **2006**, *21*, 49–73. [CrossRef] [PubMed]
59. Li, B.; Shen, Y.; Hu, W.; Luo, L. Surface modification of Ti-6Al-4V alloy via friction-stir processing: Microstructure evolution and dry sliding wear performance. *Surf. Coat. Technol.* **2014**, *239*, 160–170. [CrossRef]
60. Oonishi, H.; Clarke, I.C.; Good, V.; Amino, H.; Ueno, M.; Masuda, S.; Oomamiuda, K.; Ishimaru, H.; Yamamoto, M.; Tsuji, E. Needs of bioceramics to longevity of total joint arthroplasty. *Key Eng. Mater.* **2003**, *240–242*, 735–754. [CrossRef]
61. Sedel, L. Evolution of alumina-on-alumina implants—A review. *Clin. Orthop. Relat. Res.* **2000**, *379*, 48–54. [CrossRef]
62. Sedel, L. Clinical applications of ceramic-ceramic combinations in joint replacement. In *Bioceramics and Their Clinical Applications*; Kokubo, T., Ed.; Woohhead: Cambridge, UK; CRC: Boca Raton, FL, USA, 2008; pp. 688–698.
63. Jarrett, C.A.; Ranawat, A.S.; Bruzzone, M.; Blum, Y.C.; Rodriguez, J.A.; Ranawat, C.S. The squeaking hip: a phenomenon of ceramic-on-ceramic total hip arthroplasty. *J. Bone Jt. Surg. Am.* **2009**, *91*, 1344–1349. [CrossRef] [PubMed]
64. Hannouche, D.; Zaoui, A.; Zadegan, F.; Sedel, L.; Nizard, R. Thirty years of experience with alumina-on-alumina bearings in total hip arthroplasty. *Int. Orthop.* **2011**, *35*, 207–213. [CrossRef] [PubMed]
65. Clarke, I.C.; Manley, M.T. How do alternative bearing surfaces influence wear behavior? *J. Am. Acad. Orthop. Surg.* **2008**, *16*, S86–S93. [CrossRef] [PubMed]
66. Tipper, J.L.; Hatton, A.; Nevelos, J.E.; Ingham, E.; Doyle, C.; Streicher, R.; Nevelos, A.B.; Fisher, J. Alumina-alumina artificial hip joints. Part II: Characterisation of the wear debris from *in vitro* hip joint simulations. *Biomaterials* **2002**, *23*, 3441–3448. [CrossRef]
67. Bizot, P.; Banallec, L.; Sedel, L.; Nizard, R. Alumina-on-alumina total hip prostheses in patients 40 years of age or younger. *Clin. Orthop. Relat. Res.* **2000**, *379*, 68–76. [CrossRef]
68. Bizot, P.; Larrouy, M.; Witvoet, J.; Sedel, L.; Nizard, R. Press-fit metal-backed alumina sockets: A minimum 5-year followup study. *Clin. Orthop. Relat. Res.* **2000**, *379*, 134–142. [CrossRef]
69. Kim, Y.H.; Kim, J.S.; Cho, S.H. A comparison of polyethylene wear in hips with cobalt-chrome or zirconia heads. A prospective, randomised study. *J. Bone Jt. Surg. Br.* **2001**, *83*, 742–750. [CrossRef]
70. Wroblewski, M.; Siney, P.D.; Nagai, H.; Fleming, P.A. Wear of ultra-high-molecular-weight polyethylene cup articulating with 22.225 mm zirconia diameter head in cemented total hip arthroplasty. *J. Orthop. Sci.* **2004**, *9*, 253–255. [CrossRef] [PubMed]
71. Skyrme, A.D.; Richards, S.; John, A.; Chia, M.; Walter, W.K.; Walter, W.L.; Zicat, B. Polyethylene wear rates with Zirconia and cobalt chrome heads in the ABG hip. *Hip Int.* **2005**, *15*, 63–70. [CrossRef] [PubMed]
72. Piconi, C.; Maccauro, G.; Pilloni, L.; Burger, W.; Muratori, F.; Richter, H.G. On the fracture of a zirconia ball head. *J. Mater. Sci. Mater. Med.* **2006**, *17*, 289–300. [CrossRef] [PubMed]
73. Kurtz, S.M.; Kocagoz, S.; Arnholt, C.; Huet, R.; Ueno, M.; Walter, W.L. Advances in zirconia toughened alumina biomaterials for total joint replacement. *J. Mech. Behav. Biomed. Mater.* **2014**, *31*, 107–116. [CrossRef] [PubMed]

74. Green, D.; Pezzotti, G.; Sakakura, S.; Ries, M.; Clarke, I. Zirconia ceramic femoral heads in the USA-retrieved zirconia heads-2 to 10 years out. In Proceedings of the 49th Annual Meeting of the Orthopaedic Research Society, New Orleans, LA, USA, 2–5 February 2003.
75. Bal, B.S.; Rahaman, M.N. Orthopedic applications of silicon nitride ceramics. *Acta Biomater.* **2012**, *8*, 2889–2898. [CrossRef] [PubMed]
76. McEntire, B.J.; Bal, B.S.; Rahaman, M.N.; Chevalier, J.; Pezzotti, G. Ceramics and ceramic coatings in orthopaedics. *J. Eur. Ceram. Soc.* **2015**, *35*, 4327–4369. [CrossRef]
77. Chen, F.C.; Ardell, A.J. Fracture toughness of ceramics and semi-brittle alloys using a miniaturized disk-bend test. *Mater. Res. Innov.* **2000**, *3*, 250–262. [CrossRef]
78. Roebben, G.; Sarbu, C.; Lube, T.; Van der Biest, O. Quantitative determination of the volume fraction of intergranular amorphous phase in sintered silicon nitride. *Mater. Sci. Eng. A* **2004**, *370*, 453–458. [CrossRef]
79. Olofsson, J.; Pettersson, M.; Teuscher, N.; Heilmann, A.; Larsson, K.; Grandfield, K.; Persson, C.; Jacobson, S.; Engqvist, H. Fabrication and evaluation of Si_xN_y coatings for total joint replacements. *J. Mater. Sci. Mater. Med.* **2012**, *23*, 1879–1889. [CrossRef] [PubMed]
80. Pettersson, M.; Berlind, T.; Schmidt, S.; Jacobson, S.; Hultman, L.; Persson, C.; Engqvist, H. Structure and composition of silicon nitride and silicon carbon nitride coatings for joint replacements. *Surf. Coat. Technol.* **2013**, *235*, 827–834. [CrossRef]
81. Bal, B.S.; Khandkar, A.; Lakshminarayanan, R.; Clarke, I.; Hoffman, A.A.; Rahaman, M.N. Fabrication and testing of silicon nitride bearings in total hip arthroplasty winner of the 2007 "HAP" PAUL Award. *J. Arthroplast.* **2009**, *24*, 110–116. [CrossRef] [PubMed]
82. McEntire, B.J.; Lakshminarayanan, R.; Ray, D.A.; Clarke, I.C.; Puppulin, L.; Pezzotti, G. Silicon nitride bearings for total joint arthroplasty. *Lubricants* **2016**, *4*, 35. [CrossRef]
83. Green, D.; Donaldson, T.; Williams, P.; Pezzotti, G.; Clarke, I. Long term strip wear rates of 3rd and 4th generation ceramic-on-ceramic under microseparation. In Proceedings of the 53rd Annual Meeting of the Orthopaedic Research Society, San Diego, CA, USA, 11–14 February 2007; p. 1776.
84. Clarke, I.; Gustafson, A. The design of ceramics for joint replacement. In *Bioceramics and Their Clinical Applications*; Kokubo, T., Ed.; Woodhead: Cambridge, UK, 2008; pp. 106–132.
85. Begand, S.; Oberbach, T.; Glien, W. ATZ—A new material with a high potential in joint replacement. *Key Eng. Mater.* **2005**, *284–286*, 983–986. [CrossRef]
86. Kremers, H.M.; Larson, D.R.; Crowson, C.S.; Kremers, W.K.; Washington, R.E.; Steiner, C.A.; Jiranek, W.A.; Berry, D.J. Prevalence of total hip and knee replacement in the United States. *J. Bone Jt. Surg. Am.* **2015**, *97*, 1386–1397. [CrossRef] [PubMed]
87. Abu-Amer, Y.; Darwech, I.; Clohisy, J.C. Aseptic loosening of total joint replacements: Mechanisms underlying osteolysis and potential therapies. *Arthritis Res. Ther.* **2007**, *9*, S6. [CrossRef] [PubMed]
88. Narayan, R.J. Nanostructured diamondlike carbon thin films for medical applications. *Mater. Sci. Eng. C* **2005**, *25*, 405–416. [CrossRef]
89. Pappas, M.J.; Makris, G.; Buechel, F.F. Titanium nitride ceramic film against polyethylene: A 48-million cycle wear test. *Clin. Orthop. Relat. Res.* **1995**, *317*, 64–70.
90. Hauert, R.; Falub, C.V.; Thorwarth, G.; Thorwarth, K.; Affolter, C.; Stiefel, M.; Podleska, L.E.; Taeger, G. Retrospective lifetime estimation of failed and explanted diamond-like carbon coated hip joint balls. *Acta Biomater.* **2012**, *8*, 3170–3176. [CrossRef] [PubMed]
91. Choudhury, D.; Lackner, J.M.; Major, L.; Morita, T.; Sawae, Y.; Bin Mamat, A.; Stavness, I.; Roy, C.K.; Krupka, I. Improved wear resistance of functional diamond like carbon coated Ti-6Al-4V alloys in an edge loading conditions. *J. Mech. Behav. Biomed. Mater.* **2016**, *59*, 586–595. [CrossRef] [PubMed]
92. Catledge, S.A.; Vohra, Y.K. Effect of nitrogen addition on the microstructure and mechanical properties of diamond films grown using high-methane concentrations. *J. Appl. Phys.* **1999**, *86*, 698–700. [CrossRef]
93. Catledge, S.A.; Vaid, R.; Diggins, P.; Weimer, J.J.; Koopman, M.; Vohra, Y.K. Improved adhesion of ultra-hard carbon films on cobalt-chromium orthopaedic implant alloy. *J. Mater. Sci. Mater. Med.* **2011**, *22*, 307–316. [CrossRef] [PubMed]
94. Papo, M.J.; Catledge, S.A.; Vohra, Y.K. Mechanical wear behavior of nanocrystalline and multilayer diamond coatings on temporomandibular joint implants. *J. Mater. Sci. Mater. Med.* **2004**, *15*, 773–777. [CrossRef] [PubMed]

95. Amaral, M.; Abreu, C.S.; Oliveira, F.J.; Gomes, J.R.; Silva, R.F. Tribological characterization of NCD in physiological fluids. *Diam. Relat. Mater.* **2008**, *17*, 848–852. [CrossRef]
96. Vila, M.; Amaral, M.; Oliveira, F.J.; Silva, R.F.; Fernandes, A.J.S.; Soares, M.R. Residual stress minimum in nanocrystalline diamond films. *Appl. Phys. Lett.* **2006**, *89*, 093109. [CrossRef]
97. Ries, M.D.; Salehi, A.; Widding, K.; Hunter, G. Polyethylene wear performance of oxidized zirconium and cobalt-chromium knee components under abrasive conditions. *J. Bone Jt. Surg. Am.* **2002**, *84*, 129–135. [CrossRef]
98. Evangelista, G.T.; Fulkerson, E.; Kummer, E.; Di Cesare, P.E. Surface damage to an Oxinium femoral head prosthesis after dislocation. *J. Bone Jt. Surg. Br.* **2007**, *89*, 535–537. [CrossRef] [PubMed]
99. Jaffe, W.L.; Strauss, E.J.; Cardinale, M.; Herrera, L.; Kummer, F.J. Surface oxidized zirconium total hip arthroplasty head damage due to closed reduction. *J. Arthroplast.* **2009**, *24*, 898–902. [CrossRef] [PubMed]
100. Khanna, R.; Matsushita, T.; Kokubo, T.; Takadama, H. Formation of alumina layer on Ti alloy for artificial hip joint. *Key Eng. Mater.* **2014**, *614*, 200–205. [CrossRef]
101. Khanna, R.; Kokubo, T.; Matsushita, T.; Nomura, Y.; Nose, N.; Oomori, Y.; Yoshida, T.; Wakita, K.; Takadama, H. Novel artificial hip joint: A layer of alumina on Ti–6Al–4V alloy formed by micro-arc oxidation. *Mater. Sci. Eng. C* **2015**, *55*, 393–400. [CrossRef] [PubMed]
102. Khanna, R.; Kokubo, T.; Matsushita, T.; Takadama, H. Fabrication of dense α-alumina layer on Ti–6Al–4V alloy hybrid for bearing surfaces of artificial hip joint. *Mater. Sci. Eng. C* **2016**, *69*, 1229–1239. [CrossRef] [PubMed]
103. Khanna, R.; Rajeev, G.; Takadama, H.; Rao Bakshi, S. Fabrication of dense alumina layer on Ti alloy hybrid by cold metal transfer and micro-arc oxidation methods. *J. Mater. Res.* **2017**. [CrossRef]
104. Angadji, A.; Royle, M.; Collins, S.N.; Shelton, J.C. Influence of cup orientation on the wear performance of metal-on-metal hip replacements. *Proc. Inst. Mech. Eng. H* **2009**, *223*, 449–457. [CrossRef] [PubMed]
105. Elkins, J.M.; O'Brien, M.K.; Stroud, N.J.; Pedersen, D.R.; Callaghan, J.J.; Brown, T.D. Hard-on-hard total hip impingement causes extreme contact stress concentrations. *Clin. Orthop. Relat. Res.* **2011**, *469*, 454–463. [CrossRef] [PubMed]
106. Barrack, R.L.; Burak, C.; Skinner, H.B. Concerns about ceramics in THA. *Clin. Orthop. Relat. Res.* **2004**, *429*, 73–79. [CrossRef]
107. Langton, D.J.; Jameson, S.S.; Joyce, T.J.; Gandhi, J.N.; Sidaginamale, R.; Mereddy, P.; Lord, J.; Nargol, A.V. Accelerating failure rate of the ASR total hip replacement. *J. Bone Jt. Surg. Br.* **2011**, *93*, 1011–1106. [CrossRef] [PubMed]
108. Mao, X.; Tay, G.H.; Godbolt, D.B.; Crawford, R.W. Pseudotumor in a well-fixed metal-on-polyethylene uncemented hip arthroplasty. *J. Arthroplast.* **2012**, *27*, 493.e13–493.e17. [CrossRef] [PubMed]
109. So, K.; Kaneuji, A.; Matsumoto, T.; Matsuda, S.; Akiyama, H. Is the bone-bonding ability of a cementless total hip prosthesis enhanced by alkaline and heat treatments? *Clin. Orthop. Relat. Res.* **2013**, *471*, 3847–3855. [CrossRef] [PubMed]
110. Briggs, E.P.; Walpole, A.R.; Wilshaw, P.R.; Karlsson, M.; Palsgard, E. Formation of highly adherent nano-porous alumina on Ti-based substrates: A novel bone implant coating. *J. Mater. Sci. Mater. Med.* **2004**, *15*, 1021–1029. [CrossRef] [PubMed]
111. Varlese, F.A.; Tului, M.; Sabbadini, S.; Pellissero, F.; Sebastiani, M.; Bemporad, E. Optimized coating procedure for the protection of TiAl intermetallic alloy against high temperature oxidation. *Intermetallics* **2013**, *37*, 76–82. [CrossRef]
112. Zhang, K.; Wang, Q.M.; Sun, C.; Wang, F.H. Preparation and oxidation resistance of a crack-free Al diffusion coating on $Ti_{22}Al_{26}Nb$. *Corros. Sci.* **2007**, *49*, 3598–3609. [CrossRef]
113. Chu, M.S.; Wu, S.K. The improvement of high temperature oxidation of Ti-50Al by sputtering Al film and subsequent interdiffusion treatment. *Acta Mater.* **2003**, *51*, 3109–3120. [CrossRef]
114. Novoselova, T.; Celotto, S.; Morgan, R.; Fox, P.; O'Neill, W. Formation of TiAl intermetallics by heat treatment of cold-sprayed precursor deposits. *J. Alloy. Compd.* **2007**, *436*, 69–77. [CrossRef]
115. Balani, K.; Laha, T.; Agarwal, A.; Karthikeyan, J.; Munroe, N. Effect of carrier gases on microstructural and electrochemical behavior of cold-sprayed 1100 aluminum coating. *Surf. Coat. Technol.* **2005**, *195*, 272–279. [CrossRef]
116. Kang, K.; Won, J.; Bae, G.; Ha, S.; Lee, C. Interfacial bonding and microstructural evolution of Al in kinetic spraying. *J. Mater. Sci.* **2012**, *47*, 4649–4659. [CrossRef]

117. Yerokhin, A.L.; Nie, X.; Leyland, A.; Matthews, A.; Dowey, S.J. Plasma electrolysis for surface engineering. *Surf. Coat. Technol.* **1999**, *122*, 73–93. [CrossRef]
118. Nie, X.; Leyland, A.; Song, H.W.; Yerokhin, A.L.; Dowey, S.J.; Matthews, A. Thickness effects on the mechanical properties of micro-arc discharge oxide coatings on aluminium alloys. *Surf. Coat. Technol.* **1999**, *116*, 1055–1060. [CrossRef]
119. Sundararajan, G.; Krishna, L.R. Mechanisms underlying the formation of thick alumina coatings through the MAO coating technology. *Surf. Coat. Technol.* **2003**, *167*, 269–277. [CrossRef]
120. Hussein, R.O.; Nie, X.; Northwood, D.O. Influence of process parameters on electrolytic plasma discharging behaviour and aluminum oxide coating microstructure. *Surf. Coat. Technol.* **2010**, *205*, 1659–1667. [CrossRef]
121. Yerokhin, A.L.; Shatrov, A.; Samsonov, V.; Shashkov, P.; Pilkington, A.; Leyland, A.; Matthews, A. Oxide ceramic coatings on aluminium alloys produced by a pulsed bipolar plasma electrolytic oxidation process. *Surf. Coat. Technol.* **2005**, *199*, 150–157. [CrossRef]

© 2017 by the authors. Licensee MDPI, Basel, Switzerland. This article is an open access article distributed under the terms and conditions of the Creative Commons Attribution (CC BY) license (http://creativecommons.org/licenses/by/4.0/).

MDPI
St. Alban-Anlage 66
4052 Basel
Switzerland
www.mdpi.com

Coatings Editorial Office
E-mail: coatings@mdpi.com
www.mdpi.com/journal/coatings

Disclaimer/Publisher's Note: The statements, opinions and data contained in all publications are solely those of the individual author(s) and contributor(s) and not of MDPI and/or the editor(s). MDPI and/or the editor(s) disclaim responsibility for any injury to people or property resulting from any ideas, methods, instructions or products referred to in the content.

www.ingramcontent.com/pod-product-compliance
Lightning Source LLC
LaVergne TN
LVHW070735100526
838202LV00013B/1241